comparison methods and
stability theory

PURE AND APPLIED MATHEMATICS

A Program of Monographs, Textbooks, and Lecture Notes

LECTURE NOTES IN PURE AND APPLIED MATHEMATICS

1. *N. Jacobson,* Exceptional Lie Algebras
2. *L.-Å. Lindahl and F. Poulsen,* Thin Sets in Harmonic Analysis
3. *I. Satake,* Classification Theory of Semi-Simple Algebraic Groups
4. *F. Hirzebruch, W. D. Newmann, and S. S. Koh,* Differentiable Manifolds and Quadratic Forms
5. *I. Chavel,* Riemannian Symmetric Spaces of Rank One
6. *R. B. Burckel,* Characterization of C(X) Among Its Subalgebras
7. *B. R. McDonald, A. R. Magid, and K. C. Smith,* Ring Theory: Proceedings of the Oklahoma Conference
8. *Y.-T. Siu,* Techniques of Extension on Analytic Objects
9. *S. R. Caradus, W. E. Pfaffenberger, and B. Yood,* Calkin Algebras and Algebras of Operators on Banach Spaces
10. *E. O. Roxin, P.-T. Liu, and R. L. Sternberg,* Differential Games and Control Theory
11. *M. Orzech and C. Small,* The Brauer Group of Commutative Rings
12. *S. Thomier,* Topology and Its Applications
13. *J. M. Lopez and K. A. Ross,* Sidon Sets
14. *W. W. Comfort and S. Negrepontis,* Continuous Pseudometrics
15. *K. McKennon and J. M. Robertson,* Locally Convex Spaces
16. *M. Carmeli and S. Malin,* Representations of the Rotation and Lorentz Groups: An Introduction
17. *G. B. Seligman,* Rational Methods in Lie Algebras
18. *D. G. de Figueiredo,* Functional Analysis: Proceedings of the Brazilian Mathematical Society Symposium
19. *L. Cesari, R. Kannan, and J. D. Schuur,* Nonlinear Functional Analysis and Differential Equations: Proceedings of the Michigan State University Conference
20. *J. J. Schäffer,* Geometry of Spheres in Normed Spaces
21. *K. Yano and M. Kon,* Anti-Invariant Submanifolds
22. *W. V. Vasconcelos,* The Rings of Dimension Two
23. *R. E. Chandler,* Hausdorff Compactifications
24. *S. P. Franklin and B. V. S. Thomas,* Topology: Proceedings of the Memphis State University Conference
25. *S. K. Jain,* Ring Theory: Proceedings of the Ohio University Conference
26. *B. R. McDonald and R. A. Morris,* Ring Theory II: Proceedings of the Second Oklahoma Conference
27. *R. B. Mura and A. Rhemtulla,* Orderable Groups
28. *J. R. Graef,* Stability of Dynamical Systems: Theory and Applications
29. *H.-C. Wang,* Homogeneous Branch Algebras
30. *E. O. Roxin, P.-T. Liu, and R. L. Sternberg,* Differential Games and Control Theory II
31. *R. D. Porter,* Introduction to Fibre Bundles
32. *M. Altman,* Contractors and Contractor Directions Theory and Applications
33. *J. S. Golan,* Decomposition and Dimension in Module Categories
34. *G. Fairweather,* Finite Element Galerkin Methods for Differential Equations
35. *J. D. Sally,* Numbers of Generators of Ideals in Local Rings
36. *S. S. Miller,* Complex Analysis: Proceedings of the S.U.N.Y. Brockport Conference
37. *R. Gordon,* Representation Theory of Algebras: Proceedings of the Philadelphia Conference
38. *M. Goto and F. D. Grosshans,* Semisimple Lie Algebras
39. *A. I. Arruda, N. C. A. da Costa, and R. Chuaqui,* Mathematical Logic: Proceedings of the First Brazilian Conference
40. *F. Van Oystaeyen,* Ring Theory: Proceedings of the 1977 Antwerp Conference
41. *F. Van Oystaeyen and A. Verschoren,* Reflectors and Localization: Application to Sheaf Theory
42. *M. Satyanarayana,* Positively Ordered Semigroups
43. *D. L Russell,* Mathematics of Finite-Dimensional Control Systems
44. *P.-T. Liu and E. Roxin,* Differential Games and Control Theory III: Proceedings of the Third Kingston Conference, Part A
45. *A. Geramita and J. Seberry,* Orthogonal Designs: Quadratic Forms and Hadamard Matrices
46. *J. Cigler, V. Losert, and P. Michor,* Banach Modules and Functors on Categories of Banach Spaces

47. *P.-T. Liu and J. G. Sutinen,* Control Theory in Mathematical Economics: Proceedings of the Third Kingston Conference, Part B
48. *C. Byrnes,* Partial Differential Equations and Geometry
49. *G. Klambauer,* Problems and Propositions in Analysis
50. *J. Knopfmacher,* Analytic Arithmetic of Algebraic Function Fields
51. *F. Van Oystaeyen,* Ring Theory: Proceedings of the 1978 Antwerp Conference
52. *B. Kadem,* Binary Time Series
53. *J. Barros-Neto and R. A. Artino,* Hypoelliptic Boundary-Value Problems
54. *R. L. Sternberg, A. J. Kalinowski, and J. S. Papadakis,* Nonlinear Partial Differential Equations in Engineering and Applied Science
55. *B. R. McDonald,* Ring Theory and Algebra III: Proceedngs of the Third Oklahoma Conference
56. *J. S. Golan,* Structure Sheaves Over a Noncommutative Ring
57. *T. V. Narayana, J. G. Williams, and R. M. Mathsen,* Combinatorics, Representation Theory and Statistical Methods in Groups: YOUNG DAY Proceedings
58. *T. A. Burton,* Modeling and Differential Equations in Biology
59. *K. H. Kim and F. W. Roush,* Introduction to Mathematical Consensus Theory
60. *J. Banas and K. Goebel,* Measures of Noncompactness in Banach Spaces
61. *O. A. Nielson,* Direct Integral Theory
62. *J. E. Smith, G. O. Kenny, and R. N. Ball,* Ordered Groups: Proceedings of the Boise State Conference
63. *J. Cronin,* Mathematics of Cell Electrophysiology
64. *J. W. Brewer,* Power Series Over Commutative Rings
65. *P. K. Kamthan and M. Gupta,* Sequence Spaces and Series
66. *T. G. McLaughlin,* Regressive Sets and the Theory of Isols
67. *T. L. Herdman, S. M. Rankin III, and H. W. Stech,* Integral and Functional Differential Equations
68. *R. Draper,* Commutative Algebra: Analytic Methods
69. *W. G. McKay and J. Patera,* Tables of Dimensions, Indices, and Branching Rules for Representations of Simple Lie Algebras
70. *R. L. Devaney and Z. H. Nitecki,* Classical Mechanics and Dynamical Systems
71. *J. Van Geel,* Places and Valuations in Noncommutative Ring Theory
72. *C. Faith,* Injective Modules and Injective Quotient Rings
73. *A. Fiacco,* Mathematical Programming with Data Perturbations I
74. *P. Schultz, C. Praeger, and R. Sullivan,* Algebraic Structures and Applications: Proceedings of the First Western Australian Conference on Algebra
75. *L Bican, T. Kepka, and P. Nemec,* Rings, Modules, and Preradicals
76. *D. C. Kay and M. Breen,* Convexity and Related Combinatorial Geometry: Proceedings of the Second University of Oklahoma Conference
77. *P. Fletcher and W. F. Lindgren,* Quasi-Uniform Spaces
78. *C.-C. Yang,* Factorization Theory of Meromorphic Functions
79. *O. Taussky,* Ternary Quadratic Forms and Norms
80. *S. P. Singh and J. H. Burry,* Nonlinear Analysis and Applications
81. *K. B. Hannsgen, T. L. Herdman, H. W. Stech, and R. L. Wheeler,* Volterra and Functional Differential Equations
82. *N. L. Johnson, M. J. Kallaher, and C. T. Long,* Finite Geometries: Proceedings of a Conference in Honor of T. G. Ostrom
83. *G. I. Zapata,* Functional Analysis, Holomorphy, and Approximation Theory
84. *S. Greco and G. Valla,* Commutative Algebra: Proceedings of the Trento Conference
85. *A. V. Fiacco,* Mathematical Programming with Data Perturbations II
86. *J.-B. Hiriart-Urruty, W. Oettli, and J. Stoer,* Optimization: Theory and Algorithms
87. *A. Figa Talamanca and M. A. Picardello,* Harmonic Analysis on Free Groups
88. *M. Harada,* Factor Categories with Applications to Direct Decomposition of Modules
89. *V. I. Istrăţescu,* Strict Convexity and Complex Strict Convexity
90. *V. Lakshmikantham,* Trends in Theory and Practice of Nonlinear Differential Equations
91. *H. L. Manocha and J. B. Srivastava,* Algebra and Its Applications
92. *D. V. Chudnovsky and G. V. Chudnovsky,* Classical and Quantum Models and Arithmetic Problems
93. *J. W. Longley,* Least Squares Computations Using Orthogonalization Methods
94. *L. P. de Alcantara,* Mathematical Logic and Formal Systems
95. *C. E. Aull,* Rings of Continuous Functions
96. *R. Chuaqui,* Analysis, Geometry, and Probability
97. *L. Fuchs and L. Salce,* Modules Over Valuation Domains

98. *P. Fischer and W. R. Smith,* Chaos, Fractals, and Dynamics
99. *W. B. Powell and C. Tsinakis,* Ordered Algebraic Structures
100. *G. M. Rassias and T. M. Rassias,* Differential Geometry, Calculus of Variations, and Their Applications
101. *R.-E. Hoffmann and K. H. Hofmann,* Continuous Lattices and Their Applications
102. *J. H. Lightbourne III and S. M. Rankin III,* Physical Mathematics and Nonlinear Partial Differential Equations
103. *C. A. Baker and L, M. Batten,* Finite Geometrics
104. *J. W. Brewer, J. W. Bunce, and F. S. Van Vleck,* Linear Systems Over Commutative Rings
105. *C. McCrory and T. Shifrin,* Geometry and Topology: Manifolds, Varieties, and Knots
106. *D. W. Kueker, E. G. K. Lopez-Escobar, and C. H. Smith,* Mathematical Logic and Theoretical Computer Science
107. *B.-L. Lin and S. Simons,* Nonlinear and Convex Analysis: Proceedings in Honor of Ky Fan
108. *S. J. Lee,* Operator Methods for Optimal Control Problems
109. *V. Lakshmikantham,* Nonlinear Analysis and Applications
110. *S. F. McCormick,* Multigrid Methods: Theory, Applications, and Supercomputing
111. *M. C. Tangora,* Computers in Algebra
112. *D. V. Chudnovsky and G. V. Chudnovsky,* Search Theory: Some Recent Developments
113. *D. V. Chudnovsky and R. D. Jenks,* Computer Algebra
114. *M. C. Tangora,* Computers in Geometry and Topology
115. *P. Nelson, V. Faber, T. A. Manteuffel, D. L. Seth, and A. B. White, Jr.,* Transport Theory, Invariant Imbedding, and Integral Equations: Proceedings in Honor of G. M. Wing's 65th Birthday
116. *P. Clément, S. Invernizzi, E. Mitidieri, and I. I. Vrabie,* Semigroup Theory and Applications
117. *J. Vinuesa,* Orthogonal Polynomials and Their Applications: Proceedings of the International Congress
118. *C. M. Dafermos, G. Ladas, and G. Papanicolaou,* Differential Equations: Proceedings of the EQUADIFF Conference
119. *E. O. Roxin,* Modern Optimal Control: A Conference in Honor of Solomon Lefschetz and Joseph P. Lasalle
120. *J. C. Díaz,* Mathematics for Large Scale Computing
121. *P. S. Milojević,* Nonlinear Functional Analysis
122. *C. Sadosky,* Analysis and Partial Differential Equations: A Collection of Papers Dedicated to Mischa Cotlar
123. *R. M. Shortt,* General Topology and Applications: Proceedings of the 1988 Northeast Conference
124. *R. Wong,* Asymptotic and Computational Analysis: Conference in Honor of Frank W. J. Olver's 65th Birthday
125. *D. V. Chudnovsky and R. D. Jenks,* Computers in Mathematics
126. *W. D. Wallis, H. Shen, W. Wei, and L. Zhu,* Combinatorial Designs and Applications
127. *S. Elaydi,* Differential Equations: Stability and Control
128. *G. Chen, E. B. Lee, W. Littman, and L. Markus,* Distributed Parameter Control Systems: New Trends and Applications
129. *W. N. Everitt,* Inequalities: Fifty Years On from Hardy, Littlewood and Pólya
130. *H. G. Kaper and M. Garbey,* Asymptotic Analysis and the Numerical Solution of Partial Differential Equations
131. *O. Arino, D. E. Axelrod, and M. Kimmel,* Mathematical Population Dynamics: Proceedings of the Second International Conference
132. *S. Coen,* Geometry and Complex Variables
133. *J. A. Goldstein, F. Kappel, and W. Schappacher,* Differential Equations with Applications in Biology, Physics, and Engineering
134. *S. J. Andima, R. Kopperman, P. R. Misra, J. Z. Reichman, and A. R. Todd,* General Topology and Applications
135. *P Clément, E. Mitidieri, B. de Pagter,* Semigroup Theory and Evolution Equations: The Second International Conference
136. *K. Jarosz,* Function Spaces
137. *J. M. Bayod, N. De Grande-De Kimpe, and J. Martínez-Maurica,* p-adic Functional Analysis
138. *G. A. Anastassiou,* Approximation Theory: Proceedings of the Sixth Southeastern Approximation Theorists Annual Conference
139. *R. S. Rees,* Graphs, Matrices, and Designs: Festschrift in Honor of Norman J. Pullman
140. *G. Abrams, J. Haefner, and K. M. Rangaswamy,* Methods in Module Theory

141. *G. L. Mullen and P. J.-S. Shiue*, Finite Fields, Coding Theory, and Advances in Communications and Computing

142. *M. C. Joshi and A. V. Balakrishnan*, Mathematical Theory of Control: Proceedings of the International Conference

143. *G. Komatsu and Y. Sakane*, Complex Geometry: Proceedings of the Osaka International Conference

144. *I. J. Bakelman*, Geometric Analysis and Nonlinear Partial Differential Equations

145. *T. Mabuchi and S. Mukai*, Einstein Metrics and Yang–Mills Connections: Proceedings of the 27th Taniguchi International Symposium

146. *L. Fuchs and R. Göbel*, Abelian Groups: Proceedings of the 1991 Curaçao Conference

147. *A. D. Pollington and W. Moran*, Number Theory with an Emphasis on the Markoff Spectrum

148. *G. Dore, A. Favini, E. Obrecht, and A. Venni*, Differential Equations in Banach Spaces

149. *T. West*, Continuum Theory and Dynamical Systems

150. *K. D. Bierstedt, A. Pietsch, W. Ruess, and D. Vogt*, Functional Analysis

151. *K. G. Fischer, P. Loustaunau, J. Shapiro, E. L. Green, and D. Farkas*, Computational Algebra

152. *K. D. Elworthy, W. N. Everitt, and E. B. Lee*, Differential Equations, Dynamical Systems, and Control Science

153. *P.-J. Cahen, D. L. Costa, M. Fontana, and S.-E. Kabbaj*, Commutative Ring Theory

154. *S. C. Cooper and W. J. Thron*, Continued Fractions and Orthogonal Functions: Theory and Applications

155. *P. Clément and G. Lumer*, Evolution Equations, Control Theory, and Biomathematics

156. *M. Gyllenberg and L. Persson*, Analysis, Algebra, and Computers in Mathematical Research: Proceedings of the Twenty-First Nordic Congress of Mathematicians

157. *W. O. Bray, P. S. Milojević, and Č. V. Stanojević*, Fourier Analysis: Analytic and Geometric Aspects

158. *J. Bergen and S. Montgomery*, Advances in Hopf Algebras

159. *A. R. Magid*, Rings, Extensions, and Cohomology

160. *N. H. Pavel*, Optimal Control of Differential Equations

161. *M. Ikawa*, Spectral and Scattering Theory: Proceedings of the Taniguchi International Workshop

162. *X. Liu and D. Siegel*, Comparison Methods and Stability Theory

163. *J.-P. Zolésio*, Boundary Control and Variation

164. *M. Křížek, P. Neittaanmäki, and R. Stenberg*, Finite Element Methods: Fifty Years of the Courant Element

165. *G. Da Prato and L. Tubaro*, Control of Partial Differential Equations

Additional Volumes in Preparation

comparison methods and stability theory

edited by

Xinzhi Liu
David Siegel

University of Waterloo
Waterloo, Ontario
Canada

Marcel Dekker, Inc.　　　　　**New York • Basel • Hong Kong**

Library of Congress Cataloging-in-Publication Data

Comparison methods and stability theory / edited by Xinzhi Liu, David Siegel.
    p.   cm. — (Lecture notes in pure and applied mathematics; v. 162)
    Proceedings from the Symposium on Comparison Methods and Stability Theory,
held in Waterloo, Ont., Canada, June 3-6, 1993.
    Includes bibliographical references and index.
    ISBN 0-8247-9270-X (acid-free)
    1. Differential equations—Congresses. 2. Numerical analysis—Congresses.
3. Stability—Congresses. I. Liu, X. Z. (Xinzhi Z.) II. Siegel, David.
III. Symposium on Comparison Methods and Stability Theory (1993 : Waterloo,
Ont.) IV. Series.
QA370.C62 1994
515'.35—dc20   94-20391
                                                      CIP

The publisher offers discounts on this book when ordered in bulk quantities. For more information, write to Special Sales/Professional Marketing at the address below.

This book is printed on acid-free paper.

MARCEL DEKKER, INC.
270 Madison Avenue, New York, New York 10016

Current printing (last digit):
10 9 8 7 6 5 4 3 2 1

PRINTED IN THE UNITED STATES OF AMERICA

Preface

The Symposium on Comparison Methods and Stability Theory was held in Waterloo, Ontario, Canada, June 3-6, 1993. These proceedings consist of articles that were delivered at the conference as invited one-hour lectures as well as half-hour reports. Over 80 mathematicians, pure and applied, scientists and engineers from several countries attended the conference. The published proceedings reflect the contemporary achievements and topics in comparison methods and stability theory in a wide range of nonlinear problems.

Comparison methods and stability theory form the common core of a variety of branches of differential equations and their applications, where there has been substantial progress in recent years. This conference brought together distinguished experts in various fields as well as young scholars from different parts of the globe, which gave rise to exciting interactions and communications among researchers. We do hope that this conference will stimulate new important research and bring more cooperation and collaboration to the world community.

This book includes some of the most recent developments in the areas of ordinary differential equations, functional differential equations, impulsive differential equations, integro-differential equations, partial differential equations and differential equations with uncertainties as well as their applications. This book discusses progress that has occurred in the direct method of Lyapunov, monotone iterative techniques, numerical methods, monotone flows, large-scale systems, the method of upper-lower solutions, boundary value problems and control theory. It also features numerous applications to real world problems such as population growth models, semiconductor equations, Schrödinger equations, Hamilton equations, neural networks, chemical kinetics and stochastic processes. This book may be useful to pure and applied mathematicians, numerical analysts, researchers in differential equations and dynamical systems, and graduate students in these disciplines.

Finally, we wish to thank the University of Waterloo and the Fields Institute for their financial support without which this conference would not have been possible. We also wish to thank the staff of Marcel Dekker, Inc. for their patience and cooperation.

Xinzhi Liu
David Siegel

Contents

Preface   *iii*

Contributors   *ix*

On 2-Layer Free-Boundary Problems with Generalized Joining Conditions:
Convexity and Successive Approximation of Solutions   1
A. Acker

Nonisothermal Semiconductor Systems   17
W. Allegretto and H. Xie

A Model for the Growth of the Subpopulation of Lawyers   25
John V. Baxley and Peter A. Cummings

Differential Inequalities and Existence Theory for Differential, Integral,
and Delay Equations   35
T. A. Burton

Monotone Iterative Algorithms for Coupled Systems of Nonlinear Parabolic
Boundary Value Problems   57
Ying Chen and Xinzhi Liu

Steady State Bifurcation Hypersurfaces of Chemical Mechanisms   67
Bruce L. Clarke

Stability Problems for Volterra Functional Differential Equations   87
C. Corduneanu

Persistence (Permanence), Compressivity, and Practical Persistence
in Some Reaction–Diffusion Models from Ecology   101
Chris Cosner

Perturbing Vector Lyapunov Functions and Applications to Large–Scale
Dynamic Systems   117
Zahia Drici

On the Existence of Multiple Positive Solutions of Nonlinear Boundary
Value Problems   127
L. H. Erbe and Shouchuan Hu

Gradient and Gauss Curvature Bounds for *H*-Graphs   137
Robert Finn

Some Applications of Geometric Methods in Mechanics   151
Zhong Ge and W. F. Shadwick

Comparison of Even-Order Elliptic Equations   159
Velmer B. Headley

Positive Equilibria and Convergence in Subhomogeneous Monotone Dynamics   169
Morris W. Hirsch

Blowup of Solution for the Heat Equation with a Nonlinear Boundary Condition   189
Bei Hu and Hong-Min Yin

On the Existence of Extremal Solutions for Impulsive Differential Equations
with Variable Time   199
Saroop Kaul

Global Asymptotic Stability of Competitive Neural Networks   203
Semen Koksal

A Graph Theoretical Approach to Monotonicity with Respect to Initial Conditions   207
H. Kunze and D. Siegel

On the Stabilization of Uncertain Differential Systems   217
A. B. Kurzhanski

Comparison Principle for Impulsive Differential Equations with Variable Times   227
V. Lakshmikantham

The Relationship Between the Boundary Behavior of and the Comparison
Principles Satisfied by Approximate Solutions of Elliptic Dirichlet Problems   237
Kirk E. Lancaster

Numerical Solutions for Linear Integro-Differential Equations of Parabolic Type
with Weakly Singular Kernels   261
Yanping Lin

Impulsive Stabilization   269
Xinzhi Liu and Allan R. Willms

Comparison Methods and Stability Analysis of Reaction Diffusion Systems 277
C. V. Pao

Some Applications of the Maximum Principle to a Free Stekloff Eigenvalue Problem
and to Spatial Gradient Decay Estimates 293
G. A. Philippin

Comparison Methods in Control Theory 303
Emilio O. Roxin

The Self-Destruction of the Perfect Democracy 309
Rudolf Starkermann

A Nonlinear Stochastic Process for Quality Growth 319
Chris P. Tsokos

An Extension of the Method of Quasilinearization for Reaction–Diffusion Equations 331
A. S. Vatsala

Geometric Methods in Population Dynamics 339
M. L. Zeeman

Uniform Asymptotic Stability in Functional Differential Equations with Infinite Delay 349
Bo Zhang

Index 363

Contributors

A. Acker   The Wichita State University, Wichita, Kansas

W. Allegretto   University of Alberta, Edmonton, Alberta, Canada

John V. Baxley   Wake Forest University, Winston-Salem, North Carolina

T. A. Burton   Southern Illinois University at Carbondale, Carbondale, Illinois

Ying Chen   University of Waterloo, Waterloo, Ontario, Canada

Bruce L. Clarke   University of Alberta, Edmonton, Alberta, Canada

C. Corduneanu   University of Texas at Arlington, Arlington, Texas

Chris Cosner   University of Miami, Coral Gables, Florida

Peter A. Cummings   Wake Forest University, Winston-Salem, North Carolina

Zahia Drici   Florida Institute of Technology, Melbourne, Florida

L. H. Erbe   University of Alberta, Edmonton, Alberta, Canada

Robert Finn   Stanford University, Stanford, California

Zhong Ge   The Fields Institute for Research in Mathematical Sciences, Waterloo, Ontario, Canada

Velmer B. Headley   Brock University, St. Catharines, Ontario, Canada

Morris W. Hirsch   University of California, Berkeley, Berkeley, California

Bei Hu   University of Notre Dame, Notre Dame, Indiana

Shouchuan Hu   Southwest Missouri State University, Springfield, Missouri

Saroop Kaul   University of Regina, Regina, Saskatchewan, Canada

Semen Koksal   Bradley University, Peoria, Illinois

H. Kunze   University of Waterloo, Waterloo, Ontario, Canada

A. B. Kurzhanski   Faculty of Computational Mathematics and Cybernetics (VMK), Moscow State University, Moscow, Russia

V. Lakshmikantham   Florida Institute of Technology, Melbourne, Florida

Kirk E. Lancaster   Wichita State University, Wichita, Kansas

Yanping Lin   University of Alberta, Edmonton, Alberta, Canada

Xinzhi Liu   University of Waterloo, Waterloo, Ontario, Canada

C. V. Pao   North Carolina State University, Raleigh, North Carolina

G. A. Philippin   Laval University, Quebec, Canada

Emilio O. Roxin   College of Arts and Sciences, The University of Rhode Island, Kingston, Rhode Island

W. F. Shadwick   The Fields Institute for Research in Mathematical Sciences, Waterloo, Ontario, Canada

D. Siegel   University of Waterloo, Waterloo, Ontario, Canada

Rudolf Starkermann   Nd'Rohrdorf, Switzerland

Chris P. Tsokos   University of South Florida, Tampa, Florida

A. S. Vatsala   College of Sciences, The University of Southwestern Louisiana, Lafayette, Louisiana

Allan R. Willms   University of Waterloo, Waterloo, Ontario, Canada

H. Xie   University of Alberta, Edmonton, Alberta, Canada

Hong Min-Yin   University of Notre Dame, Notre Dame, Indiana

M. L. Zeeman   Division of Mathematics, Computer Science and Statistics, University of Texas at San Antonio, San Antonio, Texas

Bo Zhang   Fayetteville State University, Fayetteville, North Carolina

On 2-Layer Free-Boundary Problems with Generalized Joining Conditions: Convexity and Successive Approximation of Solutions

A. ACKER, Department of Mathematics and Statistics, The Wichita State University, Wichita, Kansas 67260-0033.

1 INTRODUCTION

This work is in the context of a generalized 2-layer free-boundary problem, of which the 2-layer fluid problem is the best known special case. In the 2-layer fluid problem, one seeks an interface between the fluid layers (the free boundary) such that the derivatives of the stream functions in the layers satisfy a joining condition across the interface which corresponds to Bernoulli's law (see [11], [5]). In our generalized problem (in arbitrary space dimensions), the derivatives of several distinct capacitary potentials defined in the first layer are required to be related to the derivatives of several distinct capacitary potentials defined in the second layer by means of a given, nonlinear joining condition across the unknown layer-interface. We obtain uniqueness, starlikeness, and monotonicity results for this problem (in §2), give a trial free-boundary method for the successive approximation of the free-boundary solution (in §3), establish the convexity of the solution under suitable conditions (in §4), and study a related convex variational problem (in §5). In all these cases, except §2, our results follow from the study of certain free-boundary perturbation operators, which the author originally introduced in the context of one-layer problems (see [1], [2], [3], [10]), but has also applied in various 2-layer and multilayer free-boundary problems (see [4], [5], [6], [7], [8]). One of the main objectives of this paper is to generalize the operator method to the case of general nonlinear joining conditions, since all previous applications have been restricted to the specific form motivated by Bernoulli's law. For a general discussion of operator methods and convexity results in the context of multilayer problems, we refer the reader to the author's recent survey article [9].

2   TWO-LAYER PROBLEM: UNIQUENESS

2.1   Problem   In $\mathbb{R}^N$, $N \geq 2$, let be given an annular domain Ω of the form $\Omega = D_*^+ \setminus Cl(D_*^-)$, where $D_*^\pm$ are fixed, bounded, simply connected, nested domains with C^1-boundaries $\Gamma_*^\pm = \partial D_*^\pm$. Given the values $p_i^\pm > 1, i \in I^\pm$, and the continuously differentiable function $F(x,s,t): \mathbb{R}^N \times \mathbb{R}_+^k \times \mathbb{R}_+^m \to \mathbb{R}$ (where $\mathbb{R}_+ := (0,\infty)$, $I^- = \{1,\cdots,k\}$, and $I^+ = \{1,\cdots,m\}$), we seek a domain D (with C^1-boundary $\Gamma = \partial D$) such that $Cl(D_*^-) \subset D \subset Cl(D) \subset D_*^+$, and such that

$$F\big(x, Q^-(x), Q^+(x)\big) = 0 \text{ on } \Gamma. \tag{1}$$

Here, we define $\Omega^\pm$ to be the annular domain whose boundary is $\partial \Omega^\pm = \Gamma \cup \Gamma_*^\pm$, and we define $Q^\pm = (Q_i^\pm)_{i \in I^\pm}$ and $U^\pm(x) = (U_i^\pm)_{i \in I^\pm}$, where, for each $i \in I^\pm$, $Q_i^\pm = |\nabla U_i^\pm|$ on Γ and $U_i^\pm(x)$ denotes the $p_i^\pm$–capacitary potential in $\Omega^\pm$.

2.2   Remark   For any $p > 1$, the p-Laplacian is defined by $\Delta_p U = \nabla \cdot (|\nabla U|^{p-2} \nabla U)$. The p-capacitary potential $U^\pm$ in $\Omega^\pm$ is the solution of the boundary value problem:

$$\Delta_p U^\pm = 0 \text{ in } \Omega^\pm, \ U^\pm(\Gamma) = 0, \ U^\pm(\Gamma_*^\pm) = 1. \tag{2}$$

2.3   Theorem   Assume that the domains $D_*^\pm$ are starlike (relative to the origin), and that the function $F(x,s,t): \mathbb{R}^N \times \mathbb{R}_+^k \times \mathbb{R}_+^m \to \mathbb{R}$ is strictly decreasing in the components of $s = (s_1,\cdots,s_k)$ and strictly increasing in the components of $t = (t_1,\cdots,t_m)$ for all x. Also assume that if $F(x,s,t) \geq 0$ for a given point $(x,s,t) \in \mathbb{R}^N \times \mathbb{R}_+^k \times \mathbb{R}_+^m$, then $F(\lambda x, s/\lambda, t/\lambda) \geq 0$ for any value $\lambda \geq 1$. Then Problem 2.1 has at most one classical solution D such that $\Gamma := \partial D$ is a C^2-surface. Moreover, D is starlike.

Proof   We will prove the following more general assertion: If D and $\tilde{D}$ are classical lower and upper solutions of Problem 2.1, respectively, and if Γ and $\tilde{\Gamma}$ are C^2-surfaces, then $\tilde{D} \subset D$. By assumption, we have that

$$F\big(x, Q^-(x), Q^+(x)\big) \geq 0 \text{ on } \Gamma; \ F\big(x, \tilde{Q}^-(x), \tilde{Q}^+(x)\big) \leq 0 \text{ on } \tilde{\Gamma}, \tag{3 a,b}$$

where $\tilde{U}^\pm$ and $\tilde{Q}^\pm$ are defined relative to $\tilde{\Gamma} = \partial \tilde{D}$, and where the derivatives in (3 a,b) are assumed to exist. Choose $\lambda > 0$ to be minimum subject to the requirement that $\tilde{D} \subset tD$ for all $t \geq \lambda$, where $tD = \{tx : x \in D\}$. We assert that $\lambda \leq 1$. Assume (for the purpose of obtaining a contradiction) that $\lambda > 1$. Set $\hat{\Gamma} = \lambda\Gamma$, $\hat{\Gamma}_*^\pm = \lambda\Gamma_*^\pm$, and $\hat{\Omega}^\pm = \lambda\Omega^\pm$, and define $\hat{U}^\pm(x) = U^\pm(x/\lambda)$ in $\hat{\Omega}^\pm$. Then $\hat{D} \supset \tilde{D}$, and one concludes by applications of the comparison principle for $p_i^\pm$–harmonic functions that $\pm(\tilde{U}_i^\pm - \hat{U}_i^\pm) > 0$ in $\hat{\Omega}^\pm \cap \tilde{\Omega}^\pm$ for $i \in I^\pm$. By the assumed smoothness of Γ and $\tilde{\Gamma}$, we conclude that

$$\pm\big(\tilde{Q}_i^\pm(\hat{x}_0) - \hat{Q}_i^\pm(\hat{x}_0)\big) > 0 \tag{4}$$

for $i \in I^\pm$ at any point $\hat{x}_0 = \lambda x_0 \in \tilde{\Gamma} \cap \hat{\Gamma}$. By change of scale, we also have

$$\hat{Q}^\pm(\hat{x}_0) = \big(Q^\pm(x_0)/\lambda\big). \tag{5}$$

We conclude from (4), (5), (3 a), and the assumptions regarding F that

$$F\big(\hat{x}_0, \tilde{Q}^-(\hat{x}_0), \tilde{Q}^+(\hat{x}_0)\big) > F\big(\hat{x}_0, \hat{Q}^-(\hat{x}_0), \hat{Q}^+(\hat{x}_0)\big)$$

$$= F\big(\lambda x_0, (Q^-(x_0)/\lambda), (Q^+(x_0)/\lambda)\big) \geq F\big(x_0, Q^-(x_0), Q^+(x_0)\big) \geq 0.$$

This contradiction of (3 b) proves that $0 < \lambda \leq 1$, and thus that $\tilde{D} \subset D$. The asserted uniqueness follows in the case where $D, \tilde{D}$ are both solutions of Problem 2.1. The asserted starlikeness follows from the same steps in the case where $D = \tilde{D}$.

2.4 Remark To the author's knowledge, the literature regarding the existence question for Problem 2.1 has been mainly restricted to the case where $p = 2$, $k = m = 1$, and $F(x, s, t) = A(x) + t^2 - s^2$ for a given function $A(x) : \mathbb{R}^N \to \mathbb{R}$ (see [11] and [7],§3). For this case, the requirements on $F(x, s, t)$ in Theorem 2.3 are met if $A(x) = a^2(x)$, where $a(x) > 0$ and $\lambda a(\lambda x)$ is nondecreasing in $\lambda > 0$ for each $x \in \mathbb{R}^N$. (See Remark 3.17(b).)

3 SUCCESSIVE APPROXIMATION OF SOLUTIONS

3.1 Additional assumptions for Problem 2.1 Define the continuously differentiable function $f(x, s, t) : \mathbb{R}^N \times \mathbb{R}^k_+ \times \mathbb{R}^m_+ \to \mathbb{R}$ such that $f(x, s, t) = F(x, \sigma, \tau)$, where $\sigma_i = 1/s_i$ for $i = 1, \cdots, k$ and $\tau_j = 1/t_j$ for $j = 1, \cdots, m$. In the context of Problem 2.1, we make the following additional assumptions throughout this section (and the following section):

(A1) We assume that the given domains $D^\pm_*$ are starlike relative to all points in a given ball $B_\delta(0)$ of radius δ centered at the origin, where $\delta > 0$ is sufficiently small.

(A2) We assume that the given surfaces $\Gamma^\pm_* = \partial D^\pm_*$ are C^2-surfaces.

(A3) For any constant $\rho_0 > 0$ and compact set $K \subset \mathbb{R}^N$, we assume there exist constants $\delta_0, C_0 > 0$ such that throughout $K \times (0, \rho_0]^k \times (0, \rho_0]^m$, we have $\partial f / \partial s_i \geq \delta_0$ for $i \in I^-$, $\partial f / \partial t_j \leq -\delta_0$ for $j \in I^+$, and $|\nabla_x f| \leq C_0 \max \{\partial f / \partial s_i, |\partial f / \partial t_j| : i \in I^-, j \in I^+\}$.

(A4) For any given constants $0 < \delta_0 < \rho_0$ and compact set $K \subset \mathbb{R}^N$, we have $f(x, s, t) < 0$ whenever $(x, s, t) \in K \times (0, \rho_0]^k \times [\delta_0, \rho_0]^m$ and $s_i > 0$ is sufficiently small for some $i \in I^-$, and $f(x, s, t) > 0$ whenever $(x, s, t) \in K \times [\delta_0, \rho_0]^k \times (0, \rho_0]^m$, and $t_j > 0$ is sufficiently small for some $j \in I^+$.

(A5) We assume that if $f(x_0 + v, s, t) \geq 0$ for any particular choice of $(v, s, t) \in \mathbb{R}^N \times \mathbb{R}^k_+ \times \mathbb{R}^m_+$ and $x_0 \in B_\delta(0)$, then $f(x_0 + \lambda v, \lambda s, \lambda t) \geq 0$ for any $\lambda \geq 1$.

3.2 Definitions for the operator method Let $\mathbb{X}$ denote the family of all $(N-1)$-dimensional surfaces Γ of the form $\Gamma = \partial D$, where $Cl(D^-_*) \subset D \subset Cl(D) \subset D^+_*$, and where D is starlike relative to all points in $B_\delta(0)$. For each $\Gamma \in \mathbb{X}$, let $D(\Gamma)$ denote the interior complement of Γ, let $\Omega^\pm(\Gamma)$ denote the annular domain with boundary $\partial \Omega^\pm(\Gamma) = \Gamma \cup \Gamma^\pm_*$, and let $U^\pm(\Gamma; x) = (U^\pm_i(\Gamma; x))_{i \in I^\pm}$ in $Cl(\Omega^\pm(\Gamma))$, where $U^\pm_i(\Gamma; x)$ solves (2) with $p = p_i^{\pm}$. (The functions $U_i^{\pm}(\Gamma; x)$ are classical solutions of (2) if $\Gamma \in X \cap C^2$; see [16], Prop. 2.4.1.) For $\Gamma_1, \Gamma_2 \in X$, we define $\Gamma_1 \leq \Gamma_2$ (resp. $\Gamma_1 < \Gamma_2$) if $D_1 \subset D_2$ (resp. $Cl(D_1) \subset D_2$), where $D_i := D(\Gamma_i)$, $i = 1, 2$. We also define the metric

$$\Delta(\Gamma_1, \Gamma_2) = \max\{|\ln(\lambda)| : x \in \Gamma_1, \lambda x \in \Gamma_2, \lambda > 0\}$$

in X. For any positive integer $n \in \mathbb{N}$, and for any multi-surfaces $\Gamma_1 := (\Gamma_{1,1}, \cdots, \Gamma_{n,1})$, $\Gamma_2 := (\Gamma_{1,2}, \cdots, \Gamma_{n,2})$ both in X^n, we write $\Gamma_1 \leq \Gamma_2$ (resp. $\Gamma_1 < \Gamma_2$) if $\Gamma_{i,1} \leq \Gamma_{i,2}$ (resp. $\Gamma_{i,1} < \Gamma_{i,2}$) for all $i = 1, \cdots, n$. For multisurfaces $\Gamma_1 := (\Gamma_{1,1}, \cdots, \Gamma_{k,1}) \in X^k$ and $\Gamma_2 := (\Gamma_{1,2}, \cdots, \Gamma_{m,2}) \in X^m$, we write $\Gamma_1 \ll \Gamma_2$ if $\Gamma_{i,1} < \Gamma_{j,2}$ for all $i \in I^-$ and $j \in I^+$. For any $0 < \varepsilon < 1$, we define the mappings $\Phi_\varepsilon^- : X \to X^k$ and $\Phi_\varepsilon^+ : X \to X^m$ componentwise such that

$$\Phi_{\varepsilon,i}^{\pm}(\Gamma) = \{x \in \Omega^{\pm}(\Gamma) : U_i^{\pm}(\Gamma; x) = \varepsilon\} \in X$$

for $i \in I^{\pm}$. Clearly $\Phi_\varepsilon^-(\Gamma) \ll \Gamma \ll \Phi_\varepsilon^+(\Gamma)$ for $\Gamma \in X$, $0 < \varepsilon < 1$. For any multi-surfaces $\Gamma^- := (\Gamma_1^-, \cdots, \Gamma_k^-) \in X^k$ and $\Gamma^+ := (\Gamma_1^+, \cdots, \Gamma_m^+) \in X^m$ such that $\Gamma^- \ll \Gamma^+$, we define

$$\Psi_\varepsilon(\Gamma^-, \Gamma^+) = \{x \in \omega : f\big(x, d(x, \Gamma_1^-)/\varepsilon, \cdots, d(x, \Gamma_k^-)/\varepsilon, d(x, \Gamma_1^+)/\varepsilon, \cdots, d(x, \Gamma_m^+)/\varepsilon\big) = 0\} \in X,$$

where $\omega = \omega(\Gamma^-, \Gamma^+)$ denotes the annular domain which is in the exterior complement of all the surfaces Γ_i^-, $i \in I^-$, and in the interior complement of all the surfaces Γ_i^+, $i \in I^+$, and where we define $d(x, S) = \inf\{|x - y| : y \in S\}$ for any $x \in \mathbb{R}^N$ and set $S \subset \mathbb{R}^N$. Finally, for any $0 < \varepsilon < 1$, we define the mapping $T_\varepsilon : X \to X$ such that

$$T_\varepsilon(\Gamma) = \Psi_\varepsilon(\Phi_\varepsilon^-(\Gamma), \Phi_\varepsilon^+(\Gamma)).$$

3.3 Theorem For any $0 < \varepsilon < 1$, we have that:
(a) $\Phi_{\varepsilon,i}^{\pm} : X \to X$ for each $i \in I^{\pm}$.
(b) $\Psi_\varepsilon(\Gamma^-, \Gamma^+) \in X$ for any multi-surfaces $\Gamma^- \in X^k$, $\Gamma^+ \in X^m$ such that $\Gamma^- \ll \Gamma^+$.
(c) $T_\varepsilon : X \to X$.

3.4   Proof of Theorem 3.3(a)   Let $U_i^{\pm}(x) = U_i^{\pm}(\Gamma, x)$ in $\Omega^{\pm} = \Omega^{\pm}(\Gamma)$. Comparison principles show that $0 < U_i^{\pm}(x) < 1$ in $\Omega^{\pm}$, and hence that $\pm(U_i^{\pm}(\lambda x)) - U_i^{\pm}(x)) \geq 0$ throughout $\Omega^{\pm} \cap (\Omega^{\pm}/\lambda)$ for any $\lambda \geq 1$. The starlikeness of $\Phi_{\varepsilon,i}^{\pm}(\Gamma)$ relative to the origin follows directly from this. The starlikeness of $\Phi_{\varepsilon,i}^{\pm}(\Gamma)$ relative to any point $x_0 \in B_\delta(0)$ follows by the same argument with x replaced by $y := x - x_0$.

3.5 Lemma Let $\phi(\lambda) = (d(\lambda e, \Gamma)/\lambda) = d(e, (1/\lambda)\Gamma)$ for all $\lambda > 0$, where e is a given vector in $\mathbb{R}^N$, and where $\Gamma = \partial D$ for a given bounded, starlike domain D. Then $\phi(\lambda)$ is strictly decreasing for $\lambda e \in D$ and strictly increasing for $\lambda e \notin Cl(D)$.

Proof   For $0 < \alpha < \beta$, we have $d(\alpha e, \Gamma) > d(\alpha e, (\alpha/\beta)\Gamma) = (\alpha/\beta) d(\beta e, \Gamma)$ for $\beta e \in D$, whereas $d(\beta e, \Gamma) > d(\beta e, (\beta/\alpha)\Gamma) = (\beta/\alpha) d(\alpha e, \Gamma)$ for $\alpha e \notin Cl(D)$.

3.6 Lemma Under assumptions (A3), (A4), and (A5) (with $x_0 = 0$ in (A5)), if the multi-surfaces $\Gamma^- := (\Gamma_1^-, \cdots, \Gamma_k^-)$ and $\Gamma^+ := (\Gamma_1^+, \cdots, \Gamma_m^+)$ are componentwise starlike closed surfaces such that $\Gamma^- \ll \Gamma^+$ (in the sense defined in §3.2), then $\Psi_\epsilon(\Gamma^-,\Gamma^+)$ is a starlike closed surface such that $\Gamma^- \ll \Psi_\epsilon(\Gamma^-,\Gamma^+) \ll \Gamma^+$.

Proof Given a unit vector e, we define $g(r)=g(\Gamma^-,\Gamma^+;r)=f(re,r\phi^-(r),r\phi^+(r))$ for $r \in J = J(\Gamma^-,\Gamma^+) := \{r>0: re \in \omega\}$, where $\phi^\pm(r)$ is defined componentwise such that $\phi_i^\pm(r)=(d(re,\Gamma_i^\pm)/re)$ for $i \in I^\pm$, and where $\omega=\omega(\Gamma^-,\Gamma^+)$. Observe that $\pm\phi_i^\pm(r)$ is strictly decreasing in J, due to Lemma 3.5. By assumption (A4), we have $g(r)<0$ if $\phi_i^-(r)$ is sufficiently small for some $i \in I^-$, and $g(r)>0$ if $\phi_i^+(r)$ is sufficiently small for some $i \in I^+$. By continuity, there exists a value $\alpha \in J$ such that $g(\alpha)=0$ and $g(r)<0$ for $r<\alpha$ in J. For any $\beta \in J$ with $\beta>\alpha$, we have that

$$g(\beta) > f(\beta e,\beta\phi^-(\alpha),\beta\phi^+(\alpha)) = f(\mu\alpha e,\mu\alpha\phi^-(\alpha),\mu\alpha\phi^+(\alpha)) \geq g(\alpha) = 0,$$

due to assumptions (A3) and (A5), where $\mu=(\beta/\alpha)>1$.

3.7 Proof of Theorem 3.3 (b), (c) In view of assumption (A5), one can apply coordinate translations to Lemma 3.6 to show that if Γ^- and Γ^+ are both componentwise starlike relative to $x_0 \in B_\delta(0)$, then so is $\Psi_\epsilon(\Gamma^-,\Gamma^+)$, completing the proof of part (b). Part (c) follows from parts (a) and (b).

3.8 Lemma Let the four multisurfaces $\Gamma_1^-,\Gamma_2^- \in \mathsf{X}^k$ and $\Gamma_1^+,\Gamma_2^+ \in \mathsf{X}^m$ be such that $\Gamma_1^\pm \leq \Gamma_2^\pm$, $\Gamma_1^- \ll \Gamma_1^+$, and $\Gamma_2^- \ll \Gamma_2^+$. Then $\Psi_\epsilon(\Gamma_1^-,\Gamma_1^+) \leq \Psi_\epsilon(\Gamma_2^-,\Gamma_2^+)$ for any $0<\epsilon<1$.

Proof It suffices to show (for fixed $0<\epsilon<1$) that

$$\Psi_\epsilon(\Gamma_1^-,\Gamma_1^+) \leq \Psi_\epsilon(\Gamma_1^-,\Gamma_2^+); \quad \Psi_\epsilon(\Gamma_1^-,\Gamma_2^+) \leq \Psi_\epsilon(\Gamma_2^-,\Gamma_2^+). \tag{6 a,b}$$

We will prove (6 a) only. In terms of notation introduced in the proof of Lemma 3.6, we define $g_i(r)=g(\Gamma_1^-,\Gamma_i^+;r)$ on the interval $J_i = J(\Gamma_1^-,\Gamma_i^+)$ for $i=1,2$. Assume for a fixed unit vector e that $\alpha e \in \Psi_\epsilon(\Gamma_1^-,\Gamma_1^+)$ for some $\alpha>0$. Therefore $g_1(\alpha)=0$. Since $d(\alpha e,\Gamma_{1,j}^+) \leq d(\alpha e,\Gamma_{2,j}^+)$ for $j=1,\cdots,m$, it follows that $g_2(\alpha) \leq 0$. In view of the monotonicity of $g_2(r)$ (see proof of Lemma 3.6), we conclude that $\alpha \leq \beta$, where $\beta e \in \Psi_\epsilon(\Gamma_1^-,\Gamma_2^+)$ and $g_2(\beta)=0$.

3.9 Lemma For any $0<\epsilon<1$, we have $\Phi_\epsilon^\pm(\Gamma_1) \leq \Phi_\epsilon^\pm(\Gamma_2)$ and $T_\epsilon(\Gamma_1) \leq T_\epsilon(\Gamma_2)$ whenever $\Gamma_1 \leq \Gamma_2$ in X.

Proof Let $U_{j,i}^\pm(x)=U_i^\pm(\Gamma_j,x)$ in $\Omega_j^\pm=\Omega^\pm(\Gamma_j)$ for $i \in I^\pm$ and $j=1,2$. Comparison principles show that $\pm(U_{2,i}^\pm - U_{1,i}^\pm) \leq 0$ in $\Omega_1^\pm \cap \Omega_2^\pm$ for each $i \in I^\pm$. Also, $\pm U_{j,i}^\pm$ is an increasing function of $|x|$ on radial lines in $\Omega_j^\pm$ for $j=1,2$ (see §3.4)). This implies the monotonicity of $\Phi_\epsilon^\pm$. The monotonicity of T_ϵ follows from this, in view of Lemma 3.8.

3.10 Lemma If $\Gamma^- \ll \Gamma^+$ for $\Gamma^- \in \mathsf{X}^k$ and $\Gamma^+ \in \mathsf{X}^m$, then

$$\Psi_\epsilon(\lambda\Gamma^-,\lambda\Gamma^+) \geq \lambda\Psi_\epsilon(\Gamma^-,\Gamma^+) \text{ for } 0<\lambda\leq 1,$$

$$\Psi_\epsilon(\lambda\Gamma^-,\lambda\Gamma^+) \leq \lambda\Psi_\epsilon(\Gamma^-,\Gamma^+) \text{ for } \lambda\geq 1.$$

Proof For simplicity, we consider only the case where $\lambda \geq 1$. Assume for fixed $0 < \varepsilon < 1$ and for a fixed unit vector e that $\alpha e \in \Psi_\varepsilon(\Gamma^-, \Gamma^+)$ and $\beta e \in \Psi_\varepsilon(\lambda \Gamma^-, \lambda \Gamma^+)$, where $\alpha, \beta > 0$. In terms of notation introduced in the proof of Lemma 3.6, we have $g(\alpha) = 0$ and $g(\lambda, \beta) = 0$, where $g(\lambda, r) := g(\lambda \Gamma^-, \lambda \Gamma^+; r)$. We need to show that $\beta \leq \lambda \alpha$. Assume to the contrary that $\beta > \lambda \alpha$. Then

$$\pm \left((\beta/\alpha) \, d(\alpha e, \Gamma_i^\pm) - d(\beta e, \lambda \Gamma_i^\pm) \right) > 0 \tag{7}$$

for $i \in I^\pm$, by Lemma 3.5. It follow from (7) and assumptions (A3), (A5) that

$$g(\lambda, \beta) > g(\mu \alpha e, \mu \alpha \phi^-(\alpha), \mu \alpha \phi^+(\alpha)) \geq g(\alpha e, \alpha \phi^-(\alpha), \alpha \phi^+(\alpha)) = g(\alpha) = 0,$$

where $\mu = (\beta/\alpha) > 1$, contradicting the requirement that $g(\lambda, \beta) = 0$.

3.11 Lemma Let $\tilde{\Gamma} \in \mathbb{X}$ be a classical solution of Problem 2.1, and let $\tilde{D} = D(\tilde{\Gamma})$. For $1 < r \leq r_0$ (where $r_0 > 1$ is such that $r_0^2 \Gamma_*^- < \Gamma_*^+$), let $\tilde{\Gamma}_{\lambda, r} = \partial((\lambda \tilde{D}) \cup (r D_*^-)) \in \mathbb{X}$ for $0 < \lambda \leq 1$ and $\tilde{\Gamma}_{\lambda, r} = \partial((\lambda \tilde{D}) \cap ((1/r) D_*^+)) \in \mathbb{X}$ for $\lambda \geq 1$. Then for each $0 < \varepsilon < 1$, we have

$$\Phi_\varepsilon^\pm(\tilde{\Gamma}_{\lambda, r}) \geq \lambda E(\lambda, \varepsilon) \Phi_\varepsilon^\pm(\tilde{\Gamma}) \text{ for } a_0 < \lambda \leq 1,$$

$$\Phi_\varepsilon^\pm(\tilde{\Gamma}_{\lambda, r}) \leq \lambda E(\lambda, \varepsilon) \Phi_\varepsilon^\pm(\tilde{\Gamma}) \text{ for } 1 \leq \lambda \leq b_0,$$

$$(\lambda - 1)(\lambda E(\lambda, \varepsilon) - 1) > 0 \text{ for } a_0 \leq \lambda \leq b_0, \lambda \neq 1,$$

where $E(\lambda, \varepsilon) = 1 + C_1(1 - \lambda)\varepsilon$ for some sufficiently small constant $C_1 \in (0, (1/b_0)]$, and where $0 < a_0 < 1 < b_0$ are constants chosen such that $\tilde{\Gamma}_{\lambda, r} = r \Gamma_*^-$ for $0 < \lambda \leq a_0$ and $\tilde{\Gamma}_{\lambda, r} = (1/r) \Gamma_*^+$ for $\lambda \geq b_0$ (independent of $r \in (0, r_0]$).

Proof For the case where $p_i^\pm = 2$, these assertions (taken componentwise) summarize the conclusions of Definition 3.4, Remark 3.5, Lemma 3.6, and Remark 3.7 in [5]. The proof there was based primarily on maximum principles for harmonic functions. It is shown in [16], Lemma 6.6.1 that these arguments extend to p-harmonic functions (for any $p > 1$).

3.12 Lemma Let $\tilde{\Gamma} \in \mathbb{X}$ denote a classical solution of Problem 2.1. Then, using notation from Lemma 3.11, we have

$$\alpha E(\alpha, \varepsilon) \, T_\varepsilon(\tilde{\Gamma}) \leq T_\varepsilon(\Gamma) \leq \beta E(\beta, \varepsilon) \, T_\varepsilon(\tilde{\Gamma})$$

for any $0 < \varepsilon < 1$ and for all $\Gamma \in \mathbb{X}$ satisfying $\alpha \tilde{\Gamma} \leq \Gamma \leq \beta \tilde{\Gamma}$, where $a_0 \leq \alpha \leq 1 \leq \beta \leq b_0$.

Proof We treat only the first inequality. If $\Gamma \geq \alpha \tilde{\Gamma}$, then $\Gamma \geq \tilde{\Gamma}_{\alpha, r}$ for sufficiently small $r \in (0, r_0]$. It then follows from Lemmas 3.8, 3.9, 3.10, and 3.11 that

$$T_\varepsilon(\Gamma) \geq T_\varepsilon(\tilde{\Gamma}_{\alpha, r}) = \Psi_\varepsilon \left(\Phi_\varepsilon^-(\tilde{\Gamma}_{\alpha, r}), \Phi_\varepsilon^+(\tilde{\Gamma}_{\alpha, r}) \right) \geq$$

$$\Psi_\varepsilon \left(\alpha E(\alpha, \varepsilon) \Phi_\varepsilon^-(\tilde{\Gamma}), \alpha E(\alpha, \varepsilon) \Phi_\varepsilon^+(\tilde{\Gamma}) \right) \geq$$

$$\alpha E(\alpha, \varepsilon) \Psi_\varepsilon \left(\Phi_\varepsilon^-(\tilde{\Gamma}), \Phi_\varepsilon^+(\tilde{\Gamma}) \right) = \alpha E(\alpha, \varepsilon) \, T_\varepsilon(\tilde{\Gamma}).$$

3.13 Lemma Let $\tilde{\Gamma} \in X$ be a classical solution of Problem 2.1 such that $\nabla \tilde{U}_i^{\pm}(x) \neq 0$ on $\tilde{\Gamma}$ for $i \in I^{\pm}$, where we define $\tilde{U}_i^{\pm}(x) = U_i^{\pm}(\tilde{\Gamma}, x)$. Then $(1/\varepsilon)\Delta(\tilde{\Gamma}, T_\varepsilon(\tilde{\Gamma})) \to 0$ as $\varepsilon \to 0+$.

Proof For fixed $x \in \tilde{\Gamma}$, let $x(\alpha) = x + \alpha\nu$ for $\alpha \in \mathbb{R}$, where ν denotes the exterior normal to $\tilde{\Gamma}$ at x. For $i \in I^{\pm}$, choose $\alpha_{\varepsilon,i}^{\pm}$ such that $|\alpha_{\varepsilon,i}^{\pm}|$ is minimum subject to the requirement that $x_{\varepsilon,i}^{\pm} = x(\alpha_{\varepsilon,i}^{\pm}) \in \tilde{\Gamma}_{\varepsilon,i}^{\pm} := \Phi_{\varepsilon,i}^{\pm}(\tilde{\Gamma})$. Define $\phi_\varepsilon(\alpha) = f(x(\alpha), g^-(\varepsilon, \alpha), g^+(\varepsilon, \alpha))$ for $\alpha \in J_\varepsilon :=$ $(\alpha_\varepsilon^-, \alpha_\varepsilon^+)$, where $\alpha_\varepsilon^{\pm} = \pm\min\{\pm\alpha_{\varepsilon,i}^{\pm}: i \in I^{\pm}\}$, and where the functions $g^{\pm}(\varepsilon, \alpha)$ are defined componentwise such that $g_i^{\pm}(\varepsilon, \alpha) = |x(\alpha) - x_{\varepsilon,i}^{\pm}|$ for $i \in I^{\pm}$. In view of the assumed regularity properties of $\tilde{\Gamma}$ and $\tilde{U}_i^{\pm}(x)$, Taylor's theorem implies that

$$\varepsilon = \tilde{U}_i^{\pm}(x_{\varepsilon,i}^{\pm}) = \left(|\nabla \tilde{U}_i^{\pm}(x)| + \zeta_i^{\pm}(g_i^{\pm}(\varepsilon, 0))\right)g_i^{\pm}(\varepsilon, 0),$$

for $i \in I^{\pm}$, where $\zeta_i^{\pm}(\tau) \to 0$ as $\tau \to 0+$. It follows that $g_i^{\pm}(\varepsilon, 0) = O(\varepsilon)$, and hence that $(g_i^{\pm}(\varepsilon, 0)/\varepsilon) \to (1/|\nabla \tilde{U}_i^{\pm}(x)|)$ for $i \in I^{\pm}$ as $\varepsilon \to 0+$. In view of (1) and the continuity of $f(x, s, t)$, it follows that $\phi_\varepsilon(0) \to 0$ as $\varepsilon \to 0+$. A similar argument shows that $\phi_\varepsilon(\alpha(\varepsilon)) \to 0$ as $\varepsilon \to 0+$, where we choose $\alpha(\varepsilon) \in J_\varepsilon$ such that $x(\alpha(\varepsilon)) \in T_\varepsilon(\tilde{\Gamma})$. Using the fact that $\partial g_i^{\pm}(\varepsilon, \alpha)/\partial\alpha = \pm 1$ for $i \in I^{\pm}$, and the fact that $\max\{g^{\pm}(\varepsilon, \alpha)/\varepsilon : \alpha \in J_\varepsilon\} = O(1)$ as $\varepsilon \to 0+$, we conclude from assumption (A3) that $\phi_\varepsilon'(\alpha) \geq (C_1/\varepsilon)$ uniformly over all $\alpha \in J_\varepsilon$, where C_1 is a positive constant. It follows that $(\alpha(\varepsilon)/\varepsilon) \to 0$ as $\varepsilon \to 0+$. These estimates are independent of $x \in \Gamma$.

3.14 Theorem Let $\tilde{\Gamma} \in X$ denote a classical solution of Problem 2.1 such that $\nabla \tilde{U}_i^{\pm}(x) \neq 0$ on $\tilde{\Gamma}$ for $i \in I^{\pm}$. Given any surface $\Gamma \in X$, let $(\Gamma_n)_{n=1}^{\infty}$ be the sequence of surfaces in X defined such that

$$\Gamma_1 = \Gamma, \quad \Gamma_{n+1} = T_{\varepsilon_n}(\Gamma_n), \quad n = 1, 2, 3, \cdots,$$

where $(\varepsilon_n)_{n=1}^{\infty}$ is a nullsequence of values in the interval $(0,1)$ such that $\sum\limits_{n=1}^{\infty}\varepsilon_n = \infty$. Then $\Gamma_n \to \tilde{\Gamma}$ as $n \to \infty$ in the sense that $\Delta(\Gamma_n, \tilde{\Gamma}) \to 0$ as $n \to \infty$.

Proof Apart from the more general context and the more general definition of the operators $T_\varepsilon : X \to X, 0 < \varepsilon < 1$, Theorem 3.14 is identical to [5], Theorem 3.1. Lemmas 3.12 and 3.13 in this section are also formally identical to Lemmas 3.8 and 3.9 of [5]. In fact the proof of Theorem 3.14 based on Lemmas 3.12 and 3.13 is identical to the proof of [5], Theorem 3.1 based on [5], Lemmas 3.8 and 3.9.

3.15 Lemma Let $\tilde{X} = \{\Gamma \in X : \tilde{\Gamma}_*^- \leq \Gamma \leq \tilde{\Gamma}_*^+\}$, where $\tilde{\Gamma}_*^{\pm} \in X$ are chosen such that for sufficiently small $\varepsilon_0 > 0$, we have that $T_\varepsilon : \tilde{X} \to \tilde{X}$ for $0 < \varepsilon \leq \varepsilon_0$. Then for any $\varepsilon \in (0, \varepsilon_0]$, there exists a value $\alpha = \alpha(\varepsilon) < 1$ such that $\Phi_\varepsilon^{\pm}(\lambda\Gamma) \leq \lambda^{\alpha}\Phi_\varepsilon^{\pm}(\Gamma)$ uniformly for all $\Gamma \in \tilde{X}$ and all $1 \leq \lambda \leq \tilde{\lambda} := \Delta(\tilde{\Gamma}_*^-, \tilde{\Gamma}_*^+)$.

Proof A proof for the " $-$ " case (generalizing [2], §3 and [5], §3), is given in [16], §5.1.

3.16 Theorem For any $\varepsilon \in (0, \varepsilon_0]$, the mapping $T_\varepsilon : \tilde{\mathbb{X}} \to \tilde{\mathbb{X}}$ is a contraction in the sense that

$$\Delta(T_\varepsilon(\Gamma_1), T_\varepsilon(\Gamma_1)) \leq \alpha \Delta(\Gamma_1, \Gamma_2) \text{ for all } \Gamma_1, \Gamma_2 \in \tilde{\mathbb{X}},$$

where $\alpha = \alpha(\varepsilon) < 1$. It therefore has a unique fixed point $\tilde{\Gamma}_\varepsilon \in \tilde{\mathbb{X}}$ such that $T_\varepsilon(\tilde{\Gamma}_\varepsilon) = \tilde{\Gamma}_\varepsilon$, and such that $\Delta(T_\varepsilon^n(\Gamma), \tilde{\Gamma}_\varepsilon) \leq (\alpha^n / (1 - \alpha)) \Delta(T_\varepsilon(\Gamma), \Gamma)$ for any $\Gamma \in \tilde{\mathbb{X}}$ and $n \in \mathbb{N}$.

Proof If $\Delta(\Gamma_1, \Gamma_2) = ln(\lambda)$ for $\Gamma_1, \Gamma_2 \in \tilde{\mathbb{X}}$ and $\lambda > 1$, then $\Gamma_1 \leq \lambda \Gamma_2$ and $\Gamma_2 \leq \lambda \Gamma_1$. It follows from Lemmas 3.8, 3.9, 3.10, and 3.15 that

$$T_\varepsilon(\Gamma_i) \leq T_\varepsilon(\lambda \Gamma_j) = \Psi_\varepsilon\left(\Phi_\varepsilon^-(\lambda \Gamma_j), \Phi_\varepsilon^+(\lambda \Gamma_j)\right)$$
$$\leq \Psi_\varepsilon\left(\lambda^\alpha \Phi_\varepsilon^-(\Gamma_j), \lambda^\alpha \Phi_\varepsilon^+(\Gamma_j)\right) \leq \lambda^\alpha \Psi_\varepsilon\left(\Phi_\varepsilon^-(\Gamma_j), \Phi_\varepsilon^+(\Gamma_j)\right) = \lambda^\alpha T_\varepsilon(\Gamma_j)$$

for $i = 1$, $j = 2$ and for $i = 2$, $j = 1$. Therefore $\Delta(T_\varepsilon(\Gamma_1), T_\varepsilon(\Gamma_1)) \leq \alpha \, ln(\lambda) = \alpha \Delta(\Gamma_1, \Gamma_2)$.

3.17 Remarks (a) If Problem 2.1 has a classical solution $\tilde{\Gamma} \in \mathbb{X}$ such that $\nabla \tilde{U}_i^\pm(x) \neq 0$ on $\tilde{\Gamma}$ for $i \in I^\pm$, then $\Delta(\tilde{\Gamma}, \tilde{\Gamma}_\varepsilon) \to 0$ as $\varepsilon \to 0+$. (b) The author conjectures that Theorem 3.16 can be made the basis for an existence proof for Problem 2.1 (under essentially the assumptions of this section). The idea is that any sequence of fixed points $(\tilde{\Gamma}_{\varepsilon_n})$ has a convergent subsequence (by Ascoli-Arzela), whose limit $\tilde{\Gamma} \in \tilde{\mathbb{X}}$ is (in some appropriate sense) a weak solution of Problem 2.1. It remains to show that $\tilde{\Gamma}$ is a classical solution.

4 CONVEXITY OF SOLUTIONS

4.1 Assumptions In this section, we continue to assume (A1)-(A5). Also, the following assumptions apply in Theorems 4.3, 4.4, and 4.5:

(A6) We set $k = m = 1$ (and simplify notation so that $p_1^\pm$, $U_1^\pm(x)$, $\Phi_{\varepsilon,1}^\pm(\Gamma)$, etc. become $p^\pm$, $U^\pm(x)$, $\Phi_\varepsilon^\pm(\Gamma)$, etc.).

(A7) We assume the given domains $D_*^\pm$ (and the surfaces $\tilde{\Gamma}_*^\pm$ defining $\tilde{\mathbb{X}}$) are convex.

(A8) We assume the function $f(x, s, t) := F(x, 1/s, 1/t)$ has the property that $\max\{\psi(\lambda): 0 \leq \lambda \leq 1\} \leq 0$ for any linear function $\phi(\lambda): [0,1] \to \Omega \times \mathbb{R}_+ \times \mathbb{R}_+$ such that $\psi(0) = \psi(1) = 0$, where $\psi(\lambda) := f(\phi(\lambda))$.

4.2 Definition We use $\mathbb{X}_C$ to denote the family of all surfaces $\Gamma \in \mathbb{X}$ such that the interior complement $D(\Gamma)$ is convex.

4.3 Theorem For any $0 < \varepsilon < 1$, we have that:
(a) $\Phi_\varepsilon^\pm : \mathbb{X}_C \to \mathbb{X}_C$.
(b) $\Psi_\varepsilon(\Gamma^-, \Gamma^+) \in \mathbb{X}_C$ for any (convex) surfaces $\Gamma^- < \Gamma^+$ in $\mathbb{X}_C$.
(c) $T_\varepsilon : \mathbb{X}_C \to \mathbb{X}_C$.

Proof of part (a) See J.L. Lewis [15]. (Related results are given in [12], [14].)

Proof of parts (b), (c) Part (c) follows from parts (a) and (b). Turning to the proof of part (b), let ε and $\Gamma^\pm$ be fixed, let $D^\pm = D(\Gamma^\pm)$, and let $\omega = D^+ \backslash Cl(D^-)$. Let

$\Gamma = \Psi_\varepsilon(\Gamma^-, \Gamma^+)$, so that $g(x) := f\big(x, (d(x, \Gamma^-)/\varepsilon), (d(x, \Gamma^+)/\varepsilon)\big) = 0$ for all $x \in \Gamma$. Then $\Gamma = \partial D$, where we define $D = Cl(D^-) \cup \{x \in \omega : g(x) < 0\}$. Given points $x_1, x_2 \in \Gamma$, let L denote the straight line-segment joining x_1 to x_2. To prove that D is convex, we will show that $L \subset Cl(D)$ whenever $x_1, x_2 \in \Gamma$ are sufficiently close to guarantee that $L \subset \omega$. It suffices to prove that $g(x) \leq 0$ on $L \subset \omega$, where $g(x_1) = g(x_2) = 0$. Given $x_1, x_2 \in \Gamma$, choose points $x_1^-, x_2^- \in \Gamma^-$ such that $r_i^- := |x_i - x_i^-| = d(x_i, \Gamma^-)$ for $i = 1, 2$, and let L^- denote the straight line-segment joining these points. Also choose maximum values r_1^+ and r_2^+ such that for $i = 1, 2$, the (open) ball of radius r_i^+ centered at x_i is contained in D^+, and let $S = \partial H$, where H denotes the convex hull of the union of these two balls. Clearly, we have $H \subset D^+$ and $L^- \subset Cl(D^-)$, due to the assumed convexity of $D^\pm$. Therefore, $d(x, \Gamma^-) \leq d(x, L^-)$ and $d(x, S^+) \leq d(x, \Gamma^+)$ for $x \in L$, so that

$$g(x) \leq \tilde{g}(x) := f\big(x, (d(x, L^-)/\varepsilon), (d(x, S^+)/\varepsilon)\big)$$

on L by assumption (A3), where $\tilde{g}(x_1) = \tilde{g}(x_2) = 0$. It is easily seen that

$$d((1-\lambda)x_1 + \lambda x_2, S^+) = (1-\lambda)r_1^+ + \lambda r_2^+,$$

$$d((1-\lambda)x_1 + \lambda x_2, L^-) \leq (1-\lambda)r_1^- + \lambda r_2^-,$$

both for all $0 \leq \lambda \leq 1$ (see [5], §4). Therefore, $\tilde{g}((1-\lambda)x_1 + \lambda x_2) \leq h(\lambda)$ for $0 \leq \lambda \leq 1$, due to assumption (A3), where

$$h(\lambda) := f\big([(1-\lambda)x_1 + \lambda x_2], ([(1-\lambda)r_1^- + \lambda r_2^-]/\varepsilon), ([(1-\lambda)r_1^+ + \lambda r_2^+]/\varepsilon)\big).$$

Observe that $h(0) = h(1) = 0$. We now conclude, by applying Assumption (A8), that $g((1-\lambda)x_1 + \lambda x_2) \leq h(\lambda) \leq 0$ for $0 \leq \lambda \leq 1$, so that D is convex.

4.4   Theorem   Let $\tilde{\Gamma}$ denote any classical solution of Problem 2.1 such that $\nabla \tilde{U}_i^\pm(x) \neq 0$ on $\tilde{\Gamma}$. Then $\tilde{\Gamma}$ is convex.

Proof Define the sequence $(\Gamma_n)_{n=1}^\infty$ as in Theorem 3.14, where $\Gamma \in \mathbb{X}_c$. Then $\Gamma_n \in \mathbb{X}_c$ for $n = 1, 2, 3, \cdots$, due to Theorem 4.3. Since $\Gamma_n \to \tilde{\Gamma}$ as $n \to \infty$ (due to Theorem 3.14), we conclude that $\tilde{\Gamma} \in \mathbb{X}_c$.

4.5 Theorem For any $\varepsilon \in (0, \varepsilon_0]$, the fixed point $\tilde{\Gamma}_\varepsilon$ of the mapping $T_\varepsilon : \tilde{\mathbb{X}} \to \tilde{\mathbb{X}}$ is convex.

4.6 Remark Let the restriction to $\Omega \times \mathbb{R}_+ \times \mathbb{R}_+$ of the equation $f(x, s, t) = 0$ be solved by $s = g(x, t) : \Omega \times \mathbb{R}_+ \to \mathbb{R}_+$. Then assumption (A8) is satisfied if for any linear mapping $\phi(\lambda) : [0, 1] \to \Omega$, the function $\psi(\lambda, t) := g(\phi(\lambda), t)$ satisfies $\psi_{\lambda\lambda} \leq 0$, $\psi_{tt} \leq 0$, and $\psi_{\lambda t}^2 \leq \psi_{\lambda\lambda}\psi_{tt}$.

4.7 Example Consider Problem 2.1, where we assume (A6) and (A7), where we choose $F(x, s, t) = A(x) + t^q - s^p$ (with $p, q > 0$), and where the function $A(x) : \mathbb{R}^N \to \mathbb{R}_+$ is strictly positive and twice continuously differentiable. The requirements (A3) and (A4) are satisfied automatically. A direct calculation based on the criteria in Remark 4.6 (where

$g(x,t) = \left(A(x) + t^{-q} \right)^{-(1/p)}$ shows that assumption (A8) is satisfied if $q \leq p$, and if the directional derivatives of $A(x)$ satisfy the condition: $(p+1) A_\nu^2 \leq p A A_{\nu\nu}$ in Ω for any vector ν. Assumption (A5) is satisfied if, for any $x_0 \in B_\delta(0)$ and $v \in \mathbb{R}^N$, the function $\phi(\lambda) := \lambda^p A(x_0 + \lambda v)$ is weakly increasing in $\lambda \geq 1$. Under these assumptions on $p, q,$ and $A(x)$, Theorem 4.4 shows that any classical solution $\tilde{\Gamma} \in X$ of Problem 2.1 (such that $\nabla \tilde{U}_i^{\pm}(x) \neq 0$ on $\tilde{\Gamma}$) must be convex, and Theorem 4.5 shows that the fixed points $\tilde{\Gamma}_\varepsilon \in \tilde{X}$, $\varepsilon \in (0, \varepsilon_0]$, are convex. (No relation is assumed between p and p^-, or between q and p^+.)

4.8 Remark Assume in Problem 2.1 that $k=1$, but m is arbitrary, and that $D_*^{\pm}$ are convex. Let $F(x,s,t) = \alpha + \sum_{i=1}^{m} \alpha_i t_i^{p_i} - s^p$, where $\alpha > 0, \alpha_i > 0$, and $p \geq p_i > 0$ for $i=1,\cdots,m$. (This automatically satisfies (A3),(A4),(A5).) Then any classical solution $\tilde{\Gamma} \in X$ of Problem 2.1 (with non-vanishing derivatives along $\tilde{\Gamma}$) or fixed point $\tilde{\Gamma}_\varepsilon$ of $T_\varepsilon \colon \tilde{X} \to \tilde{X}, 0 < \varepsilon < \varepsilon_0$, must be convex, due to Theorems 3.14, 3.16, and 5.11.

4.9 Remark In [6], the author proved that Problem 2.1 does not have any convex classical solution in a certain case in which the given domains $D_*^{\pm}$ are convex and $F(x,s,t) = t^2 - s^2 - \lambda^2$, with $\lambda > 0$ sufficiently large. This example shows that the condition $A(x) > 0$ plays a crucial role in the result in Example 4.7.

5 A VARIATIONAL APPROACH TO CONVEX SOLUTIONS

The primary purpose of this (largely self-contained) section is to discuss an alternative approach to Problem 2.1 in the context of convex variational inequalities, which could become the basis for an existence theorem. The application of these ideas is (at present) restricted to the following particular case of Problem 2.1:

5.1 Free-Boundary Problem Given the closed, convex surfaces $\Gamma_*^{\pm}$ (in $\mathbb{R}^N$, $N \geq 2$) such that $\Gamma_*^- < \Gamma_*^+$, let X_c denote the family of all convex closed surfaces Γ such that $\Gamma_*^- < \Gamma < \Gamma_*^+$ (where "$<$" refers to strict inclusion of the interior complements). For each $\Gamma \in X_c$, let $\Omega^{\pm}(\Gamma)$ denote the annular domains such that $\partial \Omega^{\pm}(\Gamma) = \Gamma \cup \Gamma_*^{\pm}$. For given constants $p, p_1, p_2, \cdots, p_m > 1$, and for any $\Gamma \in X_c$, let $U^-(\Gamma; x)$ denote the p-capacitary potential in $\Omega^-(\Gamma)$ and let $U_i^+(\Gamma; x)$ denote the p_i-capacitary potential in $\Omega^+(\Gamma)$ for $i=1,\cdots,m$ (oriented such that $U^-(\Gamma; x) = U_i^+(\Gamma; x) = 0$ on Γ). Given the constants $\alpha, \alpha_1, \alpha_2, \cdots, \alpha_m > 0$, we seek a (convex) surface $\Gamma \in X_c$ such that for all $x \in \Gamma$, we have

$$|\nabla U^-(\Gamma; x)|^p = \alpha + \sum_{i=1}^{m} \alpha_i |\nabla U_i^+(\Gamma; x)|^{p_i}. \qquad (8)$$

5.2 Extremal Problem Minimize the functional $I(\Gamma) \colon X_c \to \mathbb{R}$, where

$$I(\Gamma) = \int_{\Omega^-(\Gamma)} \left([|\nabla U^-(\Gamma; x)|^p/(p-1)] + \alpha \right) dx$$
$$+ \sum_{i=1}^{m} (\alpha_i/(p_i - 1)) \int_{\Omega^+(\Gamma)} |\nabla U_i^+(\Gamma; x)|^{p_i} dx. \qquad (9)$$

5.3 Remark Problem 5.1 has at most one classical solution $\Gamma \in \mathbb{X}_c$ such that Γ is a C^2-surface, due to Theorem 2.3.

5.4 Conjecture Assume in Problems 5.1, 5.2 that $p_1, \cdots, p_m \leq p$. Then
(a) Problem 5.2 has at least one solution $\Gamma \in \mathbb{X}_c$.
(b) Any solution $\Gamma \in \mathbb{X}_c$ of Problem 5.2 is a classical solution of Problem 5.1.

5.5 Discussion The author has proved (a multilayer generalization of) Conjecture 5.4 in the special case where $m=1$, $p=p_1=2$, and Γ_*^- is a C^2-surface (see [8] and [9], §5). The purpose of this section is to present a formal proof of Conjecture 5.4(b), valid under the additional assumption that the solution of 5.4(a) be sufficiently regular. This argument is based on a variational approach to the generalized operator method introduced in §3. (We ignore the relatively straightforward proof of 5.4(a).) The author's work in [8] gives indications of how to adapt the argument given here in order to rigorously prove the conjecture. In summary, we will prove the following:

5.6 Theorem Assume in Problems 5.1, 5.2 that $p_1, \cdots, p_m \leq p$. Then any sufficiently regular (see §5.12) solution Γ of Problem 5.2 is a classical solution of Problem 5.1.

5.7 Operator definitions For any $\Gamma \in \mathbb{X}_c$ and $0 < \varepsilon < 1$, let

$$\Gamma_\varepsilon^- = \Phi_\varepsilon^-(\Gamma) = \{x \in \Omega^-(\Gamma): U^-(\Gamma; x) = \varepsilon\} \tag{10}$$

and let

$$\Gamma_{\varepsilon,i}^+ = \Phi_{\varepsilon,i}^+(\Gamma) = \{x \in \Omega^+(\Gamma): U_i^+(\Gamma; x) = \varepsilon\}, \ i=1,\cdots,m. \tag{11}$$

In terms of (10) and (11), let

$$T_\varepsilon(\Gamma) = \Psi_\varepsilon(\Gamma_\varepsilon^-, \Gamma_{\varepsilon,1}^+, \cdots, \Gamma_{\varepsilon,m}^+) = \{x \in \omega_\varepsilon: (\varepsilon/d(x, \Gamma_\varepsilon^-))^p = \alpha + \sum_{i=1}^m \alpha_i (\varepsilon/d(x, \Gamma_{\varepsilon,i}^+))^{p_i}\}$$

for $\Gamma \in \mathbb{X}_c$ and $0 < \varepsilon < 1$, where $d(x, S) = \inf\{|x-y|: y \in S\}$ for any $x \in \mathbb{R}^N$ and any set $S \subset \mathbb{R}^N$. Here, $\omega_\varepsilon = \omega(\Gamma_\varepsilon^-, \Gamma_{\varepsilon,1}^+, \cdots, \Gamma_{\varepsilon,m}^+)$ denotes the annular domain which lies outside the surface Γ_ε^-, but inside all the surfaces $\Gamma_{\varepsilon,i}^+, i=1,\cdots,m$.

5.8 Lemma $\Phi_\varepsilon^-: \mathbb{X}_c \to \mathbb{X}_c$ and $\Phi_{\varepsilon,i}^+: \mathbb{X}_c \to \mathbb{X}_c$ for $0 < \varepsilon < 1$ and $i=1,\cdots,m$.

Proof This follows from the work of J. Lewis [15], which establishes the convexity of level surfaces of p-capacitary potentials (for arbitrary $p>1$) under convex conditions.

5.9 Lemma Assume $\Gamma^- < \Gamma_i^+, i=1,\cdots,m$, where $\Gamma^-, \Gamma_i^+ \in \mathbb{X}_c$. Then $\hat{\Gamma} := \Psi_\varepsilon(\Gamma^-, \Gamma_1^+, \cdots, \Gamma_m^+) \in \mathbb{X}_c$ for any $0 < \varepsilon < 1$.

Proof sketch For fixed $0 < \varepsilon < 1$, let

$$f(x) = \left((\varepsilon/d(x, \Gamma^-))^p - \sum_{i=1}^m \alpha_i (\varepsilon/d(x, \Gamma_i^+))^{p_i}\right)$$

in $\mathbb{R}^N$. Then $f(x) = \alpha$ on $\hat{\Gamma}$ and $f(x) > \alpha$ in the interior complement of $\hat{\Gamma}$ relative to

$\omega = \omega(\Gamma^-, \Gamma_1^+, \cdots, \Gamma_m^+)$. Choose points $x_1, x_2 \in \hat{\Gamma}$, and let L denote the straight line-segment joining them. We need to show that $f(x) \geq \alpha$ on L whenever the points $x_1, x_2 \in \hat{\Gamma}$ are sufficiently close together (so that $L \subset \omega$). But then

$$g(x_1) = g(x_2) = \alpha \text{ and } f(x) \geq g(x) \text{ for all } x \in L,$$

where we define

$$g(x) = \left((\varepsilon/d(x, L^-)) \right)^p - \sum_{i=1}^m \alpha_i (\varepsilon/d(x, S_i^+))^{p_i} \tag{12}$$

for all $x \in L$. In (12), L^- denotes the straight line-segment joining points $x_1^-, x_2^- \in \Gamma^-$, which are chosen such that $r_j^- := |x_j^- - x_j| = d(x_j, \Gamma^-), j = 1, 2$. Also, $S_i^+ = \partial H_i^+$ for each $i = 1, \cdots, m$. Here, H_i^+ denotes the convex hull of the union of two balls, a ball of radius $r_{i,1}^+$ centered at x_1, and a ball of radius $r_{i,2}^+$ centered at x_2, where $r_{i,1}^+$ and $r_{i,2}^+$ are maximum subject to the requirement that the balls both be contained in $D_i^+ := D(\Gamma_i^+)$. Also, for $0 \leq t \leq 1$ and $i = 1, \cdots, m$, we have

$$d(tx_1 + (1-t)x_2, S_i^+) = tr_{1,i}^+ + (1-t) r_{2,i}^+,$$

$$\rho(t) := d(tx_1 + (1-t)x_2, L^-) \leq tr_1^- + (1-t) r_2^- \text{ (since } \rho'' \geq 0\text{).}$$

Thus, if $x = tx_1 + (1-t)x_2$, then

$$h(0) = h(1) = \alpha \text{ and } f(x) \geq g(x) \geq h(t) \text{ for } 0 \leq t \leq 1,$$

where

$$h(t) = \left(\varepsilon/(tr_1^- + (1-t) r_2^-) \right)^p - \sum_{i=1}^m \alpha_i \left(\varepsilon/(tr_{1,i}^+ + (1-t) r_{2,i}^+) \right)^{p_i}.$$

At this point, the assertion can be obtained from the following lemma:

5.10 Lemma Let $\phi(t) = (At+B)^{-p} - \sum_{i=1}^m (A_i t + B_i)^{-p_i}$ for $0 \leq t \leq 1$, where $p \geq p_i > 1$ and $A, (A+B), A_i, (A_i + B_i) > 0$. Assume for a constant $\eta > 0$ that $\phi(0), \phi(1) \geq \eta$. Then:

(a) We have $\phi(t) > 0$ for $0 \leq t \leq 1$.

(b) We have $\phi''(t_0) < 0$ at any critical point $0 < t_0 < 1$. Thus, $\phi(t) \geq \eta$ for $0 \leq t \leq 1$.

Proof (part (a)) We have $\phi(t) > 0$ if and only if $\psi(S) < 1$. Here, $\psi(S) = \sum_{i=1}^m S^{-\lambda_i} (\alpha_i S + \beta_i)^{-p_i}$, where $S = (1/(At+B))$, $\lambda_i = p - p_i \geq 0$, $\alpha_i = ((AB_i - BA_i)/A)$, and $\beta_i = (A_i/A)$. A direct calculation shows that $\psi''(S) \geq 0$, from which the assertion follows.

Proof (part (b)) For fixed $0 < t < 1$, we have that

$$S^p \phi(t) = \left(1 - \sum_{i=1}^m (\theta_i/\lambda_i) \right), \tag{13}$$

$$S^{p+2} \phi''(t) = p(p+1) A^2 \left(1 - \sum_{i=1}^m \theta_i \lambda_i r_i \right), \tag{14}$$

where $S = (At+B)$, $S_i = (A_i t + B_i)$, $\Theta_i = (A_i p_i S^{p+1}/A p S_i^{p_i+1})$, $\theta_i = |\Theta_i|$, $\lambda_i = |A_i p_i S/A p S_i|$, and $r_i = [p(p_i+1)/p_i(p+1)]$ for $i = 1, \cdots, m$. One can show that if $\phi'(t) = 0$, then $\sum_{i=1}^m \Theta_i = 1$, whence $\sum_{i=1}^{m} \theta_i \geq 1$. Also, it follows from $1 < p_i \leq p$ that $r_i \geq 1$. Therefore, if $\phi'(t) = 0$, then

$$\sum_{i=1}^{m} (\theta_i/\lambda_i) \sum_{i=1}^{m} \theta_i \lambda_i r_i \geq \sum_{i=1}^{m} (\theta_i/\lambda_i) \sum_{i=1}^{m} \theta_i \lambda_i$$

$$= \sum_{i=1}^{m} \theta_i^2 + \sum_{i<j} \theta_i \theta_j \big((\lambda_i/\lambda_j) + (\lambda_j/\lambda_i)\big) \geq \sum_{i=1}^{m} \theta_i^2 + 2\sum_{i<j} \theta_i \theta_j = \Big(\sum_{i=1}^{m} \theta_i\Big)^2 \geq 1, \qquad (15)$$

since $\tau + (1/\tau) \geq 2$ for all $\tau > 0$. Since $\phi(t) > 0$, the assertion follows from (13), (14), (15).

5.11 Theorem $T_\varepsilon \colon \mathbb{X}_c \to \mathbb{X}_c$ for any $0 < \varepsilon < 1$.

5.12 Infinitesimal variations generated by T_ε Assume that $\Gamma \in C^2$, $\nabla U^- \in C^1(\Omega^- \cup \Gamma)$, $\nabla U^- \neq 0$ on Γ, $\nabla U_i^+ \in C^1(\Omega^+ \cup \Gamma)$, and $\nabla U_i^+ \neq 0$ on Γ, where $i = 1, \cdots, m$. For $0 < \varepsilon < 1$, we define the function $f_\varepsilon(x) \colon \Gamma \to \mathbb{R}$ such that

$$x + f_\varepsilon(x)\nu(x) \in T_\varepsilon(\Gamma), \qquad (16)$$

for each $x \in \Gamma$, where $\nu(x)$ denotes the exterior normal to Γ at $x \in \Gamma$, and where $|f_\varepsilon(x)|$ is minimum subject to (16). Then at each $x \in \Gamma$, we have

$$(\varepsilon/(b\varepsilon + f_\varepsilon))^p = \alpha + \sum_{i=1}^{m} \alpha_i(\varepsilon/(a_i\varepsilon - f_\varepsilon))^{p_i} + o(\varepsilon)$$

as $\varepsilon \to 0+$, where $B = B(x) = |\nabla U^-|$, $b(x) = 1/B(x)$, $A_i = A_i(x) = |\nabla U_i^+|$, and $a_i(x) = 1/A_i(x)$. It follows that

$$(f_\varepsilon(x)/\varepsilon) \to \phi(x) \qquad (17)$$

for each $x \in \Gamma$ as $\varepsilon \to 0+$, where

$$g(x, \phi(x)) = \alpha, \qquad (18)$$

and where we define

$$g(x, t) = (b(x) + t)^{-p} - \sum_{i=1}^{m} \alpha_i(a_i(x) - t)^{-p_i} \qquad (19)$$

for $-b(x) < t < a(x) := \min\{a_i(x) : i = 1, \cdots, m\}$.

5.13 Lemma Under the assumptions of §5.12, we have

$$\phi(x) = \Big(B^p(x) - \sum_{i=1}^{m} \alpha_i A_i^{p_i}(x) - \alpha\Big)/r(x), \qquad (20)$$

for each $x \in \Gamma$, where the value $r(x)$ satisfies at least one of the following two inequalities:

$$0 < r(x) \leq p\Big(\alpha + \sum_{i=1}^{m} \alpha_i A_i^{p_i}\Big)^{((p+1)/p)} + \sum_{i=1}^{m} \alpha_i p_i A_i^{p_i+1}, \qquad (21)$$

$$0 < r(x) \leq pB^{p+1} + \sum_{i=1}^{m} p_i (1/\alpha_i)^{1/p_i} B^{(p(p_i+1)/p_i)}. \qquad (22)$$

Proof For fixed $x \in \Gamma$ (which we suppress), it follows from (18) and the theorem of the mean that the value $\phi = \phi(x)$ satisfies the equation

$$|g'(\xi)|\phi = g(0) - g(\phi) = B^p - \sum_{i=1}^{m} \alpha_i A_i^{p_i} - \alpha, \qquad (23)$$

where ξ is a value between 0 and ϕ, and where

$$|g'(t)| = p(b+t)^{-p-1} + \sum_{i=1}^{m} \alpha_i p_i (a_i - t)^{-p_i - 1} \tag{24}$$

for $-b < t < a := \min\{a_i\}$. This verifies (20), where we set $r = r(x) = |g'(\xi)|$. Suppose that $-b < \phi \leq \xi \leq 0$. By applying (18) and (19), we obtain the inequality:

$$(b+\xi)^{-p-1} \leq (b+\phi)^{-p-1} = [(b+\phi)^{-p}]^{((p+1)/p)}$$

$$= \left(\alpha + \sum_{i=1}^{m} \alpha_i (a_i - \phi)^{-p_i}\right)^{((p+1)/p)} \leq \left(\alpha + \sum_{i=1}^{m} \alpha_i A_i^{p_i}\right)^{((p+1)/p)}. \tag{25}$$

Now (21) follows from (24), (25), and the fact that $(a_i - \xi)^{-p_i - 1} \leq A_i^{p_i + 1}$ for $i = 1, \cdots, m$. Now suppose, alternatively, that $0 \leq \xi \leq \phi < a$. Then $a_i - \xi \geq a_i - \phi > 0$, and it follows from (18) and (19) that

$$\alpha_i (a_i - \phi)^{-p_i} \leq (b+\phi)^{-p} - \alpha \leq B^p$$

for $i = 1, \cdots, m$. Therefore

$$\alpha_i p_i (a_i - \xi)^{-p_i - 1} \leq \alpha_i p_i (a_i - \phi)^{-p_i - 1}$$

$$= p_i (1/\alpha_i)^{1/p_i} \left(\alpha_i (a_i - \phi)^{-p_i}\right)^{((p_i+1)/p_i)} \leq p_i (1/\alpha_i)^{1/p_i} B^{(p(p_i+1)/p_i)} \tag{26}$$

for $i = 1, \cdots, m$. Now (22) follows from (24), (26), and the trivial inequality:

$$p(b+\xi)^{-p-1} \leq pB^{p+1}.$$

5.14 Lemma (variational formula) Let $\Gamma \in \mathbb{X}_c$ satisfy the assumptions in §5.12, and let the functions $A_i(x), B(x): \Gamma \to \mathbb{R}$ be as defined in §5.12. Then there exists a constant $M > 0$ such that for $\varepsilon \to 0+$, we have that

$$I(T_\varepsilon(\Gamma)) \leq I(\Gamma) - (\varepsilon/M) \int_\Gamma \left(\alpha + \sum_{i=1}^{m} \alpha_i A_i^{p_i}(x) - B^p(x)\right)^2 d\sigma + o(\varepsilon). \tag{27}$$

Proof sketch For small, smooth variations in a C^2-surface $\Gamma \in \mathbb{X}_c$, the corresponding first order variation in the functional $I(\Gamma)$ is given by $\delta I = \int_\Gamma \psi(x) \delta \nu(x) d\sigma$, where

$$\psi(x) = \left(\alpha + \sum_{i=1}^{m} \alpha_i A_i^{p_i}(x) - B^p(x)\right),$$

and where $\delta \nu(x)$ denotes the variation in Γ in the direction of the exterior normal. For the normal variation in Γ generated by T_ε, we have $\delta \nu = -(\psi(x)/r(x))\varepsilon$ to first order in ε, as follows from (17) and (20). Thus $\delta I = -\varepsilon \int_\Gamma (\psi^2(x)/r(x)) d\sigma$ to first order in ε. The assertion now follows from (21) and (22), which provide an upper bound for $r(x) > 0$.

5.15 Proof of Theorem 5.6 Let $\Gamma \in \mathbb{X}_c$ (having the regularity properties stated in §5.12) minimize the functional $I(\Gamma): \mathbb{X}_c \to \mathbb{R}$. Then $I(T_\varepsilon(\Gamma)) \geq I(\Gamma)$, due to Theorem 5.11. Since the functions $A_i(x), B(x): \Gamma \to \mathbb{R}$ are continuous, it follows from Lemma 5.14 that (8) holds pointwise on Γ.

REFERENCES

1. A. Acker, Free-boundary optimization – a constructive iterative method, *J. Appl. Math. Phys. (ZAMP) 30*: 886-900 (1979).

2. A. Acker, How to approximate the solutions of a certain free-boundary problem for the Laplace equation by using the contraction principle, *J. Appl. Math. Phys. (ZAMP) 32*: 22-33 (1981).

3. A. Acker, Interior free-boundary problems for the Laplace equation, *Arch. Rat'l. Mech. Anal. 75*: 157-168 (1981).

4. A. Acker, On the convexity of equilibrium plasma configurations, *Math. Meth. Appl. Sci. 3*: 435-443 (1981).

5. A. Acker, On the convexity and on the successive approximation of solutions in a free boundary problem with two fluid phases, *Commun. in P.D.E. 14*: 1635-1652 (1989).

6. A. Acker, On the nonconvexity in free boundary problems arising in plasma physics and fluid dynamics, *Comm. Pure Appl. Math. 42*: 1165-1174 (1989). Addendum, *44*: 869-872 (1991).

7. A. Acker, On the multi-layer fluid problem: regularity, uniqueness, convexity, and successive approximation of solutions, *Commun. in P.D.E. 16*: 647-666 (1991).

8. A. Acker, On the existence of convex classical solutions to multilayer fluid problems in arbitrary space dimensions, *Pacific J. Math.* (to appear).

9. A. Acker, Convex free boundaries and the operator method, in: *Variational Problems (A. Friedman and J. Spruck, Editors), IMA Volumes in Mathematics and its Applications #53, Springer Verlag* (1993).

10. A. Acker, L.E. Payne, G. Philippin, On the convexity of level curves of the fundamental mode in the clamped membrane problem and the existence of convex solutions in a related free-boundary problem. *J. Appl. Math. Phys. (ZAMP) 32*: 683-694 (1981).

11. H.W. Alt, L.A. Caffarelli, A. Friedman, Variational problems with two fluid phases and their free boundaries, *T.A.M.S. 282*: 431-461 (1984).

12. L.A. Caffarelli, A. Friedman, Convexity of solutions of semiliniar elliptic equations, *Duke Math, J. 52*: 431-456 (1985).

13. L.A. Caffarelli and J. Spruck, Convexity properties of some classical variational problems, *Commun. in PDE 7*: 1337-1379 (1982).

14. N. Korevaar, J. Lewis, Convex solutions to certain elliptic p.d.e.'s having constant rank hessians, *Arch. Rat'l. Mech. Anal. 97*: 19-32 (1987).

15. J.L. Lewis, Capacitary functions in convex rings, *Arch. Rat'l. Mech. Anal. 66*: 201-224 (1977).

16. R. Meyer, Approximation of solutions of free-boundary problems for the p-Laplace equation. *Doctoral dissertation, Dept. Math. and Stat., The Wichita State University* (1993).

17. A. Acker, A convex free boundary problem involving p-capacitary potentials (preliminary report). Abstract # 93T-35-138. *AMS Abstracts 90*, 701(1993).

Nonisothermal Semiconductor Systems

W. ALLEGRETTO Department of Mathematics, University of Alberta, Edmonton, Alberta T6G 2G1 Canada

H. XIE Department of Mathematics, University of Alberta, Edmonton, Alberta T6G 2G1 Canada

1. INTRODUCTION

The classical drift-diffusion semiconductor model is usually simulated under the assumption of thermal equilibrium, and consists in this case of Poisson's equation and the current-continuity equation for electrons and/or holes. In this instance, the current densities can usually be expressed as:(ref. [WAC] [COO])

$$\overrightarrow{J}_n = \mu_n n \nabla \varphi_n, \quad \overrightarrow{J}_p = \mu_p p \nabla \varphi_p \tag{1.1}$$

where

$n, p :$ electron and hole densities, respectively.

$\mu_n, \mu_p :$ mobilities of electrons and holes respectivly.

$\varphi_n, \varphi_p :$ quasi-Fermi levels, functions of the electric potential, φ; and n, p given by: $n = \exp\big((\varphi - \varphi_n)/T_n\big), \quad p = \exp\big((\varphi_p - \varphi)/T_p\big),$ where T_n, T_p are the electron and hole temperatures.

It is, however, well known that the classical model is not valid if the electric fields, current densities and concentration gradients become large, due to the fact that the

Research supported in part by the Natural Sciences and Engineering Research Council of Canada.

17

carriers can be "heated" by applied fields. Thus, the current densities (1.1) have to be supplemented by an additive term proportional to the gradient of their corresponding temperatures which act as an additional driving force ([CAL],1960):

$$\vec{J}_n = \mu_n n(\nabla \varphi_n - P_n \nabla T_n), \quad \vec{J}_p = -\mu_p p(\nabla \varphi_p - P_p \nabla T_p) \qquad (1.2)$$

The coefficients P_n, P_p are the thermoelectric power associated with the electron and hole system, respectively, and are approximated by constants in this paper.

In such cases, the temperature T has to be included as an additional dynamic state variable in the simulation of electric and thermal behavior. The new system now consists of the carrier energy transport equation (governing the temperature) along with the classical drift diffusion equations. After reducing mathematically irrelevant physical quantities to one, we arrive at the following system

$$\begin{cases} -\Delta \varphi = 1 - n + p, \\ -\nabla \big(\mu_n n(\nabla \varphi_n - P_n \nabla T_n) \big) = R, \\ -\nabla \big(\mu_p p(\nabla \varphi_p + P_p \nabla T_p) \big) = -R, \\ -\nabla \big(\kappa_n \nabla T_n \big) = H_n + c_n(T_0 - T_n), \\ -\nabla \big(\kappa_p \nabla T_p \big) = H_p + c_p(T_0 - T_p). \end{cases} \qquad (1.3)$$

Where H_n, H_p are the heat generation which take the following form:([CHR],1979)

$$\begin{cases} H_n = -\mu_n n(\nabla \varphi_n - P_n \nabla T_n) \cdot \nabla \varphi_n - RT_n, \\ H_p = -\mu_p p(\nabla \varphi_p + P_p \nabla T_p) \cdot \nabla \varphi_p + RT_p, \end{cases} \qquad (1.4)$$

and $c_n(T_0 - T_n)$ and $c_p(T_0 - T_p)$ are the heat losses to the lattice, with T_0 representing the lattice temperature, and κ_n (resp. κ_p) is the thermal conductivity which we assume equal to $nT\mu_n$ (resp. $pT\mu_p$) according to the Franz-Wiedemann law. We recall that semiconductor systems with temperature effects were considered by Seidman and Troianiello [STR] and Seidman [SEI] in the situation where no temperature gradient is involved and the mobilities are constants. In system (1.3), $R = \delta F(\varphi, \varphi_n, \varphi_p, T_n, T_p, T_0)$ denotes generation-recombination and its explicit form is not as well determined as for the case $T_n = T_p = T_0$. In particular, the commonly employed method of showing existence using constant upper-lower solutions no longer seems applicable since it depends on $T_n = T_p = T_0$ and the presence of the term $(1 - np)$ in R which is known to be the case at thermal equilibrium, but is not obvious here. We only assume R is a smooth function of all its variables. In compensation we can only deal with small δ. If the standard models for R are assumed, this corresponds to sufficiently large carrier lifetimes.

Neither system (1.3), nor the more special problem with a zero recombination-generation term, have been discussed mathematically to the best of our knowledge. The main difficulties in the analysis are the presence of the gradient of the quasi-Fermi levels and the degeneracy of the system due to the fact that the mobilities strongly depend on the temperature and go to zero as the temperature tends to infinity.

In this paper, we study the system (1.3) with the associated mixed boundary conditions:

$$
\begin{cases}
\varphi(x) = \varphi_0(x), \quad n(x) = n_0(x), \\
\qquad \text{and} \quad T_n(x) = T_p = T_0; & \text{on } \Gamma_D; \\
\frac{\partial \varphi}{\partial \nu} = 0, \quad \overrightarrow{J}_n \cdot \overrightarrow{\nu} = 0, \quad \text{and} \quad \overrightarrow{J}_p \cdot \overrightarrow{\nu} = 0; \\
nT\mu_n \frac{\partial T_n}{\partial \nu} + \gamma(T_n - T_0) = 0, \\
\qquad \text{and} \quad pT\mu_p \frac{\partial T_p}{\partial \nu} + \gamma(T_p - T_0) = 0; & \text{on } \Gamma_N;
\end{cases}
\tag{1.6}
$$

These conditions are more realistic than the standard Dirichlet or Neumann conditions. The strategy here is first to consider the zero generation-recombination case. By Degree Theory, we prove that there is a solution in a bounded open set in a suitable Banach space. We observe that the equations for electrons and holes in the system essentially decouple if $R = 0$, so that we need only treat the equations for the electrons explicitly in this case. The same procedures work for the full system. Next, for the case $R \neq 0$, Degree Theory can again be applied to obtain an existence result for the general system (1.3) and (1.6), if the parameter δ is small enough.

We assume the following throughout this paper:

Assumptions

A1: Ω is a smooth domain in R^N, $N \geq 2$, $\partial\Omega = \Gamma_D \cup \Gamma_N$ and Γ_D and Γ_N are Lipshitz with Γ_D closed and Γ_N open in $\partial\Omega$. Specific conditions on Γ_D, Γ_N, and $\overline{\Gamma}_D \cap \overline{\Gamma}_N$ may be found in [13].

A2: T_0, c_n, c_p, P_n, P_p and γ are positive constants.

A3: In the generation-recombination term, δ is a small positive parameter and the function R is Lipshitzian in its variables.

A4: The mobilities depend only on the temperatures and are smooth positive functions. Physically, the mobilities will go to zero as the temperatures tend to infinity. Without loss of generality, we may take the mobilities to be also defined for negative temperatures.

A5: All the Dirichlet boundary data admit a smooth extension to $\overline{\Omega}$.

We call a set $(\varphi, \varphi_n, \varphi_p, T_n, T_p)$ of functions in $H^1(\Omega)$ a solution of system (1.3)(1.6), if the functions satisfy the boundary conditions given in (1.6) and the equations in (1.3) hold weakly in $H^1(\Omega)$.

Our main result is the following:

EXISTENCE THEOREM. *If δ is sufficiently small, there exists at least one $C^\alpha(\overline{\Omega})$ solution of (1.3)-(1.6), for some $0 < \alpha < 1$.*

2. EXISTENCE THEOREM IN THE CASE R = 0

As mentioned in the previous section, estimates here are explicitly given only for the equations related to electrons. Similar procedures will work for the hole equations and thus for the full system.

Let

$$\widehat{\varphi}_n = \varphi_n - P_n(T_n - T_0). \tag{2.1}$$

and, let $\widehat{\varphi}_0 = \widehat{\varphi}_n|_{\Gamma_D}$. By assumption (A3), the function $\widehat{\varphi}_0$ admits a Lipshitzian extension on $\overline{\Omega}$. Furthermore, we can assume that $\widehat{\varphi}_0$ is a positive function. In fact if need be, let M be a constant which is greater than $(\max(\ln n_0) + P_n) \cdot T_0$ and replace φ, φ_n and $\widehat{\varphi}_n$ by $\varphi + M, \varphi_n + M$ and $\widehat{\varphi}_n + M$ in (2.1).

THEOREM 1. *For any bounded solution* $(\varphi, \widehat{\varphi}_n, \widehat{\varphi}_p, T_n, T_p)$ *with* $T_n, T_p > 0$*, there are two positive constants m and M, such that*

$$m \leq \varphi(x), \widehat{\varphi}_n(x), \widehat{\varphi}_p(x), T_n(x), T_p(x) \leq M \quad \text{on} \quad \overline{\Omega}. \tag{2.2}$$

Proof: Clearly, system (1.3) is uniformly elliptic for such a solution. A boot strap argument and results in [11] show that the solution is classical except at $\overline{\Gamma}_D \cap \overline{\Gamma}_N$, and we notice that the current continuity equation for electrons becomes:

$$\nabla(n\mu_n \nabla \widehat{\varphi}_n) = 0. \tag{2.3}$$

By the maximum principle, $\widehat{\varphi}_n$ can only attain its extrema on the Dirichlet boundary. Thus, there are two positive constants m_1, M_1, such that

$$m_1 \leq \widehat{\varphi}_n(x) \leq M_1, \qquad \text{on} \quad \overline{\Omega}.$$

Next in order to find an a priori bound for the temperature function T_n, we recall that T_n satisfies

$$-\nabla\left(nT_n\mu_n\nabla T_n\right) = n\mu_n\nabla\widehat{\varphi}_n \cdot \nabla\varphi_n + c(T_0 - T_n) \tag{2.4}$$

It is easy to prove that $T_n \geq m_2$ for some positive constant m_2, by the maximum principle. In order to prove the existence of a maximal bound for T_n, we replace φ_n in the above equation by $\widehat{\varphi}_n + P_nT_n$ and notice that

$$\nabla\left(n\mu_n\widehat{\varphi}_n\nabla\widehat{\varphi}_n\right) = n\mu_n\nabla\widehat{\varphi}_n \cdot \nabla\widehat{\varphi}_n \tag{2.5}$$

whence

$$-\nabla\left(n\mu_n(T_n\nabla T_n + \widehat{\varphi}_n\nabla\widehat{\varphi}_n)\right) = n\mu_nP_n\nabla\widehat{\varphi}_n\nabla T_n + c(T_0 - T_n).$$

Then the function $T_n^2 + \widehat{\varphi}_n^2$ satisfies the following boundary value problem:

$$\begin{cases} -\nabla\left(n\mu_n\nabla(T_n^2 + \widehat{\varphi}_n^2)\right) = n\mu_nP_n\nabla\widehat{\varphi}_n\nabla T_n + c(T_0 - T_n), & \text{in} \quad \Omega, \\ T_n^2 + \widehat{\varphi}_n^2 = T_0^2 + \widehat{\varphi}_0^2, & \text{on} \quad \Gamma_D, \\ n\mu_n\frac{\partial(T_n^2 + \widehat{\varphi}_n^2)}{\partial n} + 2\gamma(T_n - T_0) = 0, & \text{on} \quad \Gamma_N. \end{cases}$$

If $\max(T_n^2 + \widehat{\varphi}_n^2)$ is attained at some point $x^* \in \Omega$, then

$$-\nabla\left[n\mu_n(T_n\nabla T_n + \widehat{\varphi}_n\nabla\widehat{\varphi}_n)\right] \geq 0$$

and $T_n \nabla T_n + \widehat{\varphi}_n \nabla \widehat{\varphi}_n = 0$ there, whence it follows that

$$\nabla T_n \cdot \nabla \widehat{\varphi}_n \leq 0 \qquad \text{at} \quad x^* \in \Omega.$$

This implies that $T_0 - T_n \geq 0$ i.e $T_n \leq T_0$ at x^*, and consequently, that

$$T_n \leq (T_0^2 + M_2^2)^{1/2} \quad \text{on} \quad \overline{\Omega}.$$

If $\max (T_n^2 + \widehat{\varphi}_n^2)$ is attained at some point x^* on $\partial\Omega$, then if x^* on Γ_D, $T_n \leq (T_0^2 + \widehat{\varphi}_0^2)^{1/2}$, and if $x^* \in \Gamma_N$, $\frac{\partial(T_n^2 + \widehat{\varphi}_n^2)}{\partial\nu} \geq 0$ at x^*, which implies that $T_n \leq T_0$. Thus

$$\max(T_n) \leq M_2 = (T_0^2 + M_1^2)^{1/2}. \tag{2.6}$$

The same procedure yields that $\widehat{\varphi}_p, T_p$ are bounded above and below by two positive constants.

In order to obtain bounds for the potential function φ, one only needs to note that the right hand side of the potential equation is a monotonic decreasing function for φ. By directly applying the maximum principle, one easily gets the estimate:

$$m_3 \leq \varphi(x) \leq M_3$$

where $m_3 = \min\{\min\varphi_n, \min\varphi_0\}$ and $M_3 = \max\{\max(\varphi_n + (\ln 2)T_n), \max\varphi_p, \max\varphi_0\}$. Theorem 1 is proved by letting $m = \min\{m_1, m_2, m_3\}$, and $M = \max\{M_1, M_2, M_3\}$.

Now, let us define a smooth and monotonic increasing function g(t) as follows:

$$g(t) = \begin{cases} \frac{m}{2}, & \text{if} \quad t \leq \frac{m}{2} \\ t, & \text{if} \quad \frac{3m}{4} \leq t . \end{cases}$$

DEFINITION. Let $(\varphi', \widehat{\varphi}_n', \widehat{\varphi}_p', T_n', T_p')$ be in $L^\infty(\Omega)$ and $\lambda \in [0,1]$. We define a map $G_{\lambda,0} : L^\infty(\Omega) \to L^\infty(\Omega)$ by letting

$$(\varphi, \widehat{\varphi}_n, \widehat{\varphi}_p, T_n, T_p) = G_{\lambda,0}(\varphi', \widehat{\varphi}_n', \widehat{\varphi}_p', T_n', T_p')$$

as the unique solution of the following linearized problem P_λ:

$$\begin{cases} -\Delta\varphi = \lambda(1 - \exp((\varphi' - \varphi_n')/g(T_n')) + \exp((\varphi_p' - \varphi')/g(T_p'))), & \text{in} \quad \Omega, \\ -\nabla(n'\mu_n'\nabla\widehat{\varphi}_n) = 0, & \text{in} \quad \Omega, \\ -\nabla(n'\mu_n'g(T_n')\nabla T_n) \\ \quad = n'\mu_n'\nabla\widehat{\varphi}_n\nabla(\widehat{\varphi}_n + P_n(g(T_n') - T_0)) + c_n(T_0 - T_n), & \text{in} \quad \Omega; \\ \varphi(x) = \min\varphi_0(x) + \lambda(\varphi_0(x) - \min\varphi_0(x)), & \text{on} \quad \Gamma_D, \\ \widehat{\varphi}_n(x) = \min\widehat{\varphi}_0(x) + \lambda(\widehat{\varphi}_0(x) - \min\widehat{\varphi}_0(x)), & \text{on} \quad \Gamma_D, \\ \text{and} \quad T_n(x) = T_0, & \text{on} \quad \Gamma_D, \\ \frac{\partial\varphi}{\partial\nu} = 0, \frac{\partial\widehat{\varphi}_n}{\partial\nu} = 0, \text{and} \quad n'g(T_n')\mu_n'\frac{\partial T_n}{\partial\nu} + \gamma(T_n - T_0) = 0, & \text{on} \quad \Gamma_N. \end{cases}$$

Equations for holes are similarly defined.

where

$$n' = \exp\left((\varphi' - \varphi'_n)/g(T'_n)\right),$$
$$\mu'_n = \mu_n(g(T'_n)),$$
$$\varphi'_n = \widehat{\varphi}'_n + P_n(g(T'_n) - T_0),$$

It is not difficult to see that the map $G_{\lambda,0}$ is a well defined operator from $L^\infty(\Omega)$ to $L^\infty(\Omega)$. We are going to prove that $G_{\lambda,0}$ is a continuous and compact operator for all $\lambda \in [0,1]$ and then we may apply Degree theory to obtain an existence result for Problem (1.3)(1.6). We first observe the following connection with the original problem:

COROLLARY 1.   *Any fixed point of $G_{\lambda,0}$ satisfies the bounds given in Theorem 1. Consequently, when $\lambda = 1$, any fixed point is a bounded solution of Problem (1.3)(1.6).*

Proof: Replace T_n^2 by $2\int^{T_n} g(t)dt$ in the proof of Theorem 1, and note that $g'(T) \geq 0$. If $2\int^T g(t)dt + \widehat{\varphi}_n^2$ attains its maximum at some point x^*, we also have that $\nabla g(T) \cdot \nabla \widehat{\varphi}_n \leq 0$. The remainder is as in the proof of Theorem 1.

LEMMA 1.   *The weak solution $(\varphi, \widehat{\varphi}_n, \widehat{\varphi}_p, T_n, T_p) \in H^1(\Omega)$ of the linearized Problem P_λ satisfies the following estimates:*

$$\|\varphi\|_{H^1} \leq C(\|\Phi_0\|_{H^1} + \|1 - n' + p'\|_{L^\infty}),$$

$$\|\widehat{\varphi}_n\|_{H^1} \leq C\|\widehat{\Phi}_0\|_{H^1},$$

$$\|T_n\|_{H^1} \leq C(\|\widehat{\varphi}_n\|_{H^1} + T_0),$$

and there exit two constants δ_0, μ with $0 < \delta_0 < 1, N - 2 + \delta_0 \leq \mu < N$, such that

$$\|\nabla\varphi\|_{L^{2,\mu}} \leq C(\|\Phi_0\|_{L^{2,\mu}} + \|\varphi\|_{H^1}),$$

$$\|\nabla\widehat{\varphi}_n\|_{L^{2,\mu}} \leq C(\|\widehat{\Phi}_0\|_{L^{2,\mu}} + \|\widehat{\varphi}_n\|_{H^1}),$$

$$\|\nabla T_n\|_{L^{2,\mu}} \leq C(\|\widehat{\varphi}_n\|_{L^{2,\mu}} + T_0 + \|T_n\|_{H^1}).$$

where the constant C depends only on the bounds of $(\varphi', \widehat{\varphi}'_n, \widehat{\varphi}'_p, T'_n, T'_p)$, and is independent of λ and $\Phi_0, \widehat{\Phi}_0$ are Lipshitz extention of the function $\varphi_0, \widehat{\varphi}_0$ respectively on $\overline{\Omega}$. The same estimates hold for the variables related to holes.

Proof: First notice that the linearized system P_λ is uniformly elliptic and that the following identity holds:

$$n'\mu'_n\nabla\widehat{\varphi}_n\nabla(\widehat{\varphi}_n + P_n(g(T'_n) - T_0)) = \nabla\left(n'\mu'_n(\widehat{\varphi}_n + P_n(g(T'_n) - T_0)\nabla\widehat{\varphi}_n\right)$$

Writing the weak form of the linearized system, we can apply the standard Sobolev's estimates and $L^{2,\mu}$ estimates ([TRO], and/or [XIE]) to obtain the above estimates.

Furthermore, by employing the embedding theorem of $L^{2,\mu}$ regularity theory [TRO], one has the following $C^\alpha(\overline{\Omega})$ estimates for the system.

LEMMA 2. *There exist two constants* $K, \alpha \in (0,1)$, *such that any solution of* P_λ *will satisfy the following estimates:*

$$\|\varphi\|_{C^\alpha(\overline{\Omega})}, \quad \|\widehat{\varphi}_n, \widehat{\varphi}_p\|_{C^\alpha(\overline{\Omega})} \quad \text{and} \quad \|T_n, T_p\|_{C^\alpha(\overline{\Omega})} \le K. \tag{2.7}$$

Routine calculations then show:

LEMMA 3. *For any* $\lambda \in [0,1]$, *the map* $G_{\lambda,0}$ *maps a bounded subset of* $L^\infty(\Omega)$ *into a bounded subset of* $L^{2,\mu}(\Omega)$, *and then into a bounded subset of* $C^\alpha(\overline{\Omega})$. *Consequently the map* $G_{\lambda,0}$ *is a continous, compact map from* $[0,1] \times L^\infty(\Omega)$ *to* $L^\infty(\Omega)$.

Proof of Existence Theorem in the case $R = 0$: Let M be the constant in the Theorem 1 and let $S = \{(\varphi, \widehat{\varphi}_n, \widehat{\varphi}_p, T_n, T_p) \in L^\infty(\Omega)\}$, such that $\|(\varphi, \widehat{\varphi}_n, \widehat{\varphi}_p, T_n, T_p)\|_{L^\infty} < M+1\}$ be an open subset of $L^\infty(\Omega)$. Define $Z_\lambda = I - G_{\lambda,0}, \lambda \in [0,1]$. It follows that there is no solution of $Z_\lambda = 0$ on ∂S, since any such solution is bounded and thus must satisfy the bounds stated in Corollary 1. Now, by Lemma 2 and Lemma 3 and Degree theory [NIR], one then has:

$$\text{degree}\,(Z_\lambda, S, 0) = \text{degree}\,(Z_0, S, 0) \qquad \text{for all} \quad \lambda \in [0,1]$$

But, when $\lambda = 0$, one can easily prove the Degree is one, that is:

$$\text{degree}\,(Z_\lambda, S, 0) = \quad \text{degree}\,(Z_0, S, 0) = 1$$

The Existence Theorem follows by letting $\lambda = 1$.

3. EXISTENCE THEOREM IN THE CASE R $\neq$ 0

In the same way as in the previous section, we define a map $G_{\lambda,\delta} : R^+ \times R^+ \times L^\infty(\Omega) \to L^\infty(\Omega)$ by letting

$$(\varphi, \widehat{\varphi}_n, \widehat{\varphi}_p, T_n, T_p) = G_{\lambda,\delta}(\varphi', \widehat{\varphi}'_n, \widehat{\varphi}'_p, T'_n, T'_p)$$

as the unique solution of the following linearized problem:

$$
\begin{cases}
-\Delta\varphi = \lambda\{1 - \exp\big((\varphi' - \varphi'_n)/g(T'_n)\big) + \exp\big((\varphi'_p - \varphi')/g(T'_p)\big)\}, & \text{in} \quad \Omega, \\[4pt]
-\nabla\big(n'\mu'_n\nabla\widehat{\varphi}_n\big) = R', & \text{in} \quad \Omega, \\[4pt]
-\nabla\big(n'\mu'_n g(T'_n)\nabla T_n\big) \\[4pt]
\quad = n'\mu'_n\nabla\widehat{\varphi}_n\nabla(\widehat{\varphi}_n + P_n(g(T'_n) - T_0)) + c_n(T_0 - T_n) - R'T'_n, & \text{in} \quad \Omega; \\[4pt]
\varphi(x) = \min\varphi_0(x) + \lambda(\varphi_0(x) - \min\varphi_0(x)), & \text{on} \quad \Gamma_D, \\[4pt]
\widehat{\varphi}_n(x) = \min\widehat{\varphi}_0(x) + \lambda(\widehat{\varphi}_0(x) - \min\widehat{\varphi}_0(x)), & \text{on} \quad \Gamma_D, \\[4pt]
\text{and} \quad T_n(x) = T_0, & \text{on} \quad \Gamma_D, \\[4pt]
\frac{\partial\varphi}{\partial\nu} = 0, \frac{\partial\widehat{\varphi}_n}{\partial\nu} = 0, \text{and} \quad n'g(T'_n)\mu'_n\frac{\partial T_n}{\partial\nu} + \gamma(T_n - T_0) = 0, & \text{on} \quad \Gamma_N.
\end{cases}
$$

Equations for holes are similarly defined.

where n', μ'_n, φ'_n and the function $g(t)$ are as in the previous section, and
$R' = R(\varphi', \widehat{\varphi}'_n, \widehat{\varphi}'_p, g(T'_n), g(T'_p))$.

By Assumption (A4), the function R' is Lipshitz continuous, and one can easily prove that the mapping $G_{1,\delta}$ is a continuous, compact operator from $R^+ \times L^\infty(\Omega)$ to $L^\infty(\Omega)$.

Let $Q \subset S$ denote the set of solutions found in the previous section for the case $R = 0$, and let S' denote an ϵ neighbourhood of Q. We then have degree$(I - G_{1,0}, S', 0) \neq 0$ and without loss of generality, let ϵ be small enough so that $g(T_n) = T_n, g(T_p) = T_p$ for all $(\varphi, \varphi_n, \varphi_p, T_n, T_p) \in S'$. It follows that

$$\text{degree } (I - G_{1,\delta}, S', 0) \neq 0, \quad \text{for sufficiently small } \delta > 0.$$

since the function F is uniformly continuous in the set $\overline{S}$. This proves the main Existence Theorem given in the Section 1.

REFERENCES

1. H. B. Callen, *Thermodynamics*, Wiley, New York, London, (1960)

2. A. Chryssafis and W. Love, A computer-aided analysis of one dimensional thermal transients in $n-p-n$ power transistors, *Solid-state electronics*, 22: 249–256 (1979).

3. R. K. Cook, Numerical simulation of hot-carrier transport in silicon bipolar transistors, *IEEE* trans. on electron devices. ED-30: 1103–1110 (1983).

4. P. A. Markowich, C. A. Ringhofer and C. Schmeiser, *Semiconductor Equations*, Springer, (1990).

5. L. Nirenberg, *Topics in Nonlinear Functional Analysis*, Courant Institute of Mathematical Sciences, (1974).

6. T. I. Seidman and G. M.Troianiello, Time-dependent solutions of a nonlinear system arising in semiconductor theory, *Nonlinear Analysis, Theory, Method and Applications*, 9: 1137–1157 (1985).

7. T. I. Seidman, Time-dependent solutions of a nonlinear system arising in semiconductor theory–II. Boundedness and Periodicity, *Nonlinear Analysis, Theory, Method and Applications*, 10: 491–502 (1986).

8. S. Selberherr, *Analysis and Simulation of Semiconductor Devices*, Springer (1984).

9. R. Stratton, Diffusion of hot and cold electrons in semiconductor barries, *Physics Review*, 6: 2002–2014 (1962).

10. T. W. Tang, H. H. Ou and D. H. Navon, Prediction of velocity overshoot by a nonlocal hot-carrier transport model, Proceedings of NASECODE IV, J. Miller, Editor, Boole Press, Dublin 1985, 519–522 (1990).

11. G. M. Troianiello, *Elliptic Differential Equations and Obstacle Problem*, Plenum Press (1987).

12. G. Wachutka, Rigorous thermodynamic treatment of heat generation and conduction in semiconductor device modeling, in *Simulation of Semiconductor Devices and Processes*, G. Baccarani and M. Rudan, Editors, Tecnoprint, Bologna, Italy, 3: 26–28 (1988).

13. H. Xie, $L^{2,\mu}(\Omega)$ estimate to the mixed boundary value problem for second order elliptic equations and application in the thermistor problems, preprint (1992).

A Model for the Growth of the Subpopulation of Lawyers

JOHN V. BAXLEY and PETER A. CUMMINGS, Department of Mathematics and Computer Science, Wake Forest University, Winston-Salem, NC 27109

1 INTRODUCTION

What factors determine the percentage of lawyers in a given society at a given time? Is there some optimum portion of a population which should be in the legal profession? Comparing two different nations, say Japan and the United States, what accounts for the difference in the subpopulation of lawyers?

In this paper, we shall use the term "lawyer" in a generic way to mean any person whose livelihood depends primarily on the legal profession. So we include not only trial lawyers, civil lawyers, corporate lawyers, judges, etc., but also the paralegals, secretaries, janitors, and others whose jobs depend primarily on the existence of the lawyers. We do not include the lawmakers, for example, presidents and congressmen, although we recognize that the population of lawmakers and the population of lawyers may have a nontrivial intersection.

We propose below a somewhat abstract and stylized model for the dynamic behavior of the subpopulation of lawyers. The model is a system of two nonlinear differential equations. We seek to capture those factors which determine the number of actual lawyers in a society with a small number of functions which are assumed given and characteristic of the society in question, much like a spring constant is characteristic of a particular spring. The difference is that these functions, which play the role of parameters in our model, are not really capable of measurement, and are more like utility functions of individual consumers in the economic theory of consumer behavior.

The basic assumptions of our model can be described somewhat imprecisely as follows. We assume that some lawyers are beneficial to society, but that as the proportion of lawyers grows beyond a certain point, lawyers are less beneficial and if the numbers become larger still, the society would be better off with no lawyers. We assume that the society can support more lawyers than is ideal for the society, and that the proportion of lawyers in the society will grow towards the larger value unless some internal restraints are placed on the growth of the lawyer population.

We analyze the model with elementary methods, primarily involving linearizing about equilibria. This background material is discussed in most basic textbooks on differential equations, for example [3, Chapter 9], [4, Chapter 2], [5, Chapter 2]. We basically look for equilibria and determine (and interpret) conditions under which equilibria are stable or unstable. We are particular interested in the existence of limit cycles.

Our motivation for this work was the paper by Baggs and Freedman [1]. The question considered there concerned the chance of long term survival of a second language (such as French) in a culture dominated by a different primary language (such as English). Baggs and Freedman modeled the problem using a system of differential equations and the analysis led to rather interesting conclusions. Our problem is very different, but shares with Baggs and Freedman (1990) the interest in applying stability theory to disciplines different from the usual applied areas: the physical sciences, biology, and economics.

2 THE MODEL

Let $y(t)$ be the population density of people in a given society at time t, and let $x(t)$ be the population density of lawyers within the population at time t. Our model for the dynamic behavior of these populations is the following two dimensional system of differential equations.

$$x' = r\left[1 - \frac{x}{h(y)} - C\left(\frac{x}{g(y)}\right)\right]x \qquad (a)$$

$$y' = s\left[1 - \frac{y}{k} + f\left(\frac{x}{g(y)}\right)\right]y \qquad (b)$$

$$(1)$$

Look first at the second equation. If the function f is identically zero, we have the usual logistic equation for the general population $y(t)$. The function $g(y)$ is interpreted as the optimal population of lawyers for a given population y. This function is assumed to be independent of t, but may vary from one society to another; it is thus a characteristic of a given society and measures in some sense that society's need for lawyers. Therefore the ratio $x/g(y)$ which appears in both equations in (1) is the ratio of the actual number of lawyers to the optimal number of lawyers. The functions f and C are functions of this ratio. We thus think of f as measuring the benefit of lawyers to the general population y. We assume $f(0) = 0$, so that k is the carrying capacity of the environment in the absence of lawyers. We naturally assume that f increases from 0 to an absolute maximum at $t = 1$, and decreases thereafter. We assume the existence of $u_1 > 1$ at which $f(u_1) = 0$ and that $f(u) < 0$ for $u > u_1$.

The benefit of lawyers to the general population y can be seen more clearly as follows: assume that $u = x/g(y)$ is fixed in time. Then (1b) can be re-written as

$$y' = s(1 + f(u))\left[1 - \frac{y}{k(1 + f(u))}\right]y$$

This is just the logistic equation again, but the growth rate is now $s(1 + f(u))$ and the carrying capacity is $k(1 + f(u))$. Thus if the ratio u is between 0 and u_1, lawyers have a positive beneficial effect on society, but for $u > u_1$, the benefit is negative. The underlying idea here is that a society without lawyers is a society without "law and order" and social friction is not controlled. The presence of a moderate number of lawyers leads to a beneficial control of such social friction. But excessive numbers of lawyers leads to the need to create a demand for their services, thus "manufacturing" social friction which requires litigation.

If the function $f(u)$ falls below -1 after some value u_2, then it is clear that the population y will decrease regardless of the size of k. In this case, large lawyer populations would seem to drive the general population y to extinction.

Look now at (1a). If $C(u)$ is identically zero, we again have a logistic-like equation, where $h(y)$ plays the role of a varying carrying capacity. We interpret $h(y)$ as the maximum lawyer population that can be supported by a population level y. Certainly $h(y) < y$. We also assume that $g(y) < h(y)$; the optimal number of lawyers is surely less than the maximum number that the society can support. For small population levels where y is significantly below carrying capacity k, it seems reasonable to suppose that g and h are increasing functions of y, perhaps almost linear; with $h(y) = \alpha y$, one would interpret $\alpha \ll 1$ as the maximum portion of the population that society could support in the legal profession. For overcrowded populations with y greater than carrying capacity, it is natural to suppose that g and h are decreasing functions of y. The idea is that legal help is not as basic a need as food, clothing, and shelter; in overcrowded situations the general population must focus its attention on such basic needs and lawyers become a luxury.

We think of C as measuring the ability of the society to control the number of lawyers. This function is crucial to the model. Certainly a variety of factors may effect this function. The bar association may have interest in controlling the number of lawyers. But C should be viewed as the control which exists, not the control which may gain only lip service. If the making of the laws is in the hands of the lawyers, either because the president and congress are mostly lawyers or because they leave these decisions in the hands of staff members who are lawyers, then we can be assured that there will be an increased demand for lawyers. The income tax laws in the United States are surely a case in point. This comment should not be viewed necessarily as derogatory of lawyers; if the U.S. president and congress were all mathematicians, there would no doubt be greater need for mathematicians because society would be more mathematized. The point speaks more directly to human nature. At any rate, a wide variety of features in a society are reflected in the function C. We naturally assume that $C(1) = 0$; thus if the current population x of lawyers is optimal so that the ratio $x/g(y)$ is 1, society exerts no influence. We assume that $C(u) < 0$ for $0 \leq u < 1$ and that $C(u) > 0$ for $u > 1$; we also assume that C is an increasing function of u. The interpretation is that if there are fewer lawyers than optimal, society exerts a positive influence on the growth of the lawyer population; otherwise the influence is negative.

If u is large (in particular if x is close to y), we would assume that there would be overpowering reasons for society to reduce the number of lawyers. Thus we expect that for large value of u, $C(u)$ will be very large. Moreover, we would expect that $C(u)$ would be influenced by $f(u)$. The more damaging this large number of lawyers is for the general population (reflected in more negative values for $f(u)$), the larger $C(u)$ should be. Thus we would expect $C(u)$ to be large relative to $|f(u)|$ when u is large. A more quantitative statement will appear in the next section.

Therefore, in this over-simplified and stylized model, four functions g, h, f, and C, characterize the relationship between the general population and the subpopulation of lawyers. The interplay between these functions then determine how many lawyers there will actually be in the society at a given time. The smoothness required of these functions is minimal, but some smoothness will be needed in the analysis.

3 THE INVARIANT REGION

In order for the model to be at all reasonable, it is necessary that the region

$$T = \{(x,y) : y \geq x \geq 0$$

be an invariant subset of the phase plane. Otherwise, populations might go negative or the model allow the number of lawyers to exceed the number of people, a situation which could only be allowed in jokes. We shall see that T will be invariant under reasonable assumptions about the interplay between the four characteristic functions.

For convenience, we temporarily denote the right side of (1a) (resp. (1b)) by $G_1(x,y)$ (resp. $G_2(x,y)$). Note that our assumptions on the four characteristic functions imply that G_1 and G_2 are undefined at $(0,0)$. However, for $(x,y) \in T$, we have $0 \le x \le y$ and mild restrictions on g, h will guarantee that there exists a constant M so that

$$\| (G_1(x,y), G_2(x,y)) \| \le M \| (x,y) \|$$

in some neighborhood of $(0,0)$, where $\| \cdot \|$ denotes one's favorite norm in R^2. For example, if we assume that $g(u)$, $h(u)$ are both differentiable with non-zero derivatives near $u = 0$, then

$$\lim_{y \to 0} \frac{y}{h(y)} = \frac{1}{h'(0)},$$

and a similar result for h replaced by g, implying the desired norm inequality. Hence we may define the right sides of the equations in (4) to be zero at $(0,0)$, and not only will they be continuous on T but will satisfy a Lipschitz condition at $(0,0)$. Thus the elememtary theory for autonomous systems holds on the closed set T.

Suppose $(x(0), y(0)) \in T$. Let T' be the triangle

$$T' = \{(x,y) \in T : y \le \max\{k(1 + f(1)), y(0)\}\}.$$

Then $(x(t), y(t))$ cannot leave T' by crossing the "top" side of T' because

$$y(t) > k(1 + f(1))$$

implies

$$\begin{aligned}
y' &= s\left[1 - \frac{y}{k} + f\left(\frac{x}{g(y)}\right)\right]y \\
&\le s\left[1 - \frac{y}{k} + f(1)\right]y \\
&< 0.
\end{aligned}$$

The trajectory cannot leave T' by crossing the "left" boundary or passing through $(0,0)$ because the system is autonomous and no two distinct trajectories can intersect.

Can the trajectory escape T' by crossing the line $y = x$? Since the velocity vector (x', y') is tangent to the trajectory, we compute the dot product $(x', y') \cdot (-1, 1)$. If this dot product is positive when $x = y$, the trajectory can only cross the line $y = x$ from the outside to the inside. Substituting from (1) for x' and y', a straightforward calculation shows that this dot product for $x = y$ can be written as

$$\left[y\left(\frac{r}{h(y)} - \frac{s}{k}\right) + s - r\right] + \left[rC\left(\frac{y}{g(y)}\right) + sf\left(\frac{y}{g(y)}\right)\right]. \qquad (2)$$

The second bracketed expression in (2) is positive if $C(u)$ is large enough relative to $|f(u)|$; this is the quantitative statement mentioned in the previous section. Look at the first bracketed expression in (2). If $y \leq k$, then since also $h(y) < y$, the first bracketed expression simplifies to

$$r \frac{y}{h(y)} - s\frac{y}{k} + s - r,$$

which is surely positive. If $y > k$, then it simplifies to

$$\frac{y}{k} \left(\frac{k}{h(y)} r - s \right) + s - r.$$

Since $k/h(y)$ is much bigger than 1 and r, s have the same order of magnitude, the expression in parentheses may be assumed positive and thus the entire expression in (2) is positive.

Thus it seems reasonable that the condition for the dot product to be positive is satisfied. In any case, we assume that the four characteristic functions are related in such a way that this dot product is positive.

4   EQUILIBRIA

We now discuss existence of equilibria in the triangular region T. There are two equilibria on the boundary; they are $(0,0)$ and $(0,k)$. It is a straightforward exercise to linearize the system about $(0,k)$ and conclude that it is a saddle point which attracts solutions with initial conditions on the positive y-axis and repels solutions with other initial conditions.

Because the system (1) is not C^1 at $(0,0)$, the stability of the origin cannot be attacked by linearization. The behavior of trajectories in a neighborhood of the origin may be very complicated. We shall see later that under certain conditions, the origin must be in the omega limit set of some trajectories.

We next investigate the existence of internal equilibria in the region T. Such equilibria would be simultaneous solution pairs $(\hat{x}, \hat{y})$ of the equations

$$1 - \frac{x}{h(y)} - C\left(\frac{x}{g(y)}\right) = 0 \qquad (a)$$

$$\tag{3}$$

$$1 - \frac{y}{k} + f\left(\frac{x}{g(y)}\right) = 0 \qquad (b)$$

Recall our assumption that $f(u)$ decreases for $u > 1$ and that $f(u_1) = 0$ for some $u_1 > 1$. Choose u_2 so that $f(u_2) = -1$ if such a value exists; otherwise choose $u_2 = +\infty$, in which case we put $f(u_2) = \lim_{u \to +\infty} f(u)$.

To solve system (3), we begin by examining (3b). Let $u = \frac{x}{g(y)}$; we seek to solve

$$f(u) = \frac{y}{k} - 1 \tag{4}$$

for u where $k(1 + f(u_2)) < y < k(1 + f(1))$. Then $f(u_2) < \frac{y}{k} - 1, f(1)$. Examining the graph of f, we see that there exists a unique $\bar{u} \in (1, u_2)$ satisfying (4), and as y increases from $k(1 + f(u_2))$ to $k(1 + f(1))$, $\bar{u}$ *decreases* from u_2 to 1. For $k \leq y < k(1 + f(1))$, there is a second solution $\hat{u} \in (0,1)$ of (4), but we shall show later that this solution can be ignored.

Let $\bar{u} = F(y)$, and substitute $x = g(y)F(y)$ into equation (3a) to get

$$\frac{g(y)}{h(y)}F(y) \ + \ C(F(y)) = 1. \tag{5}$$

Consider the limiting value of the left side of (5) as $y \to k(1 + f(u_2))$. If $f(u_2) > -1$, then $u_2 = +\infty$ and $F(u) \to +\infty$ and clearly the left side of (5) tends to $+\infty$. If $f(u_2) = -1$, then L'Hôpital's rule gives the limiting value

$$\frac{g'(0)}{h'(0)}u_2 + C(u_2).$$

This value will be greater than 1 if either u_2 is sufficiently large (reflecting a situation in which a lawyer population must be significantly larger than optimal in order to be seriously damaging) or $C(u_2)$ is sufficiently large (say near 1 or larger, reflecting a situation in which society exerts reasonably powerful control over the growth of the lawyer population). We shall assume henceforth that this limiting expression is greater than 1. Otherwise, as we shall see, it is unlikely that internal equilibria exist.

Now consider the limiting value of the left side of (5) as $y \to k(1 + f(1))$. Now $F(y) \to 1$ and the left side of (5) tends to

$$\frac{g(k(1 + f(1)))}{h(k(1 + f(1)))} + C(1) < 1$$

since $C(1) = 0$ and $g(y) < h(y)$ always. Thus with our assumption above, there must be at least one solution $\hat{y}$ of (5), and hence at least one solution pair $(\hat{x}, \hat{y})$, with $\hat{x} = g(\hat{y})F(\hat{y})$ of the system (3).

Note that since $F(y)$ is monotone decreasing and $C(u)$ is monotone increasing, then $C(F(y))$ is monotone decreasing. If $g(y)/h(y)$ is non-increasing, then the left side of (5) is monotone decreasing, and we have a unique internal equilibrium. Without such assumption on the behavior of g/h, we cannot rule out the existence of multiple internal equilibria. (For a while, we were tempted to try to rationalize the belief that $g(y)/h(y)$ is essentially constant.)

Returning briefly to the case of a second solution $\hat{u}$ of (4) in $(0, 1)$, note that for such a solution the left side of (5) becomes

$$\frac{g(y)}{h(y)}\hat{u} + C(\hat{u}),$$

which is clearly less than 1 because $C(\hat{u}) < 0$; thus (5) cannot be satisfied for such a $\hat{u}$.

To summarize this discussion, internal equilibria exist if

$$1) \quad \lim_{u \to +\infty} f(u) > -1$$

$$\text{or} \quad 2) \quad \frac{g'(0)}{h'(0)}u_2 + C(u_2) > 1 \text{ where } f(u_2) = -1, \quad u_1 < u_2 \le +\infty.$$

The equilibria, of course, have no meaning unless $0 < \hat{x} < \hat{y}$. The following argument shows that this inequality is satisfied.

At the critical point,

$$\frac{g(\hat{y})}{h(\hat{y})}\bar{u} + C(\bar{u}) = 1.$$

Since $\bar{u} > 1$, $C(\bar{u}) > 0$ and we have

$$\frac{g(\hat{y})}{h(\hat{y})}\bar{u} = \frac{\hat{x}}{h(\hat{y})} < 1$$

and

$$\bar{u} = \frac{\hat{x}}{g(\hat{y})} > 1 \text{ so } \hat{x} > g(\hat{y}).$$

Combining these inequalities, we have

$$g(\hat{y}) < \hat{x} < h(\hat{y}),$$

which is a stronger inequality than $0 < \hat{x} < \hat{y}$.

We have seen that under certain conditions, at least one internal equilibrium $(\hat{x}, \hat{y})$ exists and has the following properties:

$$0 < \hat{x} < \hat{y} \qquad (a)$$
$$\hat{y} \in (0, k(1 + f(1))] \qquad (b) \qquad \qquad (6)$$
$$\hat{x} \in (g(\hat{y}), h(\hat{y})) \qquad (c)$$

Property (6a) shows that the equilibrium lies in the invariant region T. Property (6b) shows that, at the internal equilibrium, the population level must be under the "augmented" carrying capacity for the population, $k(1 + f(1))$. Property (6c) shows that the lawyer population at the internal equilibrium falls between the optimal lawyer population and the carrying capacity for the lawyers. Perhaps the interpretation for (6c) is that the equilibrium lawyer population occurs as a compromise between the population as a whole, and the lawyer population. To obtain more specific conclusions about internal equilibria, additional (perhaps quantitative) assumptions about the implicit functions (f, h, C, g) would likely be necessary.

5 STABILITY

Now we wish to analyze the qualitative behavior of solution trajectories in the invariant region T.

The Jacobian matrix may be evalutated at any internal critical point $(\hat{x}, \hat{y})$ and simplified using the fact that equations (1a) and (1b) are satisfied at the critical point. After an elementary calculation, one may conclude that the determinant of this matrix is

$$D = \frac{rs\hat{x}^2\hat{y}}{h(\hat{y})g(\hat{y})}\left[\left(\frac{g'(\hat{y})}{g(\hat{y})} - \frac{h'(\hat{y})}{h(\hat{y})}\right)f'(\bar{u}) + \frac{g(\hat{y}) + C'(\bar{u})h(\hat{y})}{k\hat{x}}\right]$$

and the trace of this matrix is

$$\text{Trace} = \frac{-s\hat{x}\hat{y}}{g(\hat{y})}\left[\frac{g'(\hat{y})}{g(\hat{y})}f'(\bar{u}) + \frac{g(\hat{y})}{k\hat{x}} + \frac{r}{s}\frac{g(\hat{y}) + C'(\bar{u})h(\hat{y})}{\hat{y}h(\hat{y})}\right],$$

where $\bar{u} = \hat{x}/g(\hat{y})$ as before.

Recall that (see e.g. [4, Section 1.5]) the eigenvalues of a the Jacobian matrix have opposite signs (and $(\hat{x}, \hat{y})$ is a saddle point) if and only if D is negative; the eigenvalues have negative real parts (and $(\hat{x}, \hat{y})$ is locally asymptotically stable) if and only if the determinant is positive and the trace is negative; and the eigenvalues have positive real parts (and $(\hat{x}, \hat{y})$ is a repellor) if and only if the determinant is positive and the trace is positive.

Now D is positive if and only if

$$-\left(\frac{g'(\hat{y})}{g(\hat{y})} - \frac{h'(\hat{y})}{h(\hat{y})}\right) f'(\bar{u}) < \frac{g(\hat{y}) + C'(\bar{u})h(\hat{y})}{k\hat{x}} \tag{7}$$

and Trace is negative if and only if

$$-\frac{g'(\hat{y})}{g(\hat{y})} f'(\bar{u}) < \frac{g(\hat{y})}{k\hat{x}} + \frac{r}{s} \frac{g(\hat{y}) + C'(\bar{u})h(\hat{y})}{\hat{y}h(\hat{y})}. \tag{8}$$

Recall that $\bar{u} > 1$ and $f'(\bar{u}) < 0$. If $g(y)/h(y)$ is non-increasing near $\hat{y}$ (a cooperative event we encountered in the previous section), then $(g/h)'$ would be non-positive at $\hat{y}$ so the left side of (7) would be non-positive and (7) would be trivially true. Again, if g/h is approximately constant then the coeffient of $f'(\bar{u})$ in (7) would be close to zero. Otherwise, for (7) to be true requires that $|f'(\bar{u})|$ be small relative to $C'(\bar{u})$.

Look at (8). If $\hat{y}$ is sufficiently large that $g'(\hat{y})$ is negative, (8) is trivially true. For such equilibria, we would have $\bar{u} < u_1$ and be in the range where the density of the lawyers was beneficial. Otherwise, (8) says that $|f'(\bar{u})|$ must be small relative to $C'(\bar{u})$.

The language of economics is suggestive here. An economist would call $f'(u)$ the marginal value of additional lawyers. In cases where $f'(u) < 0$, $|f'(u)|$ would be called the marginal damage of additional lawyers. Similarly, $C'(u)$ would be called the marginal ability of society to prevent the growth of the lawyer population. Using this language, (7) and (8) say that an equilibrium will be locally asymptotically stable if the marginal damage of additional lawyers to society is small compared to the marginal ability of society to prevent additional

lawyers from appearing, at the equilibrium. An equilibrium will be unstable in the opposite situation.

What happens if every internal equilibrium is a repellor? Since any trajectory which begins interior to the triangular region T is bounded, the Poincare-Bendixson theorem [4, Section 3.7], [3, Chapter 11] tells us that the omega limit set of the trajectory must contain a critical point or be a periodic orbit. Assuming that the trajectory does not start at an interior equilibrium, we easily conclude that the omega limit set cannot contain an interior repellor or the boundary equilibrium $(0, k)$. Thus either $(0, 0)$ is in the omega limit set (and extinction of the population results) or the omega limit set is a periodic orbit. In the latter case, the system is behaving similar to a predator- prey system with oscillating populations [2, Chapters 3-5], and the lawyers are functioning as predators. Since any periodic orbit must have an equilibrium in its interior, this situation cannot occur unless internal equilibria exist. In particular, if there are no internal equilibria, $(0, 0)$ is in the omega limit set of every trajectory starting in the interior of T and extinction surely occurs.

References

[1] I. Baggs and H. I. Freedman, A mathematical model for the dynamics of interactions between a unilingual and bilingual population: persistence versus extinction, J. Math. Soc., 16:51-75 (1990).

[2] H. I. Freedman, Deterministic Mathematical Models in Population Ecology, Marcel Dekker, Inc., New York, 1980.

[3] M. W. Hirsch and S. Smale, Differential Equations, Dynamical Systems, and Linear Algebra, Academic Press, New York, 1974.

[4] L. Perko, Differential Equations and Dynamical Systems, Springer-Verlag, New York, 1991.

[5] P. Waltman, A Second Course in Elementary Differential Equations, Academic Press, New York, 1986.

Differential Inequalities and Existence Theory for Differential, Integral, and Delay Equations

T.A. BURTON, Department of Mathematics, Southern Illinois University, Carbondale, Illinois 62901

1. Introduction. This paper is concerned with simple, concise, and unifying proofs of global existence of solutions for

$$x' = f(t, x) \tag{1}$$

where $f : [0, \infty) \times R^n \to R^n$ is continuous, for

$$x(t) = a(t) + \int_0^t D(t, s, x(s))ds \tag{2}$$

where $D : [0, \infty) \times R \times R^n \to R^n$ and $a : [0, \infty) \to R^n$ are both continuous, and for functional differential equations

$$x' = f(t, x_t) \tag{3}$$

with both finite and infinite delay. Local existence results are given as corollaries.

Classical existence theory is first local, then piecemeal, and then awkward, as we explain in the next section. In this paper we use Schaefer's fixed point theorem to prove the existence of a global solution of each equation in one step. Each theorem is proved in the same way. First we define the appropriate space and a mapping. Each theorem is then proved using three lemmas. The first lemma shows that the mapping maps bounded sets into compact sets. The second lemma shows that the mapping is continuous. The third lemma establishes a priori bounds on the solution. The results then follow from Schaefer's theorem.

Existence theory for (1) usually rests on limiting arguments with ε-approximate solutions or on careful application of Schauder's fixed point theorem after constructing an appropriate set and a mapping of that set into itself; this is usually an intricate and tedious task. Schaefer's theorem is mainly Schauder's theorem followed by a simple retract argument. Its great advantage over Schauder's theorem is that a self-mapping set need not be found.

The reader will find standard developments of existence theory for (1) in [1], [2], [8], [10], [12], and [13]. Existence theory for (2) is found in [3] and [7], while theory for (3) is found in [1], [4], [5], [9], [12], and [17].

While this paper is clearly expository in nature, Theorems 3 and 4 are new. Hale [9; p. 142] has a form of Theorem 3 in the linear case. Theorem 4 allows for unbounded initial functions, as well as larger than linear growth of f.

2. Background and motivation. Classical existence theory for (1) begins with a local result. It is shown that for each $(t_0, x_0) \in [0, \infty) \times R^n$, there is at least one solution $x(t) = x(t, t_0, x_0)$ with $x(t_0, t_0, x_0) = x_0$ and satisfying (1) on an interval $[t_0, t_1]$, where t_1 is computed from a bound on f in a closed neighborhood of (t_0, x_0). This yields a new point $(t_1, x(t_1))$ and we begin once more computing bounds on f in a closed neighborhood of $(t_1, x(t_1))$ and obtain a continuation of the solution on $[t_1, t_2]$. These are the aforementioned local and piecemeal parts.

If we continue this process on intervals $\{[t_n, t_{n+1}]\}$, can we say that $t_n \to \infty$ as $n \to \infty$? It turns out that we can unless there is an α such that $|x(t)|$ tends to infinity as t tends to

α from the left. This is the awkward part. Either implicitly or explicitly (cf. Hale [9; p. 42]) we invoke Zorn's lemma to claim that there is a solution on $[t_0, \infty)$ or one on $[t_0, \alpha)$ which can not be continued to α. Some authors call a solution on $[t_0, \infty)$ noncontinuable. We do not; for our purposes, if a solution is defined on $[t_0, \alpha)$ with $\alpha < \infty$, and if it can not be extended to α, it is said to be noncontinuable. An example of the latter case is

$$x' = x^2, \quad (t_0, x_0) = (0, 1)$$

which has the solution $x(t) = \frac{1}{1-t}$ on $[0, 1)$ and is noncontinuable. In fact, two more examples complete the range of possibilities. Solutions of $x' = -x^3$ are all continuable to $+\infty$ because they are bounded, even though the right-hand-side grows faster than in the first example. Finally, solutions of $x' = t^3 x \ell n(1 + |x|)$ are unbounded, but continuable to $+\infty$ because the right-hand-side does not grow too fast.

We come then to the question of how to rule out noncontinuable solutions of the kind mentioned above. In principle, there is a fine way of doing so. Kato and Strauss [6] prove that it always works.

DEF. A continuous function $V : [0, \infty) \times R^n \to [0, \infty)$ which is locally Lipschitz in x is said to be mildly unbounded if for each $T > 0$, $\lim_{|x| \to \infty} V(t, x) = \infty$ uniformly for $0 \leq t \leq T$.

If there is a mildly unbounded V which is differentiable, then we invoke the local existence theory and consider a solution $x(t)$ of (1) on $[t_0, \alpha)$ so that $V(t, x(t))$ is an unknown but well-defined function. The chain rule than gives

$$\frac{dV}{dt}(t, x(t)) = \sum_{i=1}^{n} \frac{\partial V}{\partial x_i} \frac{dx_i}{dt} + \frac{\partial V}{\partial t}$$
$$= \operatorname{grad} V \cdot f + \frac{\partial V}{\partial t}.$$

We can also compute V' when V is only locally Lipschitz in x (cf. Yoshizawa [17; p. 3]) and we will display such an example in a moment; in that case, one uses the upper right-hand derivative.

If V is so shrewdly chosen that it is mildly unbounded and $V' \leq 0$, then there can be no $\alpha < \infty$ with $\lim_{t \to \alpha-} |x(t)| = \infty$ because $V(t, x(t)) \leq V(t_0, x_0)$.

There is a converse theorem; if f is continuous and locally Lipschitz in x for each fixed t, then Kato and Strauss [6] show that there is a mildly unbounded V with $V' \leq 0$ if and only if all solutions can be continued for all future time. Their result is not constructive, but investigators have constructed suitable V for many important systems without any growth condition on f; we offer examples following each of our theorems. In the example $x' = -x^3$ mentioned above, $V = x^2$ yields $V' = -2x^4 \leq 0$, showing global existence.

These remarks for (1) apply in large measure to (2) and (3). In those cases we require a functional $V(t, x(\cdot))$. More importantly, as mentioned above, for (1) the only way a solution can fail to be continuable to $+\infty$ is for there to exist an α with $\lim_{t \to \alpha^-} |x(t)| = +\infty$; but for (3) we must take the limit supremum.

Wintner derived conditions on the growth of f to ensure that solutions of (1) could be continued to $+\infty$ and Conti used these to construct a suitable V. These results are most accessible in Hartman [10; pp. 29–30] for the Wintner condition and Sansone-Conti [13; p. 6] for V. A proof here will show how it works and will be a guide for an alternative proof of each of our Lemma 3 for each of our theorems.

Theorem (Conti-Wintner).   *If there are continuous functions* $\Gamma : [0, \infty) \to [0, \infty)$ *and* $W : [0, \infty) \to [1, \infty)$ *with*

$$|f(t, x)| \leq \Gamma(t) W(|x|) \text{ and } \int_0^\infty \frac{ds}{W(s)} = \infty, \text{ then}$$

$$V(t, x) = \left\{ \int_0^{|x|} \frac{ds}{W(s)} + 1 \right\} \exp - \int_0^t \Gamma(s) ds$$

is mildly unbounded and $V'(t, x(t)) \leq 0$ *along any solution of (1).*

Proof.   Let $x(t)$ be a noncontinuable solution of (1) on $[t_0, \alpha)$. By examining the difference quotient we see that $|x(t)|' \leq |x'(t)|$. Thus,

$$V'(t, x(t)) \leq \frac{|x(t)|'}{W(|x(t)|)} \exp - \int_0^t \Gamma(s) ds - \Gamma(t) V(t, x(t)) \leq 0$$

when we use $|x(t)|' \leq |x'(t)| \leq \Gamma(t) W(|x(t)|)$. This means that $V(t, x(t)) \leq V(t_0, x_0)$; since V is mildly unbounded, $\lim_{t \to \alpha^-} |x(t)| \neq \infty$. This completes the proof.

In the next two sections we will prove a result which will give a global solution in one step. But we still want a local solution as a special case. This can be accomplished by extending f over a compact set to a continuous bounded function on R^{n+1}. There are classical extension theorems which we use in Theorems 3 and 4, but the following idea of a colleague, G. Makay, makes it elementary for Theorems 1 and 2.

Remark on extension. Suppose that f is continuous on $\Delta = \{(t, x)|t_0 \leq t \leq T, |x - x_0| \leq J\}$ for $T > t_0$ and $J > 0$. We want to extend f to all of $R \times R^n$ in a bounded and continuous manner. Since Δ is convex, if we choose (t_1, y_1) as any interior point of Δ and if Q is any ray from (t_1, y_1) then Q intersects the boundary of Δ at exactly one point (t_Q, x_Q). Define $F : R \times R^n \to R^n$ by

    (i) $F(t, x) = f(t, x)$ if $(t, x) \in \Delta$, and

    (ii) $F(t, x) = f(t_Q, x_Q)$ if (t, x) is on Q and in the complement of Δ.

    Clearly, F is bounded, continuous, and agrees with f on Δ.

Our results are based on the following theorem of Schafer [14] which is discussed and proved also in Smart [16; p. 29].

Theorem (Schaefer). *Let $(C, \|\cdot\|)$ be a normed space, H a continuous mapping of C into C which is compact on each bounded subset of C. Then either*

    (i) the equation $x = \lambda H x$ has a solution for $\lambda = 1$, or

    (ii) the set of all such solutions x, for $0 < \lambda < 1$, is unbounded.

2. Existence theory for (1). Let $0 \leq \lambda \leq 1$ and consider

(4) $$x' = \lambda f(t, x), \quad x(t_0) = \lambda x_0$$

or the equivalent integral equation

(5) $$x(t) = \lambda \left[x_0 + \int_{t_0}^{t} f(s, x(s))ds \right] =: \lambda H(x)(t).$$

Theorem 1. *If either of the following conditions hold, then for each $(t_0, x_0) \in [0, \infty) \times R^n$, there is a solution of (1) on $[t_0, \infty)$.*

(I) There are continuous functions $\Gamma : [0, \infty) \to [0, \infty)$ *and* $W : [0, \infty) \to [1, \infty)$ *with*

$$|f(t, x)| \leq \Gamma(t) W(|x|) \text{ and } \int_0^\infty \frac{ds}{W(s)} = \infty.$$

(II) There is a continuous function $V : [0, \infty) \times R^n \to [0, \infty)$ *which is locally Lipschitz in* x, *mildly unbounded, and* $V'(t, x(t)) \leq 0$ *along any continuous solution of (4) defined on* $[t_0, \infty)$.

Proof. Let $T > t_0$ be given. We will show that there is a solution $x(t, t_0, x_0)$ of (1) on $[t_0, T]$.

Let $(C, \| \cdot \|)$ be the Banach space of continuous functions $\varphi : [t_0, T] \to R^n$ with the supremum norm. From (5) we consider $\varphi \in C$ and write

$$H(\varphi)(t) = x_0 + \int_{t_0}^t f(s, \varphi(s)) ds.$$

The conditions of Schaefer's theorem will be verified by three simple lemmas.

Lemma 1. $H : C \to C$ *and* H *maps bounded sets into compact sets.*

Proof. For $\varphi \in C$ we have $f(t, \varphi(t))$ continuous and so $H(\varphi)(t)$ is continuous. Thus, $H : C \to C$.

For a given $J > 0$, if $\varphi \in C$ and $\|\varphi\| \leq J$, then there is a $J^* > 0$ with $|f(t, \varphi(t))| \leq J^*$ for $t_0 \leq t \leq T$. Thus, there is a $K > 0$ with $|H(\varphi)(t)| \leq K$. Also, $|(H(\varphi)(t))'| = |f(t, \varphi(t))| \leq J^*$. By Ascoli's theorem, this set of φ is mapped into a compact set.

Lemma 2. H *is continuous.*

Proof. Let $J > 0$ be given and let $\varphi_i \in C$ with $\|\varphi_i\| \leq J$, $i = 1, 2$. Now f is uniformly continuous for $|x| \leq J$ and $t_0 \leq t \leq T$, so for each $\varepsilon > 0$ there is a $\delta > 0$ such that $[t_0 \leq t \leq T$ and $|\varphi_1(t) - \varphi_2(t)| \leq \delta]$ imply that $|f(t, \varphi_1(t)) - f(t, \varphi_2(t))| < \varepsilon$. Thus, $\|\varphi_1 - \varphi_2\| \leq \delta$ and $t_0 \leq t \leq T$ imply that

$$|H(\varphi_1)(t) - H(\varphi_2)(t)| = \left| \int_{t_0}^t [f(s, \varphi_1(s)) - f(s, \varphi_2(s))] ds \right| < \varepsilon[T - t_0]$$

and so $\|H(\varphi_1) - H(\varphi_2)\| < \varepsilon[T - t_0]$.

Lemma 3. *There is a $K > 0$ such that if $\varphi(t) = \lambda H(\varphi)(t)$ for $t_0 \leq t \leq T$, then $\|\varphi\| \leq K$.*

Proof. If (I) holds, then a suitable V for (4) is constructed in the Conti-Wintner theorem since $\lambda \leq 1$ which satisfies (II) and so (I) is a special case of (II). Thus, for the mildly unbounded V we always have $V(t_0, \varphi(t_0)) = V(t_0, \lambda x_0)$, a fixed constant for every φ; since V is continuous and x_0 is fixed, there is a $P > 0$ with $V(t_0, \lambda x_0) \leq P$ if $0 \leq \lambda \leq 1$. But by the definition of V being mildly unbounded, $V(t, \varphi) \to \infty$ as $|\varphi| \to \infty$ uniformly for $t_0 \leq t \leq T$. Hence, there is a $K > 0$ with $\|\varphi\| \leq K$ whenever φ satisfies (5).

All of the conditions of Schaefer's theorem are satisfied, his condition (ii) is ruled out by Lemma 3, and so (5) has a solution for $\lambda = 1$. That solution satisfies (1).

COR. *Let $f(t, x)$ be continuous for $|x - x_0| \leq J$ and $t_0 \leq t \leq T$, and let $|f(t, x)| \leq M$ on that set. Then (1) has a solution $x(t, t_0, x_0)$ defined for $t_0 \leq t \leq \alpha$ where $\alpha = \min[T, t_0 + J/M)]$.*

Proof. Let $\Omega = \{(t, x) | \ |x - x_0| \leq J, t_0 \leq t \leq T\}$. Since f is continuous on Ω, by the remark on extension in the previous section, we can extend f to a bounded and continuous function F on $R \times R^n$; hence, Condition (I) of Theorem 1 holds for F and there is a solution $x(t, t_0, x_0) = x(t)$ of $x' = F(t, x)$ for $t_0 \leq t < \infty$. Certainly, $x(t)$ also satisfies (1) so long as $(t, x) \in \Omega$. If $(t, x(t))$ reaches the boundary of Ω at $t_1 < T$ and $t_1 < t_0 + (J/M)$, then

$$|x(t_1) - x_0| \leq \int_{t_0}^{t} |f(s, x(s))| ds \leq M(t_1 - t_0) < J,$$

a contradiction to $(t, x(t))$ reaching the boundary at t_1. This completes the proof.

Remark. Lakshmikantham and Leela [12 (vol. I); p. 46] have a global existence theorem partially in the spirit of Theorem 1(I). But the theorem and proof fall short of ours in four ways. First, they must work up differential inequality and maximal solution theory to find an upper bound on a solution set. Next, they require theory for a locally convex topological vector space and the Tychonov fixed point theorem. Thirdly, they must find a self-mapping set. Finally, they require that the Wintner function $W(|x|)$ be monotone, a condition we do not need until we consider (2).

Example 1.   Consider the scalar equation

$$x'' + h(t, x, x')x' + g(x) = e(t)$$

where $h(t, x, y) \geq 0$, $xg(x) > 0$ if $x \neq 0$, h, g, and e are continuous. Write the equation as

$$x' = \lambda y$$

$$y' = \lambda[-h(t, x, y)y - g(x) + e(t)]$$

and define a mildly unbounded function by

$$V(t, x, y) = \left[y^2 + 2\int_0^x g(s)ds + \ln(|x| + 1) + 1\right]\exp - \int_0^t E(s)ds$$

where $E(t) = 2|e(t)| + 1$ so that

$$V'(t, x, y) \leq \left[-2\lambda h(t, x, y)y^2 + 2\lambda ye(t) + |y| - E(t)(y^2 + 1)\right]\exp - \int_0^t E(s)ds \leq 0.$$

Thus, by Theorem 1 all solutions exist on $[t_0, \infty)$.

3. An integral equation.   Consider once more

(2) $$x(t) = a(t) + \int_0^t D(t, s, x(s))ds$$

with its continuity conditions.

Theorem 2.   *If either of the following conditions hold, then (2) has a solution on* $[0, \infty)$:

(I) There are continuous increasing functions $\Gamma : [0, \infty) \to [0, \infty)$ *and* $W : [0, \infty) \to [1, \infty)$ *with*

(6) $$\int_0^\infty \frac{ds}{W(s)} = \infty \text{ and } |D(t, s, x)| \leq \Gamma(t)W(|x|) \text{ for } 0 \leq s \leq t.$$

(II) There is a differentiable scalar functional $V(t, x(\cdot))$ which is mildly unbounded along any solution of

(7) $$x(t) = \lambda\left[a(t) + \int_0^t D(t, s, x(s))ds\right] =: \lambda H(x)(t), \quad 0 \le \lambda \le 1,$$

and which satisfies $V'(t, x(\cdot)) \le 0$ along such a solution.

Proof. Let $T > 0$ and $(C, \|\cdot\|)$ be the Banach space of continuous $\varphi : [0, T] \to R^n$ with the supremum norm. We will show that there is a solution $x(t)$ of (2) on $[0, T]$.

Lemma 1. *If H is defined by (7) then $H : C \to C$ and H maps bounded sets into compact sets.*

Proof. If $\varphi \in C$, then $D(t, s, \varphi(s))$ is continuous and so $H(\varphi)$ is a continuous function of t. Let $J > 0$ be given and let $B = \{\varphi \in C | \|\varphi\| \le J\}$. Now $a(t)$ is uniformly continuous on $[0, T]$ and $D(t, s, x)$ is uniformly continuous on $\Delta = \{(t, s, x) | 0 \le s \le t \le T,$ $|x| \le J\}$. Thus, for each $\varepsilon > 0$ there is a $\delta > 0$ such that for $(t_i, s_i, x_i) \in \Delta$, $i = 1, 2$, then $|(t_1, s_1, x_1) - (t_2, s_2, x_2)| < \delta$ implies that $|D(t_1, s_1, x_1) - D(t_2, s_2, x_2)| < \varepsilon$; a similar statement holds for $a(t)$. If $\varphi \in B$ then $0 \le t_i \le T$ and $|t_1 - t_2| < \delta$ imply that

$$\begin{aligned}
|H(\varphi)(t_1) - H(\varphi)(t_2)| &\le |a(t_1) - a(t_2)| \\
&+ \left|\int_0^{t_1}\left[D(t_1, s, \varphi(s)) - D(t_2, s, \varphi(s))\right]ds\right| \\
&+ \left|\int_0^{t_1} D(t_2, s, \varphi(s))ds - \int_0^{t_2} D(t_2, s, \varphi(s))ds\right| \\
&\le \varepsilon + t_1\varepsilon + |t_1 - t_2|M \le \varepsilon(1 + T) + \delta M
\end{aligned}$$

where $M = \max_\Delta |D(t, s, x)|$. Hence, the set $A = \{H(\varphi) | \varphi \in B\}$ is equicontinuous. Moreover, $\varphi \in B$ implies that $\|H(\varphi)\| \le \|a\| + TM$. Thus, A is contained in a compact set by Ascoli's theorem.

Lemma 2. *H is continuous in φ.*

Proof. Let $J > 0$ be given, $\|\varphi_i\| \le J$ for $i = 1, 2$, and for a given $\varepsilon > 0$ find the δ of uniform continuity on the region Δ of the proof of Lemma 1 for D. If $\|\varphi_1 - \varphi_2\| < \delta$, then

$$|H(\varphi_1)(t) - H(\varphi_2)(t)| \leq \int_0^t |D(t, s, \varphi_1(s)) - D(t, s, \varphi_2(s))| ds$$

$$\leq T\varepsilon$$

so $\|H(\varphi_1) - H(\varphi_2)\| \leq T\varepsilon$.

Lemma 3.   *There is a $K > 0$ such that any solution of (7), for $0 < \lambda < 1$, satisfies* $\|\varphi\| \leq K$.

Proof.   Let (I) hold. If φ satisfies (7) on $[0, T]$, then

$$|\varphi(t)| \leq \lambda \left[A(T) + \int_0^t \Gamma(T) W(|\varphi(s)|) ds \right] \text{ for } 0 \leq t \leq T$$

where $A(T) = \max_{0 \leq t \leq T} |a(t)|$. If we define $y(t)$ by

$$y(t) = \lambda \left[A(T) + 1 + \Gamma(T) \int_0^t W(|y(s)|) ds \right]$$

then $y(t) \geq |\varphi(t)|$ on $[0, T]$; clearly, $y(0) > |\varphi(0)|$ so if there is a first t_1 with $y(t_1) = |\varphi(t_1)|$, then a contradiction is clear. But the Conti-Wintner result gives a bound K on $\|y\|$ and so the lemma is true for (I). The argument when (II) holds is identical to the proof of Lemma 3 of Theorem 1.

The conditions of Schaefer's theorem hold and, by Lemma 3, H has a fixed point for $\lambda = 1$.

Remark.   To appreciate the power of Schaefer's theorem, compare Theorem 2 with a standard treatment. For example, Corduneanu [3; pp. 95–109] arrives at Theorem 2(I) through several pages of analysis.

COR.   Let $a(t)$ be continuous for $0 \leq t \leq T$ and $D(t, s, x)$ be continuous on $U = \{(t, s, x) \mid 0 \leq s \leq t \leq T, |x - a(t)| \leq J\}$. Then there is a solution of (2) on $[0, \alpha]$ where $\alpha = \min[T, J/M]$ and $M = \max_U |D(t, s, x)|$.

We extend D to a bounded and continuous function on $R \times R \times R^n$ and apply Theorem 2, just as we did in the corollary to Theorem 1.

We now give an example which occupies a significant place in the literature (cf. Gripenberg et al [7; pp. 613–638]); and it also illustrates the change in language from Theorem 1(II) where we ask that $V(t,x)$ be mildly unbounded, and in Theorem 2(II) where we ask that $V(t,x(\cdot))$ be mildly unbounded along a solution of (7).

Example 2. Consider the scalar equation

$$(8) \qquad x(t) = b(t) - \int_{-\infty}^{t} D(t,s)g(s,x(s))ds$$

where b, D, and g are continuous, while $xg(t,x) \geq 0$. To specify a solution of (8) we require a bounded and continuous initial function $\varphi : (-\infty,0] \to R$ such that

$$(9) \qquad a(t) := b(t) - \int_{-\infty}^{0} D(t,s)g(s,\varphi(s))ds \text{ is continuous}$$

so that

$$(10) \qquad x(t) = a(t) - \int_{0}^{t} D(t,s)g(s,x(s))ds$$

is essentially of the form of (2) when φ is chosen so that $\varphi(0) = a(0)$. We also ask that there exist a continuous function M with

$$(11) \qquad -2g(t,x)[x - \lambda a(t)] \leq M(t), g(t,x) \text{ bounded for } t \leq 0 \text{ if } x \text{ is bounded .}$$

But the defining property of this example is

$$(12) \qquad D(t,s) \geq 0, \quad D_s(t,s) \geq 0, \quad D_{st}(t,s) \leq 0, \; D_t(t,0) \leq 0.$$

This is an infinite delay problem and the following convergence condition is required:

$$(13) \qquad \int_{-\infty}^{t} D(t,s)ds \quad exists.$$

It is not necessary that $\varphi(0) = a(0)$ but in that case x has a discontinuity:

$$(14) \qquad x(0) \text{ does not equal } \varphi(0).$$

For $0 \leq \lambda \leq 1$, consider the equation

(15) $$x(t) = \lambda \left[a(t) - \int_0^t D(t,s)g(s,x(s))ds \right]$$

and define

$$V(t, x(\cdot)) =$$

(16)
$$\left\{ \int_0^t D_s(t,s) \left(\int_s^t \lambda g(v, x(v))dv \right)^2 ds + D(t,0) \left(\int_0^t \lambda g(v, x(v))dv \right)^2 + 1 \right\} \exp - \int_0^t M(s)ds.$$

Proposition. *If (9) – (14) hold, then there is a solution $x(t, \varphi)$ of (8) on $[0, \infty)$.*

Proof. Let $T > 0$ be given. Because of the form of (10) and Theorem 2, we need only prove that there is a $K > 0$ such that any solution of (15) on $[0, T]$ for $0 < \lambda < 1$ satisfies $|x(t)| \leq K$ on $[0, T]$. Here are the steps:

   (i) Differentiate V.

   (ii) Integrate by parts the term in V' obtained from differentiating the inner integral in the first integral in V.

   (iii) Substitute $\lambda a(t) - x(t)$ from (15) into that last term obtained in (ii).

Since $D_{st} \leq 0$ we will now have

$$V'(t, x(\cdot)) \leq \{-2\lambda g(t, x(t))[x(t) - \lambda a(t)] - M(t)\} \exp - \int_0^t M(s)ds$$

and this is not positive by (11) since $xg(t, x) \geq 0$. Hence,

(17) $$V(t, x(\cdot)) \leq V(0, x(\cdot)).$$

Here is the surprising part. It does not appear that V is mildly unbounded; however, along a solution we have

$$(\lambda a(t) - x(t))^2 = \left(\int_0^t \lambda D(t,s)g(s,x(s))ds \right)^2$$

(from (15))

$$= \left(D(t,0) \int_0^t \lambda g(v, x(v)) dv + \int_0^t D_s(t,s) \int_s^t \lambda g(v, x(v)) dv \, ds \right)^2$$

(upon integration by parts)

$$\leq 2 \int_0^t D_s(t,s) ds \int_0^t D_s(t,s) \left(\int_s^t \lambda g(v, x(v)) dv \right)^2 ds + 2D^2(t,0) \left(\int_0^t \lambda g(v, x(v)) dv \right)^2$$

(by Schwarz's inequality) with $exp \int_0^T M(s) ds = U$

$$\leq 2[D(t,0) + D(t,t) - D(t,0)] V(t, x(\cdot)) U = 2D(t,t) V(t, x(\cdot)) U.$$

Hence, $V(t, x(\cdot)) \to \infty$ as $|x(t)| \to \infty$ uniformly for $0 \leq t \leq T$. In particular, putting this together with (17) yields

$$(\lambda a(t) - x(t))^2 \leq 2D(t,t) V(0, x(\cdot)) U$$

for $0 \leq t \leq T$. Hence, there is a K with $|x(t)| \leq K$ on $[0, T]$. This will satisfy Lemma 3 of Theorem 2 and the proposition is proved.

4. A finite delay equation. Let $(G, |\cdot|_h)$ denote the Banach space of continuous functions $\psi : [-h, 0] \to R^n$ with the supremem norm and consider the system

$$(18) \qquad\qquad x'(t) = f(t, x_t)$$

where $f : [0, \infty) \times G \to R^n$ is continuous. Here, if $x : [-h, A) \to R^n$ for $A > 0$, then $x_t(s) = x(t + s)$ for $-h \leq s \leq 0$.

To specify a solution of (18) we require a $t_0 \geq 0$ and a $\tilde{\psi} \in G$. We then seek a solution $x(t) = x(t, t_0, \tilde{\psi})$ with $x_{t_0} = \tilde{\psi}$ and $x(t)$ satisfying (18) on an interval $t_0 < t < \alpha$ for some $\alpha > 0$.

Yoshizawa [17; p. 184] shows that it is sufficient to ask that f be continuous in order to prove existence of such a solution. But if we do not at least ask that f take bounded sets into bounded sets, then a solution can have surprisingly bad behavior, as shown by Hale [9; p. 44]. On the other hand, if f is continuous and locally Lipschitz in x_t, then a contraction mapping argument will quickly lead to a unique local solution.

Hale [9; p. 142] has a version of the next result in case $W(r) = r$. His system is linear so he gets uniqueness as well.

Theorem 3.   *Suppose that either:*

(I) there are continuous functions $\Gamma : [0, \infty) \to [0, \infty)$ and $W : [0, \infty) \to [1, \infty)$ with W increasing and

$$|f(t, \psi)| \leq \Gamma(t) W(|\psi|_h) \ and \ \int_0^\infty \frac{ds}{W(s)} = \infty; \ or$$

(II) f takes bounded sets into bounded sets and there is a continuous scalar functional $V(t, x_t)$ which is locally Lipschitz in x_t, mildly unbounded in x_t, and $V' \leq 0$ along any solution of (19) (which is displayed in the following proof).
Then for each $(t_0, \tilde\psi) \in [0, \infty) \times G$, there is a solution $x(t, t_0, \tilde\psi)$ of (18) on $[t_0, \infty)$.

Proof.   Let $(t_0, \tilde\psi)$ be given and let $T > t_0$ be arbitrary. We will show that there is a solution of (18) on $[t_0, T]$ with $x_{t_0} = \tilde\psi$.

Let $(C, \|\cdot\|)$ be the Banach space of continuous functions $\varphi : [t_0 - h, T] \to R^n$ with the supremum norm. Consider the equations

$$x_{t_0} = \lambda\tilde\psi,$$
(19)
$$x(t) = \lambda\left[\tilde\psi(0) + \int_{t_0}^t f(s, x_s)ds\right]$$

and define $H : [t_0 - h, T] \to R^n$ by $\varphi \in C$ implies that

$$H(\varphi)_{t_0} = \tilde\psi,$$
$$H(\varphi)(t) = \tilde\psi(0) + \int_{t_0}^t f(s, \varphi_s)ds \ for \ t_0 \leq t \leq T.$$

Lemma 1. $H : C \to C$ and H maps bounded sets into compact sets.

Proof. Let $\varphi \in C$. We first show that $f(t, \varphi_t)$ is a continuous function of t so that $H(\varphi)(t)$ will be continuous. Let $\varepsilon > 0$ and $t_1 \in [t_0, T]$ be given; we must find $\delta > 0$ such that $t_2 \in [t_0, T]$ and $|t_1 - t_2| < \delta$ imply that $|f(t_1, \varphi_{t_1}) - f(t_2, \varphi_{t_2})| < \varepsilon$. Now φ is uniformly continuous on $[t_0 - h, T]$ and so for each $\varepsilon_1 > 0$ there is a $\delta > 0$ such that $t_i \in [t_0 - h, T]$ and $|t_1 - t_2| < \delta$ implies that $|\varphi(t_1) - \varphi(t_2)| < \varepsilon_1$; hence, $[t_i \in [t_0, T]$, $-h \le s \le 0, |t_1 - t_2| < \delta]$ imply that $|\varphi(t_1 + s) - \varphi(t_2 + s)| < \varepsilon_1$ so $|\varphi_{t_1} - \varphi_{t_2}|_h < \varepsilon_1$. But f is continuous at (t_1, φ_{t_1}) so there is an $\varepsilon_1 > 0$ such that $|t_1 - t_2| < \varepsilon_1$ and $|\varphi_{t_1} - \varphi_{t_2}| < \varepsilon_1$ imply that $|f(t_1, \varphi_{t_1}) - f(t_2, \varphi_{t_2})| < \varepsilon$. Thus, $|t_1 - t_2| < \delta < \varepsilon_1$ implies $|\varphi_{t_1} - \varphi_{t_2}|_h < \varepsilon_1$ so $|f(t_1, \varphi_{t_1}) - f(t_2, \varphi_{t_2})| < \varepsilon$.

Let $J > 0$ be given, $B = \{\varphi \in C \mid \|\varphi\| \le J\}$. Then under either (I) or (II) f takes bounded sets into bounded sets so there is a $J^* > 0$ such that $|f(t, \varphi_t)| \le J^*$ for $\varphi \in B$ and $t_0 \le t \le T$. Since $H(\varphi)_{t_0} = \tilde{\psi}$, a fixed uniformly continuous function on $[-h, 0]$ and since $t_0 < t \le T$ implies that $|(H(\varphi)(t))'| = |f(t, \varphi_t)| \le J^*$, it follows that H maps B into an equicontinuous set. Also, $\|H(\varphi)\| \le |\tilde{\psi}|_h + J^*[T - t_0]$. The lemma now follows from Ascoli's theorem.

Lemma 2. *H is continuous.*

Proof. Let $\varphi \in C$ be given. We claim that for each $\varepsilon > 0$ there is a $\delta > 0$ such that $t \in [t_0, T]$, $\psi \in C$, $\|\varphi - \psi\| < \delta$ imply that $|f(t, \varphi_t) - f(t, \psi_t)| < \varepsilon$. If this is false, then there is an $\varepsilon > 0$, $\{t_n\} \subset [t_0, T]$, $\{\psi^{(n)}\} \subset C$ such that $\|\varphi - \psi^{(n)}\| \to 0$ as $n \to \infty$, but $|f(t_n, \varphi_{t_n}) - f(t_n, \psi_{t_n}^{(n)})| \ge \varepsilon$. Now there is a subsequence, say $\{t_n\}$ again, with $\{t_n\} \to t^*$; also, $\varphi_{t_n} \to \varphi_{t^*}$ as $n \to \infty$. Thus, for large n we have

$$|f(t_n, \varphi_{t_n}) - f(t_n, \psi_{t_n}^{(n)})| \le |f(t_n, \varphi_{t_n}) - f(t^*, \varphi_{t^*})| + |f(t^*, \varphi_{t^*}) - f(t_n, \psi_{t_n}^{(n)})| < \varepsilon.$$

This is because f is continuous and the following is small:

$$|\varphi_{t^*} - \psi_{t_n}^{(n)}|_h \le |\varphi_{t^*} - \varphi_{t_n}|_h + |\varphi_{t_n} - \psi_{t_n}^{(n)}|_h.$$

This is a contradiction, so there is a $\delta > 0$ such that $\|\psi - \varphi\| < \delta$ implies that

$$\|H(\varphi) - H(\psi)\| \le \int_{t_0}^{T} |f(s, \varphi_s) - f(s, \psi_s)| ds \le \varepsilon[T - t_0],$$

proving Lemma 2.

Lemma 3.   *There is a $K > 0$ such that if $\varphi_{t_0} = \lambda\tilde{\psi}$ and $\varphi(t) = \lambda H(\varphi)(t)$ for $t_0 \le t \le T$ and for some $\lambda \in (0,1)$, then $\|\varphi\| \le K$.*

Proof. If (I) holds then we have

$$|\varphi(t)| \le |\tilde{\psi}|_h + \int_{t_0}^{t} |f(s, \varphi_s)| ds \le |\tilde{\psi}|_h + \int_{t_0}^{t} \Gamma(s)W(|\varphi_s|_h) ds$$

and, since the right-hand-side is increasing and $\varphi_{t_0} = \tilde{\psi}\lambda$, it follows that

$$|\varphi_t|_h \le |\tilde{\psi}|_h + \int_{t_0}^{t} \Gamma(s)W(|\varphi_s|_h) ds.$$

Since W is increasing, $|\varphi_t|_h$ is bounded by any solution $y(t)$ of

$$y(t) = |\tilde{\psi}|_h + 1 + \int_{t_0}^{t} \Gamma(s)W(y(s)) ds$$

or of

$$y' = \Gamma(t)W(y), \quad y(t_0) = |\tilde{\psi}|_h + 1.$$

By the Conti-Wintner argument, those solutions are bounded by some K on $[t_0, T]$. Hence, $|\varphi_t|_h \le K$ or $\|\varphi\| \le K$. This proves Lemma 3 when (I) holds. If (II) holds then a parallel argument completes the proof.

By Schaefer's theorem there is a solution of $x = \lambda H x$ for $\lambda = 1$.   □

Cor.   *Let $(t_0, \tilde{\psi}) \in [0, \infty) \times G, |\tilde{\psi}|_h = K$, and suppose there is a $J > 0, T > t_0$, and $M > 0$ such that $f(t, \psi)$ is continuous on*

$$\Omega = \{(t, \psi) \,|\, t_0 \le t \le T, |\psi|_h \le K + J\}$$

with $|f(t, \psi)| \le M$ on Ω. Then there is a solution $x(t) = x(t, t_0, \tilde{\psi})$ of (18) defined for $t_0 \le t \le \alpha$ where $\alpha = \min[T, t_0 + J/M]$.

To prove this corollary, we extend f to a bounded continuous function $F : R \times C \to R^n$ with $F(t, \psi) = f(t, \psi)$ on Ω, following Friedman [6; p. 111]. Then invoke Theorem 3 to get a global solution. The choice of α is verified exactly as in the proof of the corollary of Theorem 1.

Example 3. Consider the scalar equation

$$x' = a(t)W(x(t)) + \int_{t-h}^{t} b(s)W(x(s))ds$$

where a, b, and W are continuous, W is increasing, $|W(x)| \leq W(|x|)$, $\int_0^\infty \frac{ds}{W(s)} = \infty$. Then

$$x' = a(t)W(x(t)) + \int_{-h}^{0} b(t+s)W(x(t+s))ds$$

$$= a(t)W(x(t)) + \int_{-h}^{0} b(t+s)W(x_t(s))ds$$

$$=: f(t, x_t).$$

Let $\Gamma(t) = |a(t)| + h \max_{t-h \leq s \leq t} |b(s)|$. Then $|f(t, x_t)| \leq \Gamma(t)W(|x_t|_h)$ and the conditions of Theorem 3 are satisfied.

5. Infinite delay. Let $g : (-\infty, 0] \to [1, \infty)$ be a continuous non-increasing function and let $(G, |\cdot|_g)$ be the Banach space of all continuous functions

$$\psi : (-\infty, 0] \to R^n \text{ for which } |\psi|_g = \sup_{-\infty < s \leq 0} |\psi(s)/g(s)|$$

exists as a finite number. If $A > 0$ and if $\varphi : (-\infty, A] \to R^n$ is continuous, then for $0 \leq t \leq A$ we define $\varphi_t(s) = \varphi(t+s)$ for $-\infty < s \leq 0$.

Consider the system of functional differential equations

(20) $$x'(t) = f(t, x_t)$$

and suppose that there is a space $(G, |\cdot|_g)$ for (20) so that $f : [0, \infty) \times G \to R^n$ is continuous in the sense that if $t_1 \geq 0$ and $\psi_1 \in G$, then for each $\varepsilon > 0$ there is a $\delta > 0$ such that $[t \geq 0, \psi \in G, |t - t_1| < \delta, |\psi_1 - \psi|_g < \delta]$ imply that $|f(t_1, \psi_1) - f(t, \psi)| < \varepsilon$.

Clearly, the space $(G, |\cdot|_g)$ is chosen in view of the properties of f. The prototype is

$$x'(t) = p(t, x(t)) + \int_{-\infty}^{t} q(t, s, x(s))ds$$

where p is continuous and $|q(t, s, x)| \leq Ke^{-(t-s)}|x|^n$. In this case we can choose $g(t) = 1 + |t^j|$ for any $j > 0$, for example.

The space $(G, |\cdot|_g)$ admits $g(t) \equiv 1$ and then becomes the space of bounded continuous functions with the supremum norm, but it will not satisfy an important property encountered in many problems: when $x : (-\infty, A) \to R^n$ is bounded and continuous, then the mapping $t \to x_t$ need not be continuous, as the example $x(t) = \sin(t^2)$ shows. Virtually always we need to ask that $g(s) \to \infty$ as $s \to -\infty$ and that makes $(G, |\cdot|_g)$ a fading memory space and (20) a fading memory equation. Properties of this space are discussed extensively in Burton [1]. An example of Seifert [15]

$$x'(t) = -2x(t) + x(0)$$

with solutions

$$x(t, x(0)) = (1 + e^{-2t})x(0)/2$$

shows how disasterous the absence of a fading memory can be in asymptotic stability theory.

Notice the progression from Theorem 1 to Theorem 4 of the continuity of f. In Lemma 1 in the proofs of both Theorem 1 and 2 we point out that $f(t, x(t))$ and $D(t, s, x(s))$ are continuous functions of t and (t, s) (by the composite function theorem) when $x(t)$ is continuous. In Lemma 1 of Theorem 3 we prove that $f(t, x_t)$ is a continuous function of t when $x(t)$ is continuous. But that will not work in the infinite delay case and we are compelled to add that as a hypothesis.

For a given $t_0 \geq 0$, $\tilde{\psi}$ *in* G, and λ in $(0, 1)$, we also consider

(21) $$x' = \lambda f(t, x_t), \quad x_{t_0} = \lambda \tilde{\psi}.$$

Theorem 4.   *Suppose that if* $\varphi : R \to R^n$ *with* $\varphi_t \in G$ *for* $t \geq 0$, *then* $f(t, \varphi_t)$ *is a continuous function of* t. *Let* $(t_0, \tilde{\psi}) \in [0, \infty) \times G$ *be given and suppose that either:*

(I) there are continuous functions $\Gamma : [0, \infty) \to [0, \infty)$ *and* $W : [0, \infty) \to [1, \infty)$ *with* W *increasing,* $|f(t, \psi)| \leq \Gamma(t)W(|\psi|_g)$, *and* $\int_0^\infty \frac{ds}{W(s)} = \infty$; *or*

(II) f takes bounded sets into bounded sets and there is a continuous functional $V :$ $[0, \infty) \times G \to [0, \infty)$ *which is locally Lipschitz in* ψ *with* $V'(t, x_t) \leq 0$ *along any solution of (21) and for each* $T > 0$ *then* $V(t, \psi) \to \infty$ *as* $|\psi|_g \to \infty$ *uniformly for* $0 \leq t \leq T$.

Then there is a solution $x(t) = x(t, t_0, \tilde{\psi})$ *of (20) on* $[t_0, \infty)$ *with* $x_{t_0} = \tilde{\psi}$.

Proof.   Let $T > 0$ be given and let $(C, \|\cdot\|)$ be the Banach space of continuous functions $\varphi : (-\infty, T] \to R^n$ for which $\|\varphi\| := \sup\limits_{t_0 \leq s \leq T} |\varphi_s|_g$ exists as a finite number. Define $H : C \to C$ by $\varphi \in C$ implies that

$$H(\varphi)_{t_0} = \tilde{\psi}$$
$$H(\varphi)(t) = \tilde{\psi}(0) + \int_{t_0}^t f(s, \varphi_s)ds \text{ for } t \geq t_0.$$

Lemma 1. $H : C \to C$ *and* H *maps bounded sets into compact sets.*

Proof.   Since $\varphi \in C$ implies that $f(t, \varphi_t)$ is a continuous function of t and since $H(\varphi)(t) = \tilde{\psi}(t - t_0)$ for $t < t_0$ with $\tilde{\psi} \in G$, it follows that $H : C \to C$. For a given $J > 0$, if $t_0 \leq t \leq T$ and if $\{\varphi^{(n)}\} \subset C$ is any sequence satisfying $\|\varphi^{(n)}\| \leq J$, then there is an $M > 0$ with $|f(t, \varphi_t^{(n)})\| \leq M$ by either (I) or (II). Hence, by Ascoli's theorem there is a subsequence, say $\{\varphi^{(n)}\}$ again, such that $\{H(\varphi^{(n)})\}$ converges uniformly to a function φ for $t_0 \leq t \leq T$ and with $H(\varphi^{(n)})(t_0) = \tilde{\psi}(0)$, so that $\varphi(t_0) = \tilde{\psi}(0)$. Hence, $\{H(\varphi^{(n)})\}$ converges to φ on $[t_0, T]$ and $H(\varphi^{(n)})(t) = \tilde{\psi}(t - t_0)$ for $t < t_0$. Thus, the bounded set $\{\varphi \in C| \|\varphi\| \leq J\}$ is mapped into a compact set.

Lemma 2. H *is continuous.*

Proof.   If we replace $|.|_h$ by $|.|_g$, since we assume that $f(t, \varphi_t)$ is a continuous function of t, the proof becomes identical to that of Lemma 2 for Theorem 3.

Lemma 3.   *There is a constant K such that if $\varphi = \lambda H(\varphi)$ for $0 < \lambda < 1$, then $\|\varphi\| \leq K$.*

Proof.   If $t_0 \leq t \leq T$, then for (I) we have

$$|\varphi(t)| \leq |\tilde{\psi}|_g + \int_{t_0}^t \Gamma(s)W(|\varphi_s|_g)ds$$

and the right-hand side is increasing, while $\varphi_{t_0} = \lambda\tilde{\psi}$; hence,

$$|\varphi_t|_g \leq |\tilde{\psi}|_g + \int_{t_0}^t \Gamma(s)W(|\varphi_s|_g)ds.$$

Thus, $|\varphi_t|_g$ is bounded by any solution of

$$y(t) = |\tilde{\psi}|_g + 1 + \int_{t_0}^t \Gamma(s)W(y(s))ds.$$

By the Conti-Wintner argument, K exists. When (II) holds there is a parallel argument.

Theorem 4 now follows from Schaefer's result.

Cor.   *Let $(t_0, \tilde{\psi}) \in [0, \infty) \times G, |\tilde{\psi}|_g = K$, and suppose there are $T > t_0$ and $J > 0$ such that for $(C, \|\cdot\|)$ defined in the proof of Theorem 4,*

(i) if $t_0 \leq t \leq T$ and $\varphi \in C$ with $|\varphi_t|_g \leq J + K$, then $f(t, \varphi_t)$ is a continuous function of t, and

(ii) if $t_0 \leq t \leq T$, $\varphi \in C$, $|\varphi_t|_g \leq K + J$ then $|f(t, \varphi_t)| \leq M$.

Then (19) has a solution $x(t, t_0, \tilde{\psi})$ for $t_0 \leq t \leq \alpha$ where $\alpha = \min[T, t_0 + J/M]$.

The corollary is proved by using the extension theorem [6; p. 111] again.

The most important idea that an example can convey is that of the fading memory and how it can make $f(t, \varphi_t)$ continuous when $\varphi_t \in G$ and $\varphi(t)$ is continuous. Thus, our example is a simple one.

Example 4.   Consider the scalar equation

$$x'(t) = \int_{-\infty}^t D(t, s)x(s)ds = \int_{-\infty}^0 D(t, u + t)x_t(u)du =: f(t, x_t)$$

where D is continuous and $|D(t, s)| \leq e^{-(t-s)}$. Select $g(t) = 1 + |t|$. If $\varphi : R \to R^n$ is continuous and if for each $t \geq 0$, $\varphi_t \in G$, then this means that $|\varphi_t|_g = \sup_{-\infty < s \leq 0} |\varphi(t +$

$s)/[1 + |s|]$ exists as a finite number, say k, so that $|\varphi(t + s)| \leq k(1 + |s|)$ for $s \leq 0$. Now φ_t is certainly not continuous in the sense that $|\varphi_t - \varphi_s|_g$ is small for $|t - s|$ small and so we are not depending on a composite function theorem to make $f(t, \varphi_t)$ a continuous function of t. Instead, we rely on the fading memory. For

$$\int_{-\infty}^{0} D(t, u + t)\varphi_t(u)du = \int_{-\infty}^{-P} D(t, u + t)\varphi_t(u)du + \int_{-P}^{0} D(t, u + t)\varphi_t(u)du$$

and

$$\left| \int_{-\infty}^{-P} D(t, u + t)\varphi_t(u)du \right| \leq \int_{-\infty}^{-P} e^u k(1 + |u|)du \to 0 \text{ as}$$

$P \to \infty$, while $\int_{-P}^{0} D(t, u + t)\varphi_t(u)du$ is a continuous function of t. The tail of φ_t fades in importance. Clearly, the growth condition of (I) is satisfied.

References

1. Burton, T.A., *Stability and Periodic Solutions of Ordinary and Functional Differential Equations*, Academic Press, Orlando, Florida, 1983.

2. Coddington, E.A. and Levinson, N., *Theory of Ordinary Differential Equations*, McGraw-Hill, New York, 1955.

3. Corduneanu, C., *Integral Equations and Applications*, Cambridge University Press, Cambridge, 1991.

4. Driver, R., Existence and stability of solutions of a delay-differential system, Arch. Rational Mech. Anal. **10** (1962), 401–426.

5. Driver, R., Existence theory for a delay-differential system, Contrib. Differential Equations **1** (1963), 317–335.

6. Friedman, A., *Foundations of Modern Analysis*, Dover, New York, 1982.

7. Gripenberg, G., Londen, S.-O., and Staffans, O., *Volterra Integral and Functional Equations*, Cambridge University Press, Cambridge, 1990.

8. Hale, J.K., *Ordinary Differential Equations*, Wiley, New York, 1969.

9. Hale, J.K., *Theory of Functional Differential Equations*, Springer, New York, 1977.

10. Hartman, P., *Ordinary Differential Equations*, Wiley, New York, 1964.

11. Kato, J. and Strauss, A., On the global existence of solutions and Liapunov functions, Ann. Mat. pura appl. **77** (1967), 303–316.

12. Lakshmikantham, V. and Leela, S., *Differential and Integral Inequalities*, Vol. I and II, Academic Press, Orlando, Florida, 1969.

13. Sansone, G. and Conti, R., *Non-linear Differential Equations*, Macmillan, New York, 1964.

14. Schaefer, H., Uber die Methode der a priori Schranken, Math. Ann. **129**(1955), 415–416.

15. Seifert, G., Liapunov-Razumikhin conditions for stability and boundedness of functional differential equations of Volterra type, J. Differential Equations **14** (1972), 424–430.

16. Smart, D.R., *Fixed Point Theorems*, Cambridge University Press, Cambridge, 1980.

17. Yoshizawa, T., Stability Theory by Liapunov's Second Method, Math. Soc. Japan, Tokyo, 1966.

Monotone Iterative Algorithms for Coupled Systems of Nonlinear Parabolic Boundary Value Problems

YING CHEN   Department of Applied Mathematics University of Waterloo, Waterloo, Ontario, Canada

XINZHI LIU   Department of Applied Mathematics University of Waterloo, Waterloo, Ontario, Canada

1   INTRODUCTION

Let us consider the following coupled system of nonlinear parabolic equations

$$\partial u_s/\partial t - L^{(s)}u_s + g^{(s)}(t,x,u_s)\cdot\nabla u_s = f^{(s)}(t,x,u_1,u_2), \tag{1}$$
$$s=1,2,\ (t\in(0,T],\ x\in\Omega),$$

together with the boundary and initial conditions

$$\alpha_s(x)\partial u_s/\partial\nu + \beta_s(x)u_s = \phi_s(t,x),\quad s=1,2,\ (t\in(0,T],\ x\in\partial\Omega), \tag{2}$$

$$u_s(0,x) = \psi_s(x),\qquad s=1,2,\ (x\in\Omega), \tag{3}$$

where $L^{(s)}$ are uniform elliptic operators in Ω defined by

$$L^{(s)} = \sum_{i,j=1}^{m} a^{(s)ij}(t,x)\partial^2/\partial x_i\partial x_j,\ s=1,2$$

and

$$g^{(s)}(t, x, u_s) \cdot \nabla u_s = \sum_{i=1}^{m} g_i^{(s)}(t, x, u_s) \partial u_s / \partial x_i.$$

All functions appeared in (1) - (3) are assumed to be sufficiently smooth so that the boundary-initial value problem (1) - (3) admits a classical solution. Nonlinear systems of parabolic equations of form (1) arise in many fields of applications, e.g. fluid mechanics. One of the most powerful tools in the qualitative analysis of nonliear differential equations is the monotone iterative method which has gained much attention in recent years. The novel idea of this method is that by using suitable initial iterations, i.e. the lower and upper solutions, one can construct monotone sequences from an auxiliary linear or quasilinear system and these sequences converge to the solution of the nonlinear system. For a detailed discussion of this method see the monograghs by Ladde, Lakshmikantham and Vatsala (1985) and Pao (1992). Pao (1985,1990) adapted this method for the computation of numerical solutions of mildly nonlinear parabolic equations. But its extension to convection dominated nonlinear equations had remained unsolved since the difficulties that arise when solving such equations by other numerical methods are well known. See Wong and Xin (1989) for a discussion of this point. Recently, Liu, Wong and Ji (1992) have developed a finite difference scheme and successfully extended the monotone iterative approach to the computation of numerical solutions of convection dominated elliptic equations. In this note we shall present some partial results on its extension to coupled system of convection dominated parabolic equations.

2   DISCRETIZATION

We assume for simplicity $m = 2$ and

$$\Omega = \{(x_1, x_2) \in R^2 : \|x_i - x_{i0}\| < \delta, \delta > 0, i = 1, 2\}.$$

We set up a uniform mesh size h for $\bar{\Omega}$ and a uniform mesh size $p = \theta h^2$ for $[0, T]$ where $\theta \in (0, \frac{1}{2}]$. let $x_{ij} = (x_i, x_j)(i, j = 0, 1, \cdots, N+1)$ be an arbitrary mesh point in $\bar{\Omega}$, where N is the total number of interior points in the spatial coordinate direction. we denote by Ω_h, Λ_h, S_h the set of mesh points in Ω, $(0, T] \times \bar{\Omega}$ and $\partial\Omega$, respectively, and by $\bar{\Lambda}_h$ the set of all mesh points in $[0, T] \times \bar{\Omega}$. $u_{s(kij)} = u_s(t_k, x_{ij})(x_{ij} \in \bar{\Omega}_h)$, $f_{kij}^{(s)} = f^{(s)}(t_k, x_{ij}, u_{1(kij)}, u_{2(kij)}), i, j = 0, 1, \cdots, N+1$, $\phi_{kij} = \phi(t_k, x_{ij})$ $(x_{ij} \in S_h)$, $\psi_{ij} = \psi(x_{ij})$ $(x_{ij} \in \bar{\Omega}_h)$. $a_{kij}^{(s)12} = a_{kij}^{(s)21} \leq 0$, and there exist $\bar{a}^{(s)} > a^{(s)} > 0(s = 1, 2)$ such that

$$max\{a_{kij}^{(s)11}, a_{kij}^{(s)22}, i, j = 1, 2, \cdots, N, k = 1, 2, \cdots, K\} \leq \bar{a}^{(s)},$$

$$min\{a_{kij}^{(s)11} + a_{kij}^{(s)12}, a_{kij}^{(s)22} + a_{kij}^{(s)12}, i, j = 1, 2, \cdots, N, k = 1, 2, \cdots, K\} \geq a^{(s)}.$$

For convinience , we make the following notations:

$$g_{n(kij)}^{(s)+} = max\{0, g_{n(kij)}^{(s)}\}, \quad g_{n(kij)}^{(s)-} = g_{n(kij)}^{(s)} - g_{n(kij)}^{(s)+}, \tag{4}$$

where $g^{(s)}{}_{n(kij)} = g^{(s)}_{n(kij)}(t_k, x_{ij}, u_{s(kij)})$ with $g^{(s)}_{kij} = (g^{(s)}_{1(kij)}, g^{(s)}_{2(kij)})$, n, $s = 1, 2$, $i, j = 0, 1, 2, \cdots, N+1$, $k = 0, 1, \cdots, K$.

$$\nabla u_{s(kij)} = \frac{(1-\alpha^{(s)})}{2} u_{s(ki(j+1))} + \alpha^{(s)} u_{s(kij)} - \frac{(1+\alpha^{(s)})}{2} u_{s(ki(j-1))}; \qquad (5)$$

$$\bar{\nabla} \bar{u}_{s(kij)} = \frac{(1-\bar{\alpha}^{(s)})}{2} u_{s(k(i+1)j)} + \bar{\alpha}^{(s)} u_{s(kij)} - \frac{(1+\bar{\alpha}^{(s)})}{2} u_{s(k(i-1)j)}; \qquad (6)$$

$$\nabla_+ u_{s(kij)} = \nabla u_{s(kij)} \text{ with } \alpha^{(s)} = \alpha_1^{(s)} \in [-1, 0];$$

$$\nabla_- u_{s(kij)} = \nabla u_{s(kij)} \text{ with } \alpha^{(s)} = \alpha_2^{(s)} \in [1, 0];$$

$$\bar{\nabla}_+ u_{s(kij)} = \bar{\nabla} u_{s(kij)} \text{ with } \bar{\alpha}^{(s)} = \bar{\alpha}_1^{(s)} \in [-1, 0];$$

$$\bar{\nabla}_- u_{s(kij)} = \bar{\nabla} u_{s(kij)} \text{ with } \bar{\alpha}^{(s)} = \bar{\alpha}_2^{(s)} \in [0, 1];$$

$$\hat{c}^{(s)} = \frac{\alpha^{(s)}(x)}{\alpha^{(s)}(x) + \beta^{(s)}(x)h}; \quad \hat{d}^{(s)}(x) = \frac{h}{\alpha^{(s)}(x) + \beta^{(s)}(x)h}.$$

It is easy to see $\hat{c}^{(s)}(x) \in [0, 1], \hat{d}^{(s)}(x) \geq 0$. Then the first-order difference operator is given by

$$h^{-1}[g^{(s)+}_{1(kij)} \nabla_- u_{s(kij)} + g^{(s)-}_{1(kij)} \nabla_+ u_{s(kij)} + g^{(s)+}_{2(kij)} \bar{\nabla}_- u_{s(kij)} + g^{(s)-}_{2(kij)} \bar{\nabla}_+ u_{s(kij)}] \qquad (7)$$

where $g^{(s)\pm}_{n(kij)}, (s, n = 1, 2)$, is given by (4).

The standard second order difference operator is given by

$$Lu_{kij} = \sum_{i,j=1}^{m} (a^{11}_{kij} \frac{\partial^2 u_{kij}}{\partial^2 x_1} + a^{22}_{kij} \frac{\partial^2 u_{kij}}{\partial^2 x_2} + 2a^{12}_{kij} \frac{\partial^2 u_{kij}}{\partial x_1 \partial x_2})$$

where

$$\frac{\partial^2 u_{s(kij)}}{\partial x_1^2} = \frac{1}{h^2}[u_{s(k(i-1)j)} - 2u_{s(kij)} + u_{s(k(i+1)j)}];$$

$$\frac{\partial^2 u_{s(kij)}}{\partial x_2^2} = \frac{1}{h^2}[u_{s(ki(j-1))} - 2u_{s(kij)} + u_{s(ki(j+1))}];$$

and

$$\frac{\partial^2 u_{s(kij)}}{\partial x_1 \partial x_2} = \frac{1}{2h^2}[u_{s(ki(j+1))} - u_{s(kij)} + u_{s(ki(j-1))} - u_{s(k(i-1)(j+1))}$$

$$+ u_{s(k(i-1)j)} - u_{s(kij)} + u_{s(k(i+1)j)} - u_{s(k(i+1)(j-1))}].$$

Similarly , by letting $\phi^{(s)}_{kij} = \phi^{(s)}(t_k, x_{ij}), x_{ij} \in \partial\Omega$, $\psi^{(s)}_{ij} = \phi^{(s)}(x_{ij}), x_{ij} \in \bar{\Omega}$, $s = 1, 2, i, j = 1, 2, \cdots, N, k = 1, 2, \cdots, K$, the finite difference approximation for the boundary and initial conditions (2) and (3) becomes

$$\alpha^{(s)} |x_{ij} - \hat{x}_{ij}| |u_s(t_k, x_{ij}) - u_s(t_k, \hat{x}_{ij})| + \beta^{(s)}(x_{ij}) u_s(t_k, x_{ij}) = \phi^{(s)}_{kij}, \; x_{ij} \in \partial\Omega,$$

$$u_s(0, x_{kij}) = \psi^{(s)}_{ij}, \; x_{ij} \in \bar{\Omega}, i, j = 0, 1, \cdots, N+1, k = 1, 2, \cdots K,$$

where $B_k^{(s)}[u_{s(kij)}] = \alpha^{(s)} |x_{ij} - \hat{x}_{ij}| |u_s(t_k, x_{ij}) - u_s(t_k, \hat{x}_{ij})| + \beta^{(s)}(x_{ij})u_s(t_k, x_{ij})$ $(s = 1, 2)$
with $x_{ij} \in \partial\Omega$ and $\hat{x}_{ij}$ is a suitable point in Ω , and $|x_{ij} - \hat{x}_{ij}|$ is the distance between x_{ij}
and $\hat{x}_{ij}$.

Then an implicit finite difference approximation for the parabolic systems $(1) - (3)$ is
given by

$$A_k U_k + G_{k+}(u_{1k}, u_{2k})D_{k-}U_k + G_{k-}(u_{1k}, u_{2k})D_{k+}U_k + \bar{G}_{k+}(u_{1k}, u_{2k})\bar{D}_{k-}U_k$$

$$+ \quad \bar{G}_{k-}(u_{1k}, u_{2k})\bar{D}_{k+}U_k = F_k(u_{1k}, u_{2k}) + b_k, \quad (k = 1, 2, \cdots, K), \tag{8}$$

where

$$A_k = \begin{pmatrix} A_k^{(1)} & 0 \\ 0 & A_k^{(2)} \end{pmatrix},$$

$$G_{k\pm}(u_{1k}, u_{2k}) = \begin{pmatrix} G_{k\pm}^{(1)}(u_{1k}) & 0 \\ 0 & G_{k\pm}^{(2)}(u_{2k}) \end{pmatrix}, \quad G_{k\pm}^{(s)}(u_{sk}) = diag(g_{1(kij)}^{(s)\pm}),$$

$$\bar{G}_{k\pm}(u_{1k}, u_{2k}) = \begin{pmatrix} \bar{G}_{k\pm}^{(1)}(u_{1k}) & 0 \\ 0 & \bar{G}_{k\pm}^{(2)}(u_{2k}) \end{pmatrix}, \quad \bar{G}_{k\pm}^{(s)}(u_{sk}) = diag(g_{2(kij)}^{(s)\pm}),$$

$$D_{k\pm} = hB_{k\pm}, \bar{D}_{k\pm} = h\bar{B}_{k\pm},$$

with

$$B_{k\pm} = \begin{pmatrix} B_{k\pm}^{(1)} & 0 \\ 0 & B_{k\pm}^{(2)} \end{pmatrix}, \quad \bar{B}_{k\pm} = \begin{pmatrix} \bar{B}_{k\pm}^{(1)} & 0 \\ 0 & \bar{B}_{k\pm}^{(2)} \end{pmatrix},$$

$$F_k(u_{1k}, u_{2k}) = \begin{pmatrix} F_{1k}(u_{1k}, u_{2k}) \\ F_{2k}(u_{1k}, u_{2k}) \end{pmatrix}, \quad F_{sk}(u_{1k}, u_{2k}) = diag(h^2 f_{kij}^{(s)}),$$

$$b_k = \begin{pmatrix} b_{1k} \\ b_{2k} \end{pmatrix}, \quad u_k = \begin{pmatrix} u_{1k} \\ u_{2k} \end{pmatrix},$$

$A_k^{(s)}, B_{\pm k}^{(s)}, \bar{B}_{\pm k}^{(s)}$ are $N^2 \times N^2$ matrices , s = 1,2 .

To apply the method of upper$-$lower solutions for the construction of monotone convergent sequences we assume that the reaction function f has a mixed quasimonotone property
to be defined next.

DEFINITION 2.1 A c^1-function $f = (f^{(1)}, f^{(2)})$ is said to be mixed quasimonotone in Ω
if $\partial f^{(1)}/\partial u_2 \le 0$, $\partial f^{(2)}/\partial u_1 \ge 0$ (or vice versa) for all $(u_1, u_2) \in \Omega$.

For definiteness, we always assume in this paper $\partial f^{(1)}/\partial u_2 \leq 0$, $\partial f^{(2)}/\partial u_1 \geq 0$. When f is mixed quasimonotone, we define the concept of upper—lower solutions as follows.

DEFINITION 2.2 A pair of functions $\bar{u}$ and $\underline{u}$ with $\underline{u} \leq \bar{u}$ on $\bar{\Lambda}_h$ is called a pair of upper-lower solutions of (6) if they satisfy

$$A_k^{(1)}\bar{u}_{1k} + G_{k+}^{(1)}(\bar{u}_{1k})D_{k-}^{(1)}\bar{u}_{1k} + G_{k-}^{(1)}(\bar{u}_{1k})D_{k+}^{(1)}\bar{u}_{1k} + \bar{G}_{k+}^{(1)}(\bar{u}_{1k})\bar{D}_{k-}^{(1)}\bar{u}_{1k}$$
$$+\bar{G}_{k-}^{(1)}(\bar{u}_{1k})\bar{D}_{k+}^{(1)}\bar{u}_{1k} \geq F_k^{(1)}(\bar{u}_{1k},\underline{u}_{2k}) + b_k^{(1)}, \tag{9}$$

$$A_k^{(2)}\bar{u}_{2k} + G_{k+}^{(2)}(\bar{u}_{2k})D_{k-}^{(2)}\bar{u}_{2k} + G_{k-}^{(2)}(\bar{u}_{2k})D_{k+}^{(2)}\bar{u}_{2k} + \bar{G}_{k+}^{(2)}(\bar{u}_{2k})\bar{D}_{k-}^{(2)}\bar{u}_{2k}$$
$$+\bar{G}^{(2)k-}(\bar{u}_{2k})D_{k+}^{(2)}\bar{u}_{2k} \geq F_k^{(2)}(\bar{u}_{1k},\bar{u}_{2k}) + b_k^{(2)}, \tag{10}$$

$$A_k^{(1)}\underline{u}_{1k} + G_{k+}^{(1)}(\underline{u}_{1k})D_{k-}^{(1)}\underline{u}_{1k} + G_{k-}^{(1)}(\underline{u}_{1k})D_{k+}^{(1)}\underline{u}_{1k} + \bar{G}_{k+}^{(1)}(\underline{u}_{1k})\bar{D}_{k-}^{(1)}\underline{u}_{1k}$$
$$+\bar{G}_{k-}^{(1)}(\underline{u}_{1k})\bar{D}_{k+}^{(1)}\underline{u}_{1k} \leq F_k^{(1)}(\underline{u}_{1k},\bar{u}_{2k}) + b_k^{(1)}, \tag{11}$$

$$A_k^{(2)}\underline{u}_{2k} + G_{k+}^{(2)}(\underline{u}_{2k})D_{k-}^{(2)}\underline{u}_{2k} + G_{k-}^{(2)}(\underline{u}_{2k})D_{k+}^{(2)}\underline{u}_{2k} + \bar{G}_{k+}^{(2)}(\underline{u}_{2k})\bar{D}_{k-}^{(2)}\underline{u}_{2k}$$
$$+\bar{G}^{(2)k-}(\underline{u}_{2k})\bar{D}_{k+}^{(2)}\underline{u}_{2k} \leq F_k^{(2)}(\underline{u}_{1k},\underline{u}_{2k}) + b_k^{(2)}. \tag{12}$$

When equality holds in $(9) - (12)$, it is called a pair of quasisolutions.

3 ALGORITHMS

Let $\bar{u}$ and $\underline{u}$ be a pair of upper and lower solutions of (9) such that $\bar{u} \leq \underline{u}$ on $\bar{\Lambda}_h$. For $k = 1,2,\cdots, K, i,j = 1,2,\ldots,N$,

$$\alpha_{kij\pm}^{(s)} \geq max\{|g_{1(kij)\pm}^{(s)}|/a^{(s)} : \underline{u}_{1k} \leq u_{1(kij)} \leq \bar{u}_{1k}\}, \tag{13}$$

$$\bar{\alpha}_{kij\pm}^{(s)} \geq max\{|g_{2(kij)\pm}^{(s)}|/a^{(s)} : \underline{u}_{1k} \leq u_{1(kij)} \leq \bar{u}_{1k}\}, \tag{14}$$

then $\alpha_{kn}^{(s)}$ and $\bar{\alpha}_{kn}^{(s)}, n, s = 1, 2$, are given by

$$\alpha_{k1}^{(s)} = min\{0, -1 + \tfrac{2}{h\alpha_{kij-}^{(s)}}, i, j = 1, 2, \cdots, N\},$$

$$\alpha_{k2}^{(s)} = max\{0, 1 - \tfrac{2}{h\alpha_{kij+}^{(s)}}, i, j = 1, 2, \cdots, N\}, \tag{15}$$

$$\bar{\alpha}_{k1}^{(s)} = min\{0, -1 + \tfrac{2}{h\bar{\alpha}_{kij-}^{(s)}}, i, j = 1, 2, \cdots, N\},$$

$$\bar{\alpha}_{k2}^{(s)} = max\{0, 1 - \tfrac{2}{h\bar{\alpha}_{kij+}^{(s)}}, i, j = 1, 2, \cdots, N\}. \tag{16}$$

It is easy to see from (15) and (16) that $\alpha_{k1}^{(s)}, \bar{\alpha}_{k1}^{(s)} \in [-1,0]$ and $\alpha_{k2}^{(s)}, \bar{\alpha}_{k2}^{(s)} \in [0,1], s = 1,2, k = 1,2,\cdots,K$. Let

$$c_{kij\pm}^{(s)} = \max\{|\frac{dg_{1kij\pm}^{(s)}}{du_s}| : \underline{u}_{s(kij)} \le u_{s(kij)} \le \bar{u}_{s(kij)}\},$$

$$\bar{c}_{kij\pm}^{(s)} = \max\{|\frac{dg_{2kij\pm}^{(s)}}{du_s}| : \underline{u}_{s(kij)} \le u_{s(kij)} \le \bar{u}_{s(kij)}\},$$

$$m_{kij}^{(s)} = \max\{-\frac{\partial f_{kij}^{(s)}}{\partial u_s} : \underline{u}_{s(kij)} \le u_{s(kij)} \le \bar{u}_{s(kij)}\},$$

$$s = 1,2, \ i,j = 1,2,\cdots,N, \ k = 1,2,\cdots,K.$$

We define the operator $T_{sk}^{(m-1)}$ for a vector $u_k = (u_{1k}, u_{2k})$ with $\underline{u}_{sk} \le u_{sk} \le \bar{u}_{sk}$ $(s = 1,2)$ by

$$T_{sk}^{(m-1)}u_{sk}^{(m)} = A_k^{(s)}u_{sk}^{(m)} + G_{k+}^{(s)}(u_{1k}^{(m-1)}, u_{2k}^{(m-1)})D_{k-}^{(s)}u_{sk}^{(m)}$$

$$+ \ G_{k-}^{(s)}(u_{1k}^{(m-1)}, u_{2k}^{(m-1)})D_{k+}^{(s)}u_{sk}^{(m)} + \bar{G}_{k+}^{(s)}(u_{1k}^{(m-1)}, u_{2k}^{(m-1)})\bar{D}_{k-}^{(s)}u_{sk}^{(m)}$$

$$+ \ \bar{G}_{k-}^{(s)}(u_{1k}^{(m-1)}, u_{2k}^{(m-1)})\bar{D}_{k+}^{(s)}u_{sk}^{(m)} + Q_{sk}^{(m-1)}u_{1k}^{(m)} \quad (17)$$

and for an initial function $u_k^{(0)} = (u_{1k}^{(0)}, u_{2k}^{(0)})$, we construct the sequences $u_{sk}^{(m)}, (s = 1,2)$, such that $u_{sk}^{(m)}$ is the solution of the following system

$$T_{sk}^{(m-1)}u_{sk}^{(m)} = Q_{sk}^{(m-1)}u_{sk}^{(m-1)} + F_{sk}(u_{1k}^{(m-1)}, u_{2k}^{(m-1)}) + b_{sk}, \quad (18)$$

where $Q_{sk}^{(m-1)} = diag(q_{s(kij)}^{(m-1)})$ and for $i,j = 1,2,\cdots,N, k = 1,2,\cdots,K,$

$$q_{s(kij)}^{(m-1)} = \gamma_{s(kij)}^{(m-1)} + m_{kij}^{(s)}, \ (s = 1,2). \quad (19)$$

For simplicity, we denote the following cases by

$$(A1)_{sn} : \frac{dg_{n(kij)}^{(s)}}{du_s} \ge 0; \quad (A2)_{sn} : \frac{dg_{n(kij)}^{(s)}}{du_s} \le 0; \quad (A3)_{sn} : g_{n(kij)}^{(s)} \text{ change sign},$$

$$\underline{u}_{sk} \le u_{sk} \le \bar{u}_{sk}, \qquad i,j = 1,2,\cdots,N, \qquad k = 1,2,\cdots,K, s = 1,2.$$

$$(\lambda 1)_s^{(m-1)} = max\{0, c_{kij+}^{(s)}d_{(3.1)kij-}^{(s)(m-1)}, c_{kij-}^{(s)}d_{(3.1)kij+}^{(s)(m-1)}, c_{kij+}^{(s)}d_{(3.2)kij-}^{(s)(m-1)}, c_{kij-}^{(s)}d_{(3.2)kij+}^{(s)(m-1)}\},$$

$$(\bar{\lambda} 1)_s^{(m-1)} = max\{0, \bar{c}_{kij+}^{(s)}\bar{d}_{(3.1)kij-}^{(s)(m-1)}, \bar{c}_{kij-}^{(s)}\bar{d}_{(3.1)kij+}^{(s)(m-1)}, \bar{c}_{kij+}^{(s)}\bar{d}_{(3.2)kij-}^{(s)(m-1)}, \bar{c}_{kij-}^{(s)}\bar{d}_{(3.2)kij+}^{(s)(m-1)}\},$$

$$(\lambda 2)_s^{(m-1)} = max\{0, c_{kij+}^{(s)}e_{(3.1)kij-}^{(s)(m-1)}, c_{kij-}^{(s)}e_{(3.1)kij+}^{(s)(m-1)}, c_{kij+}^{(s)}e_{(3.2)kij-}^{(s)(m-1)}, c_{kij-}^{(s)}e_{(3.2)kij+}^{(s)(m-1)}\},$$

$$(\bar{\lambda} 2)_s^{(m-1)} = max\{0, \bar{c}_{kij+}^{(s)}\bar{e}_{(3.1)kij-}^{(s)(m-1)}, \bar{c}_{kij-}^{(s)}\bar{e}_{(3.1)kij+}^{(s)(m-1)}, \bar{c}_{kij+}^{(s)}\bar{e}_{(3.2)kij-}^{(s)(m-1)}, \bar{c}_{kij-}^{(s)}\bar{e}_{(3.2)kij+}^{(s)(m-1)}\},$$

$$(\lambda 3)_s^{(m-1)} = 2max\{(\lambda 1)_s^{(m-1)}, (\lambda 2)_s^{(m-1)}\},$$

$$(\bar{\lambda} 3)_s^{(m-1)} = 2max\{(\bar{\lambda} 1)_s^{(m-1)}, (\bar{\lambda} 2)_s^{(m-1)}\},$$

where

$$d^{(s)(m-1)}_{(3.1)kij-}, \; d^{(s)(m-1)}_{(3.1)kij+}, \; \bar{d}^{(s)(m-1)}_{(3.1)kij-}, \; \bar{d}^{(s)(m-1)}_{(3.1)kij+}, e^{(s)(m-1)}_{(3.1)kij-}, \; e^{(s)(m-1)}_{(3.1)kij+}, \; \bar{e}^{(s)(m-1)}_{(3.1)kij-}, \; \bar{e}^{(s)(m-1)}_{(3.1)kij+}$$

will be determined by Alogrithem 3.1 and

$$d^{(s)(m-1)}_{(3.2)kij-}, \; d^{(s)(m-1)}_{(3.2)kij+}, \; \bar{d}^{(s)(m-1)}_{(3.2)kij-}, \bar{d}^{(s)(m-1)}_{(3.2)kij+}, \; e^{(s)(m-1)}_{(3.1)kij+}, \; e^{(s)(m-1)}_{(3.2)kij-}, \; \bar{e}^{(s)(m-1)}_{(3.2)kij-}, \; \bar{e}^{(s)(m-1)}_{(3.2)kij+}$$

will be determined by Alogrithem 3.2.

ALGORITHM 3.1 (For construction of the monotone increasing sequences).
Set $\underline{u}^{(0)}_k = \underline{u}_k$ $(k = 1, 2, \cdots, K)$ and for $m = 1, 2, \cdots$, compute:

$$d^{s(m-1)}_{kij+} = h\left(\frac{1-\alpha^{(s)}_1}{2}\bar{u}_{s(ki(j+1))} + \alpha^{(s)}_1 u^{(m-1)}_{s(kij)} - \frac{1+\alpha^{(s)}_1}{2}u^{(m-1)}_{s(ki(j-1))}\right),$$

$$d^{s(m-1)}_{kij-} = h\left(\frac{1-\alpha^{(s)}_2}{2}\bar{u}_{s(ki(j+1))} + \alpha^{(s)}_2 \bar{u}_{s(kij)} - \frac{1+\alpha^{(s)}_2}{2}u^{(m-1)}_{s(ki(j-1))}\right),$$

$$\bar{d}^{s(m-1)}_{kij+} = h\left(\frac{1-\bar{\alpha}^{(s)}_1}{2}\bar{u}_{s(k(i+1)j)} + \bar{\alpha}^{(s)}_1 u^{(m-1)}_{s(kij)} - \frac{1+\bar{\alpha}^{(s)}_1}{2}u^{(m-1)}_{s(k(i-1)j)}\right),$$

$$\bar{d}^{s(m-1)}_{kij-} = h\left(\frac{1-\bar{\alpha}^{(s)}_2}{2}\bar{u}_{s(k(i+1)j)} + \bar{\alpha}^{(s)}_2 \bar{u}_{s(kij)} - \frac{1+\bar{\alpha}^{(s)}_2}{2}u^{(m-1)}_{s(k(i-1)j)}\right),$$

$$e^{s(m-1)}_{kij+} = h\left(\frac{1+\alpha^{(s)}_1}{2}\bar{u}_{s(ki(j-1))} - \alpha^{(s)}_1 \bar{u}_{s(kij)} - \frac{1-\alpha^{(s)}_1}{2}u^{(m-1)}_{s(ki(j+1))}\right),$$

$$e^{s(m-1)}_{kij-} = h\left(\frac{1+\alpha^{(s)}_2}{2}\bar{u}_{s(ki(j-1))} - \alpha^{(s)}_2 u^{(m-1)}_{s(kij)} - \frac{1-\alpha^{(s)}_2}{2}u^{(m-1)}_{s(ki(j+1))}\right),$$

$$\bar{e}^{s(m-1)}_{kij+} = h\left(\frac{1+\bar{\alpha}^{(s)}_1}{2}\bar{u}_{s(k(i-1)j)} - \bar{\alpha}^{(s)}_1 \bar{u}_{s(kij)} - \frac{1-\bar{\alpha}^{(s)}_1}{2}u^{(m-1)}_{s(k(i+1)j)}\right),$$

$$\bar{e}^{s(m-1)}_{kij-} = h\left(\frac{1+\bar{\alpha}^{(s)}_2}{2}\bar{u}_{s(k(i-1)j)} - \bar{\alpha}^{(s)}_2 u^{(m-1)}_{s(kij)} - \frac{1-\bar{\alpha}^{(s)}_2}{2}u^{(m-1)}_{s(k(i+1)j)}\right).$$

If $(A1)_{sn}$ hold for $s, n = 1, 2$, set

$$\gamma^{(m-1)}_{s(kij)} = (\lambda 1)^{(m-1)}_s + (\bar{\lambda}1)^{(m-1)}_s,$$

If $(A2)_{sn}$ hold for $s, n = 1, 2$, set

$$\gamma^{(m-1)}_{s(kij)} = (\lambda 2)^{(m-1)}_s + (\bar{\lambda}2)^{(m-1)}_s,$$

If $(A1)_{1n}, (A1)_{21}$, and $(A2)_{22}$ hold for $n = 1, 2.$ set

$$\gamma^{(m-1)}_{1(kij)} = (\lambda 1)^{(m-1)}_1 + (\bar{\lambda}1)^{(m-1)}_1, \; \gamma^{(m-1)}_{2(kij)} = (\lambda 1)^{(m-1)}_2 + (\bar{\lambda}2)^{(m-1)}_2,$$

If $(A3)_{sn}$ hold for $s, n = 1, 2$, set

$$\gamma_{s(kij)}^{(m-1)} = (\lambda3)_s^{(m-1)} + (\bar\lambda3)_s^{(m-1)},$$

and so on.

ALGORITHEM 3.2 (For construction of monotone decreasing sequence).
Set $\bar{u}_k^{(0)} = \bar{u}_k(k = 1, 2, \cdots, K)$. For $m = 1, 2, \cdots$, compute:

$$d_{kij+}^{(s)(m-1)} = h\left(\frac{1 - \alpha_1}{2} u_{s(ki(j+1))}^{(m-1)} + \alpha_1 \underline{u}_{s(kij)} - \frac{1 + \alpha_1}{2} u_{s(ki(j-1))}\right),$$

$$d_{kij-}^{(s)(m-1)} = h\left(\frac{1 - \alpha_2}{2} u_{s(ki(j+1))}^{(m-1)} + \alpha_2 u_{s(kij)}^{(m-1)} - \frac{1 + \alpha_2}{2} u_{s(ki(j-1))}\right),$$

$$\bar{d}_{kij+}^{(s)(m-1)} = h\left(\frac{1 - \bar\alpha_1}{2} u_{s(k(i+1)j)}^{(m-1)} + \bar\alpha_1 \underline{u}_{s(kij)} - \frac{1 + \bar\alpha_1}{2} u_{s(k(i-1)j)}\right),$$

$$\bar{d}_{kij-}^{(s)(m-1)} = h\left(\frac{1 - \bar\alpha_2}{2} u_{s(k(i+1)j)}^{(m-1)} + \bar\alpha_2 u_{s(kij)}^{(m-1)} - \frac{1 + \bar\alpha_2}{2} u_{s(k(i-1)j)}\right),$$

$$e_{kij+}^{(s)(m-1)} = h\left(\frac{1 + \alpha_1}{2} u_{s(ki(j-1))}^{(m-1)} - \alpha_1 u_{s(kij)}^{(m-1)} - \frac{1 - \alpha_1}{2} u_{s(ki(j+1))}\right),$$

$$e_{kij-}^{(s)(m-1)} = h\left(\frac{1 + \alpha_2}{2} u_{s(ki(j-1))}^{(m-1)} - \alpha_2 \underline{u}_{s(kij)} - \frac{1 - \alpha_2}{2} u_{s(ki(j+1))}\right),$$

$$\bar{e}_{kij+}^{(s)(m-1)} = h\left(\frac{1 + \bar\alpha_1}{2} u_{s(k(i-1)j)}^{(m-1)} - \bar\alpha_1 u_{s(kij)}^{(m-1)} - \frac{1 - \bar\alpha_1}{2} u_{s(k(i+1)j)}\right),$$

$$e_{kij-}^{(s)(m-1)} = h\left(\frac{1 + \alpha_2}{2} u_{s(k(i-1)j)}^{(m-1)} - \bar\alpha_2 \underline{u}_{s(kij)} - \frac{1 - \bar\alpha_2}{2} u_{s(k(i+1)j)}\right).$$

The rest are the same as in Algorithem 3.1.

Finally we remark that the sequences constructed in Algorithems $3.1 - 3.2$ are monotone
and converge to the quasisolutions $(u_{1*}, u_{2*}), (u_1^*, u_2^*)$ such that $u_{s*} \le u_s^*, s = 1, 2$, of system
(8).

To ensure that the quasisolutions $(u_{1*}, u_{2*}), (u_1^*, u_2^*)$ are true solutions of system (8) it is
necessary to show that $u_{s*} = u_s^*, s = 1, 2$, which will be discussed in a separate paper.

REFERENCES

G.S.Ladde, V.Lakshikantham, and A.Vatsala (1985). *Monotone Iterative Techniques for
Nonlinear Dfferential Equations*, Pitman.

Xinzhi Liu, Yau Shu Wang and Ji Xingzhi (1992). Monotone iterations for numerical solutions of nonlinear elliptic partial differential equations, *Appl.Math.Comp.*, **50**: 59-91.

C.V. Pao (1992). *Nonlinear Parabolic and Elliptic Equations*, Plenum Press.

C.V. Pao (1985). Monotone iterative methods for finite difference system of reaction diffusion equations, *Numer. Math.*, **46**: 153-169.

C.V. Pao (1990). Numerical methods for coupled systems of nonlinear parabolic boundary value problems, *J. Math. Anal. Appl.*, **151**: 581-608.

Y.S. Wong and X.K. Xin (1989). Eulerian-Lagrangian splitting methods for convection dominated equations, *Fluid Dynam. Rese.*, **5**: 13-28.

Steady State Bifurcation Hypersurfaces
of Chemical Mechanisms

BRUCE L. CLARKE Department of Chemistry, University of Alberta, Edmonton, Alberta, Canada

1. INTRODUCTION

Instability, oscillations, or chaos occur in a very small fraction of the chemical systems studied experimentally. Why are such phenomena so rare? An approach for answering such questions using stability analysis has been developed by Clarke *et al.* (1974-1993). Stability analysis suggests that a large fraction of all realistic chemical systems have network structures which produce a single stable steady state for all non-negative parameter values. One important role of stability analysis is to determine which network structures can produce interesting dynamics.

When a new experimental oscillator is discovered, chemists suggest molecules and reactions which are likely to be present and these become a *proposed mechanism*. Establishing the correctness of a mechanism becomes more difficult with increasing network complexity. Systems with interesting non-linear dynamics are often so complex that many species and reactions in their mechanisms are pure speculation. It is here that stability analysis becomes an important tool by predicting where steady state bifurcations should occur for comparison with experiments. Stability analysis can invalidate a mechanism by showing that its bifurcations cannot match experiments for any choice of parameters.

Stability analysis shows that instability can usually be attributed to certain kinds of positive feedback cycles. A change in the concentration of one species causes a change in another concentration around the loop. If the species in a large negative feedback cycle are involved in smaller destabilizing positive feedback cycles, Hopf bifurcations usually occur. Without the large negative cycle, saddle node bifurcations usually occur. (Clarke and Jiang, 1993)

The flow in the reactions at steady state is the sum of fundamental network flows called *extreme currents*. Different extreme currents are important in different regions of parameter space. The locations of bifurcation hypersurfaces are determined by the feedback cycles in the dominant extreme currents in each region of parameter space.

When a mechanism is missing a reaction that is needed to explain the experimental data, the network will be missing extreme currents involving this reaction. The feedback cycles in the missing currents can affect the bifurcation hypersurfaces. Discrepancies in the shape of the unstable regions can tell us much about the missing reaction.

Complex networks yield intractable systems of algebraic equations when treated exactly. These can be handled with various approximations, such as by neglecting all but the dominant or essential currents, or by using only the most important terms in polynomials. These approximations work because the range of parameters and experiments in chemistry often spans many orders of magnitude. Different approximations may be valid over various parts of the range of parameters used in experiments.

Recognizing the impossibility of obtaining an exact treatment for such complex systems, our aim is to sketch out the main features of the bifurcation surfaces when the parameters range over many orders of magnitude and the equations have a number of different approximate limiting cases which are solvable. It often turns out that such a treatment can reproduce the experimental results with surprising accuracy.

2. A MATHEMATICAL FRAMEWORK FOR CHEMICAL REACTION SYSTEMS

2.1 Definition of a Stoichiometric Network

Consider a set of n chemical species A_i, for $i = 1, \ldots, n$, and a set of r reactions R_j, for $j = 1, \ldots, r$. Let $\mathbf{N} = (\nu_{ij})$ be an arbitrary real matrix called the *stoichiometric matrix*. An element ν_{ij} is a dimensionless quantity which gives the net relative amount of species A_i produced when R_j advances by one unit. Gay-Lussac's Law states that the ratios of the elements of ν_{ij} for a given R_j are small integers. Hence the columns of $\mathbf{N}$ can be rescaled as small integers without common divisors. Since rescaling gives the same relative amounts of the species, the reaction is not considered to have changed. When scaled, $\mathbf{N}$ obeys *Gay Lussac's constraint*. We will consider $\mathbf{N}$ to be an arbitrary real matrix and occasionally assume Gay-Lussac's constraint holds.

Let $\mathbf{c} = (c_1, \ldots, c_n)$ be the concentrations of the species and $\mathbf{v} = (v_1, \ldots, v_r)$ be the rates of the reactions. Note that $\nu_{ij} v_j$ is the rate of production of A_i from R_j. Summing the rate of A_i production over all reactions gives the matrix equation

$$\mathbf{c}_t = \mathbf{N}\,\mathbf{v}. \tag{1}$$

Reaction rates depend on concentration through formulas called *rate laws*. A rate law formula is always expressed as a *rate constant* parameter k_j times a general function $u_j(\mathbf{c})$. We assume $u_j > 0$ when all $c_i > 0$. The general form

$$v_j = k_j u_j(\mathbf{c})$$

can be written

$$\mathbf{v} = \mathrm{diag}(\mathbf{k})\,\mathbf{u}(\mathbf{c}), \tag{2}$$

where $\mathrm{diag}(\mathbf{k})$ is a diagonal matrix whose diagonal is the *rate constant vector* $\mathbf{k} = (k_1, \ldots, k_r)$. Substituting in (1) gives the autonomous dynamical system

$$\mathbf{c}_t = \mathbf{N}\,\mathrm{diag}(\mathbf{k})\,\mathbf{u}(\mathbf{c}). \tag{3}$$

We will see that the trajectories lie in a space of dimension d, where d is defined as the rank of $\mathbf{N}$. If $d < n$ there is a *conservation matrix* $\mathbf{\Gamma}$ with $n - d$ linearly independent rows satisfying

$$\mathbf{\Gamma N} = \mathbf{0}. \tag{4}$$

When (1) is multiplied on the left by Γ and integrated over t, the *conservation condition*

$$\Gamma \mathbf{c} = \mathbf{C} \tag{5}$$

emerges. The vector of conservation constants $\mathbf{C} = (C_1, \ldots, C_{n-d})$ is a set of parameters required to specify the d-dimensional dynamical system in the dynamical variables $\mathbf{c}_I$.

Assume species $A_1, \ldots, A_d$ are independent and $A_{d+1}, \ldots, A_n$ and dependent. The concentration vector and other matrices split into independent and dependent blocks. Thus $\mathbf{c} = (\mathbf{c}_I, \mathbf{c}_D)$ and $\Gamma = (\Gamma_I, \Gamma_D)$. Equation (5) may be solved for the dependent concentrations

$$\mathbf{c}_D = \Gamma_D^{-1}(\mathbf{C} - \Gamma_I \mathbf{c}_I). \tag{6}$$

The general stoichiometric network is specified by a matrix $\mathbf{N}$ and functions $\mathbf{u}(\mathbf{c})$. The parameters $\mathbf{k}$ and $\mathbf{C}$ are combined into a set of $r + n - d$ parameters $(\mathbf{k}, \mathbf{C})$. The parameter domain is $k_j \geq 0$ for reactions, and $c_i \geq 0$ for species. Thus $(\mathbf{k}, \mathbf{C}) \in \bar{D}$, where $\bar{D} = \bar{R}_+^r \times \{\Gamma \mathbf{c} \mid \mathbf{c} \in \bar{R}_+^n\}$. R_+^n and $\bar{R}_+^n$ are the open and closed positive orthants of R^n.

DEFINITION: A *stoichiometric network* is a set of dynamical systems (3) with one system for each $(\mathbf{k}, \mathbf{C}) \in \bar{D}$.

Dynamical systems on the boundary of $\bar{D}$ are either missing reactions R_j because $k_j = 0$, or are missing species A_i because $c_i = 0$. Depending on the functions $u_j(\mathbf{c})$, the condition $c_i = 0$ may imply $v_j = 0$. In both cases, the boundary dynamics can be studied using a simpler network without these reactions. Hence, we frequently study the dynamics on D, the interior of $\bar{D}$.

An *open stoichiometric network* is a flow on the $(n+r)$-dimensional domain $(\mathbf{c}_I, \mathbf{k}, \mathbf{C}) \in R_+^d \times D$. It may be regarded as a d-dimensional flow of $\mathbf{c}_I \in R_+^d$ for each $(\mathbf{k}, \mathbf{C}) \in D$. The $n - d$ conservation parameters can be replaced by $\mathbf{c}_D$ using equation (6). The network is then a flow on $(\mathbf{c}, \mathbf{k}) \in R_+^{n+r}$.

Equivalent Dynamical Systems

Consider the mapping $\psi : R_+^{n+r} \to R_+^{n+r}$ which maps $(\mathbf{c}, \mathbf{k}) \mapsto (\mathbf{h}, \mathbf{v})$ defined by

$$h_i = c_i^{-1} \qquad \text{for } i = 1, \ldots, n$$
$$v_j = k_j u_j(\mathbf{c}) \qquad \text{for } j = 1, \ldots, r.$$

ψ is $1 - 1$ and the inverse map is

$$c_i = h_i^{-1} \qquad \text{for } i = 1, \ldots, n \tag{7}$$
$$k_j = v_j / u_j(\mathbf{h}^{-1}) \qquad \text{for } j = 1, \ldots, r, \tag{8}$$

where $\mathbf{h}^{-1} = (h_1^{-1}, \ldots, h_n^{-1})$. The mapped flow obeys ODE's which are equivalent to (3).

$$\mathbf{h}_t = -(\text{diag}(\mathbf{h}))^2 \mathbf{N} \mathbf{v} \tag{9}$$
$$\mathbf{v}_t = \text{diag}(\mathbf{v}) \text{diag}(\mathbf{u}^{-1})(\nabla \mathbf{u})^t \mathbf{N} \mathbf{v}. \tag{10}$$

Here $\nabla \mathbf{u}$ and $\mathbf{u}$ are functions of $\mathbf{h}$.

The steady state condition for the $(\mathbf{h}, \mathbf{v})$ flow is

$$\mathbf{N} \mathbf{v} = 0 \tag{11}$$

with no restriction on $\mathbf{h}$. Hence, the steady states are $(\mathbf{h}, \mathbf{v}) \in R_+^n \times C_v$, where

$$C_v = \{\mathbf{v} \in R_+^r \mid \mathbf{N}\mathbf{v} = \mathbf{0}\}. \tag{12}$$

This manifold is a simply connected set of dimension $n + r - d$. Also, $\psi^{-1} R_+^n \times C_v$ gives the original steady state manifold in the $(\mathbf{c}_I, \mathbf{k}, \mathbf{C})$ domain. Thus, the steady state manifold of the general stoichiometric network is simply connected.

A d-dimensional system (3) with a given $(\mathbf{k}, \mathbf{C})$ does not necessarily have a steady state. Choose $(\mathbf{k}, \mathbf{C})$ having a steady state $\mathbf{c}_{I*}$. Then $(\mathbf{c}_{I*}, \mathbf{k}, \mathbf{C})$ is a steady state of the flow on D. If $(\mathbf{h}_*, \mathbf{v}_*)$ is the corresponding steady state, equation (8) may be used to replace $\mathbf{k}$ in (3) to obtain

$$\mathbf{c}_t = \mathbf{N} \operatorname{diag}(\mathbf{v}_*) \operatorname{diag}(\mathbf{u}(\mathbf{h}_*^{-1}))^{-1} \mathbf{u}(\mathbf{c}). \tag{13}$$

The stoichiometric network ODE's have now been parametrized using the point $(\mathbf{h}_*, \mathbf{v}_*)$ on the mapped steady state manifold.

Concentrations and rate laws may be scaled relative to steady state. Define the scaled concentrations $\mathbf{x}$ and rate laws $\mathbf{w}(\mathbf{x}, \mathbf{h})$ by

$$\mathbf{x} = \operatorname{diag}(\mathbf{h}_*)\mathbf{c}$$
$$\mathbf{w}(\mathbf{x}, \mathbf{h}) = \operatorname{diag}(\mathbf{u}(\mathbf{h}_*^{-1}))^{-1}\mathbf{u}(\mathbf{c}).$$

At steady state $\mathbf{x} = \mathbf{e}$, where $\mathbf{e} = (1, \ldots, 1)^t$. Also $\mathbf{w}(\mathbf{e}, \mathbf{h}) = \mathbf{e}$. The general ODE's (13) become
$$\mathbf{x}_t = \operatorname{diag}(\mathbf{h}_*)\mathbf{N} \operatorname{diag}(\mathbf{v}_*)\mathbf{w}(\mathbf{x}, \mathbf{h}_*). \tag{14}$$

This form of the dynamics shows that $\mathbf{h}_*$ determines the time scale for the evolution of corresponding components of $\mathbf{x}$. The lower the steady state concentration of A_i, the larger h_{*i}, and the faster x_i evolves. When a low concentration species approaches an attracting pseudo-steady-state on a sufficiently fast time scale, the pseudo-steady-state approximation is often made. (Clarke 1992b.) However, this approximation sometimes eliminates instability. The approximation is safe for "flow through intermediates" in the limit $h_{*i} \to \infty$. (Clarke, 1975). An example shows that a network can be stabilized if the the intermediate X_2 is eliminated from $X_1 \to X_2 \to X_3$ when h_2 is not small enough. (Clarke, 1980, pp 199-203)

The mechanisms for experimental chemical oscillators often contain *complex reactions* which are made by combining elementary reactions using the pseudo-steady-state approximation. Complex reactions are approximations because unimportant low concentration species have been eliminated. However, certain low concentration species cannot be eliminated by this approximation because they are involved in the destabilizing feedback cycles. These rapidly evolving species are usually difficult to detect experimentally. Studying bifurcation phenomena may be one of the best means for studing their reactions.

The highest concentration species have the slowest times scales. In a real physical system, when all species including the slowest are at steady state, the system is at *thermodynamic equilibrium*. In this state, all physical processes occur in the forward and reverse directions at equal rates. A steady state of a stoichiometric network represents thermodynamic equilibrium only if the forward and reverse rates of every reaction are equal. This *detailed balance* condition is only satisfied on a subset of codimension $(r/2) - d$ of the full $(n + r - d)$-dimensional domain of the $(\mathbf{k}, \mathbf{C})$ parameters. The $(r/2) - d$ constraints on $\mathbf{k}$ are called *Wegscheider conditions* (Wegscheider, 1901).

We model non-equilibrium steady states assuming the existence of *major species*, whose time scales are so slow their concentrations are constant. The non-major species are called *intermediates*, and are usually symbolized by X. Thus, the set of species is now $A_1, \ldots, A_{n_M}, X_1, \ldots, X_{n_X}$, and the total number of species $n = n_M + n_X$. The stoichiometric matrix has an upper block $\mathbf{N}_M$ for the major species and a lower block $\mathbf{N}_X$ for the intermediates. Also $\mathbf{c} = (\mathbf{c}_M, \mathbf{c}_X)$, where $\mathbf{c}_M$ is considered constant and the dynamics of $\mathbf{c}_X$ is the focus of our attention. Thus, we will study a dynamical system of n_X intermediates with concentrations $\mathbf{c}_X$ and stoichiometric matrix $\mathbf{N}_X$. The subscript X will henceforth be dropped except when needed for clarity. An example of a major species in H^+ which appears later in Figure 9.

Extreme Currents and the $(\mathbf{h}, \mathbf{j})$ Parameters

A *current* is a steady state flow in the reactions of the network and is represented by a steady state $\mathbf{v}_*$ satisfying equation (11). A current is similar to an electrical network current which obeys Kirchhoff's Current Law. This condition says that no net charge can accumulate at a node of the network, and is a steady state condition for charge at a node. Thus, nodes in an electrical network are analogous to intermediates. Sources and sinks of charge in an electrical network are analogous to major species.

A current $\mathbf{v}_*$ produces major species at the net rate $\mathbf{N}_M \mathbf{v}_*$. Suppose this vector is rescaled by a factor v_O to obtain a dimensionless vector $\mathbf{n}_O$, so that

$$\mathbf{n}_O v_O = \mathbf{N}_M \mathbf{v}_*.$$

Then $\mathbf{n}_O$ is a possible stoichiometric vector for the *overall reaction* at steady state and v_O is its rate. This vector is usually scaled so its coefficients are integers without common divisors.

The set of all possible currents is given by C_v in equation (12). Since C_v is the intersection of a linear space and the orthant R_+^r, it is a *convex polyhedral cone* whose boundary is a subset of the orthant boundary. Transforming C_v by the mapping $\mathbf{N}_M$ produces a convex polyhedral cone of all possible overall reactions of the network.

The f edges of C_v are extreme rays, which we call *extreme currents*. One vector $\mathbf{v}_*$ in each edge is used to construct an $r \times f$ matrix $\mathbf{E}$ called the *extreme current matrix*. The ith column of $\mathbf{E}$ is a vector $\mathbf{v}_*$ which we call $\mathbf{E}_i$.

Let j_i represent the rate of the extreme current $\mathbf{E}_i$. The rates of the extreme currents form the *current rate vector* $\mathbf{j} = (j_1, \ldots, j_f)$. Given an arbitrary current $\mathbf{v}_*$, the rates are defined so

$$\mathbf{v}_* = \mathbf{E}\mathbf{j}. \tag{15}$$

C_v is a *simplicial cone* if a transversal cross-section is a simplex. The number of edges is then $f = \dim C_v = r - d$. Since $\mathbf{v}_*$ lies in the $(r - d)$-dimensional right null-space of $\mathbf{N}$, and the $r - d$ columns of $\mathbf{E}$ are a basis for this space, there is a unique set of rates $\mathbf{j}$ satisfying (15).

When C_v is not simplicial, it must be subdivided into simplicial cones. Within a simplicial subcone, the rates of the $r - d$ non-zero components of $\mathbf{j}$ are determined by (15). The convexity of C_v implies the $j_i \geq 0$.

A *simplicial decomposition* of C_v determines which set of at most $r - d$ extreme currents may have positive rates for every $\mathbf{v} \in C_v$. All other components of $\mathbf{j}$ are zero. As a result, there is a unique current rate vector $\mathbf{j}$ for every current $\mathbf{v}_* \in C_v$.

The rate vector $\mathbf{j}$ makes an excellent parameter for bifurcation theory. When (15) is substituted into (13), and the subscript $*$ is dropped from $\mathbf{h}_*$ for brevity, the equations of motion for the general stoichiometric network become

$$\mathbf{x}_t = \text{diag}(\mathbf{h})\mathbf{N}\,\text{diag}(\mathbf{E}\mathbf{j})\mathbf{w}(\mathbf{x}, \mathbf{h}). \tag{16}$$

Note that the number of parameters has not changed. Originally there were r $\mathbf{k}$-parameters and $n - d$ $\mathbf{C}$-parameters. Now there are n $\mathbf{h}$-parameters and $r - d$ possibly non-zero $\mathbf{j}$-parameters. Both cases have $n + r - d$ parameters.

3. EXAMPLE NETWORKS

Species have their own chemical symbols which are used in this section instead of $A_1, \ldots, X_1, \ldots$.

3.1 Wegscheider's Network

This network consists of three reversible reactions involving intermediates X, Y and Z. There are no major species (Wegscheider, 1901).

$$R_1 : \quad X \rightleftharpoons Y$$
$$R_2 : \quad Y \rightleftharpoons Z$$
$$R_3 : \quad Z \rightleftharpoons X$$

Each of these is actually two reactions – a forward and a reverse, which are called R_i and R_{-i} respectively. Then $n = 3$ and $r = 6$. The $n \times r$ stoichiometric matrix is

$$
\mathbf{N} = \begin{array}{c} \\ X \\ Y \\ Z \end{array}
\begin{array}{cccccc}
R_1 & R_2 & R_3 & R_{-1} & R_{-2} & R_{-3} \\
\left(\begin{array}{cccccc}
-1 & 0 & 1 & 1 & 0 & -1 \\
1 & -1 & 0 & -1 & 1 & 0 \\
0 & 1 & -1 & 0 & -1 & 1
\end{array}\right)
\end{array}
$$

This matrix is rank 2, so $d = 2$. The conservation matrix has $n - d$ rows and is

$$\mathbf{\Gamma} = (1, 1, 1).$$

There are $f = 5$ extreme currents. Three are *detailed balanced currents*, which consist of a reaction and its reverse. Another is the cycle of reactions $X \rightarrow Y \rightarrow Z \rightarrow X$, and the fifth is the reverse cycle. Note that $n + r - d = 7$. The 7 $(\mathbf{k}, \mathbf{C})$ parameters are the 6 rate constants and one conservation constant. The 7 $(\mathbf{h}, \mathbf{j})$ parameters are $\mathbf{h} = (h_X, h_Y, h_Z)$ and the rates of 4 extreme currents. C_v is 4-dimensional and is divided into two simplicial cones, each with 4 edges. Both contain the 3 detailed balanced extreme currents. The fourth current in one simplicial cone is $X \rightarrow Y \rightarrow Z \rightarrow X$. In the other simplicial cone, the fourth current is the reverse cycle.

3.2 Gray-Scott Network

The Gray-Scott network (Gray and Scott, 1983) is a very simple example with a rich bifurcation structure. It consists of two reactions R_1 and R_2 taking place in a continuously stirred tank reactor (CSTR). Reactions $R_3, \ldots, R_6$ represent the flow of A and B out of and into the reactor.

R_1 :		$A + 2B \xrightarrow{k_1} 3B$	$v_1 = k_1 A B^2$
R_2 :		$B \xrightarrow{k_2}$	$v_2 = k_2 B$
R_3 :	A exit	$A \xrightarrow{k_0}$	$v_3 = k_0 A$
R_4 :	B exit	$B \xrightarrow{k_0}$	$v_4 = k_0 B$
R_5 :	A inflow	$\xrightarrow{k_0 A_0} A$	$v_5 = k_0 A_0$
R_6 :	B inflow	$\xrightarrow{k_0 B_0} B$	$v_6 = k_0 B_0.$

There are two independent intermediates, and six reactions. Hence, there are $n + r - d = 2 + 6 - 2 = 6$ $(\mathbf{k}, \mathbf{C})$ and $(\mathbf{h}, \mathbf{j})$ parameters. In CSTR's, several reactions often have the rate constant k_0, which reduces the dimension of the parameter domain by a *CSTR constraint*. Since there are 5 *experimental parameters* k_0, k_1, k_2, A_0 and B_0, the accessible parameters domain is a co-dimension 1 subset of $\bar{D}$. The stoichiometric matrix is

$$\mathbf{N} = \begin{array}{c} \\ A \\ B \end{array} \begin{pmatrix} R_1 & R_2 & R_3 & R_4 & R_5 & R_6 \\ -1 & 0 & -1 & 0 & 1 & 0 \\ 1 & -1 & 0 & -1 & 0 & 1 \end{pmatrix}. \tag{17}$$

All chemical reactions occur via microscopic collisions between molecules. Reactions that occur as a collision are *elementary reactions* and their rate laws are monomials. Combinations of elementary reactions (*complex reactions*) frequently have monomial rate laws. Such rate laws are called *power law kinetics*. In such cases the exponents or *orders of kinetics* are the elements of the *kinetic matrix* $\mathbf{K}$, which for the Gray-Scott system is

$$\mathbf{K} = \begin{array}{c} \\ A \\ B \end{array} \begin{pmatrix} R_1 & R_2 & R_3 & R_4 & R_5 & R_6 \\ 1 & 0 & 1 & 0 & 0 & 0 \\ 2 & 1 & 0 & 1 & 0 & 0 \end{pmatrix}. \tag{18}$$

With a suitable interpretation of a vector raised to a matrix power, the monomial rate laws can be written as the matrix equation

$$\mathbf{v} = \mathrm{diag}(\mathbf{k})\mathbf{c}^{\mathbf{K}}. \tag{19}$$

The matrices $\mathbf{N}$ and $\mathbf{K}$ define the network. $\mathbf{N}$ determines the extreme current matrix

$$\mathbf{E} = \begin{pmatrix} 1 & 0 & 0 & 1 & 0 \\ 1 & 1 & 0 & 0 & 0 \\ 0 & 0 & 1 & 0 & 0 \\ 0 & 0 & 0 & 1 & 1 \\ 1 & 0 & 1 & 1 & 0 \\ 0 & 1 & 0 & 0 & 1 \end{pmatrix}$$

The columns of $\mathbf{E}$ are the edges of the current cone C_v. A transversal cross-section through C_v is the *current polytope*, whose vertices represent extreme currents. (See Figure 1.)

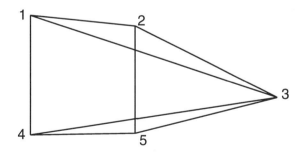

Figure 1   The current polytope of the Gray-Scott network is a three-dimensional square pyramid.

The mathematics can be interpreted in terms of chemical reactions if the network is represented by a diagram. Each species and reaction appears once on the diagram. Reactions are branched arrows that connect the species they consume and produce. Barbs and feathers on the arrows show the numerical values of $\mathbf{N}$ and $\mathbf{K}$, (provided the matrix elements are small integers). The Gray-Scott network diagram appears in Figure 2.

In each extreme current many reactions do not occur. Those that do are thick curves on the *extreme current diagram*. One current diagram for each vertex of the current polytope is shown in Figure 3.

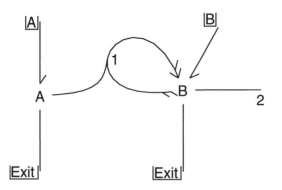

Figure 2   Network diagram of the Gray-Scott network. The reactions labelled "Exit" and "2" remove A and B. The diagram shows only the left half of the reaction and there is no right half. These "half-reactions" are analogous to the part of R_1 between A and the label "1". Barbs and feathers do not appear on them.

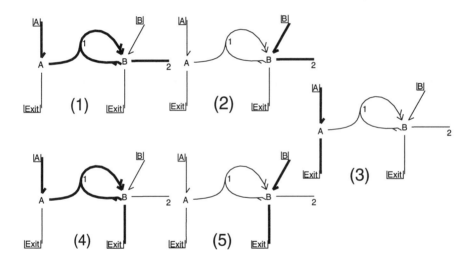

Figure 3   Extreme current diagrams for the Gray-Scott reaction network in the same arrangement as the vertices of the current polytope in Figure 1. Each current diagram represents a column of **E**.

4. STABILITY ANALYSIS OF STEADY STATES

4.1 Exact Treatment

The linearization of equation (16) about the steady state $\mathbf{x} = \mathbf{e}$ has the Jacobian

$$\mathbf{M}(\mathbf{h}, \mathbf{j}) = \operatorname{diag}(\mathbf{h})\mathbf{N} \operatorname{diag}(\mathbf{Ej})(\nabla_{\mathbf{x}}\mathbf{w}(\mathbf{x}, \mathbf{h}))^t. \tag{20}$$

When the rate laws are given by the kinetic matrix **K**,

$$\mathbf{M}(\mathbf{h}, \mathbf{j}) = \operatorname{diag}(\mathbf{h})\mathbf{N} \operatorname{diag}(\mathbf{Ej})\mathbf{K}^t\operatorname{diag}(\mathbf{h}). \tag{21}$$

The eigenvalues are solutions of the characteristic equation

$$|\lambda \mathbf{I} - \mathbf{M}(\mathbf{p})| = \lambda^n + \alpha_1(\mathbf{p})\lambda^{n-1} + \alpha_2(\mathbf{p})\lambda^{n-2} + \ldots + \alpha_n(\mathbf{p}) = 0, \tag{22}$$

whose last $n - d$ coefficients vanish because of the conservation conditions. The number of real positive eigenvalues is the number of sign changes in $\mathbf{R}$, which we define as the transpose of the first column of the Routh array (Porter, 1968)

$$\mathbf{R} = (\, 1 \quad \Delta_1 \quad \frac{\Delta_2}{\Delta_1} \quad \frac{\Delta_3}{\Delta_2} \quad \frac{\Delta_4}{\Delta_3} \quad \cdots \quad \frac{\Delta_{d-1}}{\Delta_{d-2}} \quad \alpha_d \,), \tag{23}$$

where Δ_i is the ith Hurwitz determinant, and the last element comes from the condition

$$\Delta_d = \alpha_d \Delta_{d-1}.$$

Instability occurs in two ways. A *saddle node* bifurcation occurs when one real eigenvalue crosses the imaginary axis. $\mathbf{R}$ has one sign change and α_d changes sign. The bifurcation hypersurface has the equation

$$\alpha_d(\mathbf{p}) = 0. \tag{24}$$

At a Hopf bifurcation the interior elements of $\mathbf{R}$ change sign. However, Orlando's Theorem (Orlando, 1911) says Δ_{d-1} is a product of the sums of all pairs of eigenvalues

$$\Delta_{d-1} = \prod_{\substack{j<k}}^{1,\ldots,d} (-\lambda_j - \lambda_k). \tag{25}$$

In the interior of the stable region, no pair has a sum which vanishes, but at a Hopf bifurcation a pair's sum vanishes. Hence, at every Hopf bifurcation

$$\Delta_{d-1}(\mathbf{p}) = 0. \tag{27}$$

This condition is necessary but not sufficient. Hopf bifurcations occur on a subset of this hypersurface.

Stability analysis may be carried out throughout the open $(\mathbf{h}, \mathbf{j})$ parameter domain D by constructing the polynomials α_d and Δ_{d-1} on a computer. For any given simplex of a simplicial decomposition of C_v the parameter domain is R_+^{n+r-d}. Since the parameters are positive, a polynomial can only change sign if terms with negative coefficients are present. Hence, the absence of negative coefficients proves all steady states are stable throughout D.

A very small fraction of all realistic chemical systems are capable of instability. Out of hundreds of stable ones I have studied none have negative terms in α_d and Δ_{d-1}. Hence the presence of negative terms is, in practice, necessary and *sufficient* for instability.

Each term in a polynomial α_i is a product of feedback cycles. (Clarke 1974, 1980) Positive feedback cycles produce negative factors and negative feedback cycles produce positive factors. Hence, negative terms in α_i are products with an odd number of positive feedback cycles. Most networks contain many positive feedback cycles. The negative terms from these positive feedback cycles are usually cancelled by algebraically equivalent terms from negative feedback cycles. Only under unusual conditions do the positive feedback cycles produce negative terms which remain after they are combined with algebraically equivalent positive terms.

4.2 Approximations which Work in Chemistry

Wide Parameter Ranges

Parameters in chemistry often range over more than 15 orders of magnitude. The polynomial α_d is homogeneous of order $2d$, and Δ_{d-1} is of much higher order. Unless the parameters are almost equal, only a small fraction of the terms are significant.

Newton discovered a method (*Newton's Polygon*) (Newton, 1673) for finding the leading terms in a power series expansion of a real-power polynomial function $f(x, y)$ about the singular point $(0, 0)$. We apply a generalization of this method by Clarke (1978) to polynomials with any number of variables. The general real-power polynomial has the form

$$f(\mathbf{x}) = \sum_{i=1}^{n} c_i x_1^{y_{1i}} x_2^{y_{2i}} \ldots x_n^{y_{ni}} = \sum_{i=1}^{n} c_i \mathbf{x}^{\mathbf{y}_i} = \sum_{i=1}^{n} c_i e^{(\log \mathbf{x}) \cdot \mathbf{y}_i}.$$

The convex hull of the exponents $P = \text{conv} \{\mathbf{y}_i \mid i = 1, \ldots\}$ is called the *exponent polytope*, and generalizes Newton's polygon. Almost everywhere when $\mathbf{x}$ is near the finite and infinite boundary of R_+^n, the dominant terms of $f(\mathbf{x})$ are vertices of P. If the orthant is mapped into $\log \mathbf{x}$, the boundary maps to infinity. The vertices of P are the dominant terms almost everywhere far from the origin on the domain $(\log \mathbf{x}) \in R^n$.

The dominant terms in the Hurwitz determinants come from the diagonal element

$$\Delta_i = \alpha_1 \alpha_2 \ldots \alpha_i + \ldots. \tag{28}$$

In the *α-approximation*, only this leading term is used. Then

$$\mathbf{R} = (1, \alpha_1, \alpha_2, \ldots, \alpha_d). \tag{29}$$

Hopf bifurcation hypersurfaces coincide with the sign changes of some α_i for $i < d$. Starting in the stable region, if the parameters are changed until the first α_i changes sign, for $i = 1, \ldots, d$, a Hopf bifurcation occurs if $i < d$, and a saddle node bifurcation occurs if $i = d$, except on higher codimension hypersurfaces where other bifurcations occur. The α-approximation is accurate enough to explain the experimental data for most chemical networks.

Dominant and Essential Currents

Just as chemical concentrations can range over more than 16 orders of magnitude, so do reaction rates and the rates of extreme currents. Hence the components of $\mathbf{j}$ have a very wide range. The *dominant extreme currents* are the major contributions to the important reactions and processes which affect the qualitative dynamics. Usually a good intuitive understanding of the system can be obtained with only a small number of dominant currents.

For example, Figure 4 shows the most important processes in each section of the steady state manifold for the Gray-Scott network.

The current diagrams tell us that when A is in the high steady state, it enters and washes out without reacting. Under these conditions, B enters and is either consumed by the second reaction at low k_0 or washes out at high k_0. In the lower steady state of A, reaction R_1 converts A to B, which reacts by R_2 or washes out.

The bifurcation hypersurfaces can be calculated by neglecting all extreme currents except the largest one for each reaction. The currents necessary to derive a bifurcation hypersurface equation are the *essential currents*.

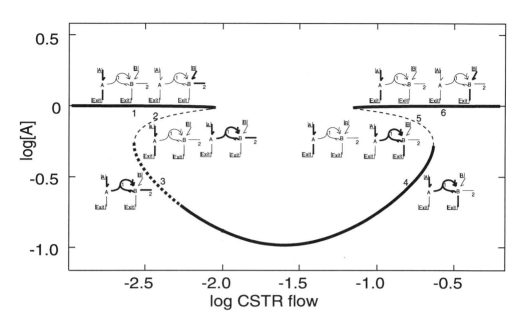

Figure 4 Dominant extreme currents in various regions of the steady state manifold of the Gray-Scott network.

5. DYNAMICAL PHASE DIAGRAMS

A "phase diagram" in chemistry shows the region where a substance has certain properties (solid, liquid or gas) versus its parameters (temperature, pressure). A *dynamical phase diagram* shows the dynamical properties throughout the $(n + r - d)$-dimensional parameter domain. A phase diagram is different from a *bifurcation diagram*, which shows how bifurcations branch from dependent system property as a parameter is varied.

We are concerned here with bifurcations of steady states and attractors, such as limit cycles, that are closely associated with these bifurcations. The regions of the diagram will be marked using the symbols in Figure 5.

$$\triangle \quad \square \quad \triangledown \quad \ominus \!\!\!\!\bigtriangleup \quad \ominus \quad \ominus\!\!\!\!\bigtriangledown \quad \ominus\!\!\!\!\bigtriangleup \quad \bigcirc \quad \bigcirc\!\!\!\!\bigtriangledown \quad \bigcirc\!\!\!\!\bigtriangleup \quad \textcircled{\bigtriangleup} \quad \textcircled{\square} \quad \textcircled{\bigtriangledown} \quad \textcircled{\ominus}$$

Figure 5 Symbols used to mark phase diagram regions. When folding of the steady state manifold produces multiple steady states, alternate steady states are always unstable. The symbols shown here are used to mark the intervening steady states. Thus, two symbols stacked on top of each other implies the existence of an unstable steady state between them, and possibly additional unstable steady states above and below. The first seven symbols classify the type of steady state without considering other attractors. They represent (left to right) a unique upper stable steady state (sss), any unique sss, unique lower sss, bistability (two sss's), unique Hopf region state (see text), upper Hopf region state with lower sss, lower Hopf region state with upper sss. The last seven symbols show the presence of limit cycles (lc) in the following situations: lc with no sss, small lc near upper unstable steady state (uss) and lower sss, small lc near lower uss and upper sss, large lc and unique upper sss, large lc and any unique sss, large lc and unique lower sss, large lc and two sss.

We will use the term *saddle node region* for the region where $\alpha_d < 0$, and *Hopf region* for the region where any $\Delta_i < 0$ for $i = 1, \ldots, d - 1$. These regions can be marked on a phase diagram using the symbols in Figure 5. Note that the Hopf region symbol does mean a limit cycle is present.

The steady state manifold of the Gray-Scott network (Figure 6) shows along the top the symbols which would classify the steady state manifold on a dynamical phase diagram.

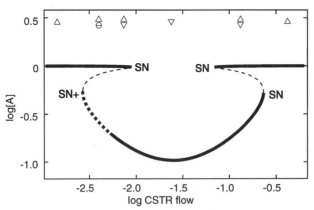

Figure 6   Steady state manifold for the Gray-Scott network as a function of the CSTR flow rate $\log k_0$. The stability and bifurcation points are indicated by line type. Thick solid lines are stable states. Dashed lines are unstable. The Hopf region is thick dashed, and the saddle node region is thin dashed. Saddle node bifurcations (SN) and saddle node plus (SN+) are marked. The symbols at the top classify the steady state manifold and would appear on a phase diagram.

Each point on a phase diagram in the $(\mathbf{h}, \mathbf{j})$-parameters represents a single steady state. There are four types of points: stable points, those in Hopf regions, those in saddle node regions, and those in both. These are shown in the left part of Figure 7, where a Hopf region overlaps a saddle node region. The four regions meet at a Takens-Bogdanov (TB) bifurcation point (center). TB points have been discussed by Guckenheimer (1986), Dumortier *et al.*(1987), Annabi *et al.*(1991) and Hale (1991). The stable region (upper right) is bounded by a Hopf bifurcation line (thick solid) and a saddle node (SN) line (thin dashed). The Hopf region (upper left) is bounded by the Hopf bifurcation line and the saddle node plus (SN$^+$) line (thick dashed). The saddle node region (lower right) is separated from the lower left region where the SN and Hopf regions overlap by a thin line obeying (27). Orlando's Theorem says that along this continuation of the Hopf bifurcation curve past the TB point, two real eigenvalues are placed symmetrically on either side of the imaginary axis. Hence we call it the *symmetric eigenvalue* (SE) curve.

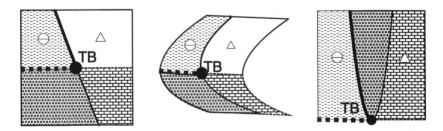

Figure 7   How the steady state manifold in the $(\mathbf{h}, \mathbf{j})$ parameters folds near a Takens-Bogdanov point. The regions below the TB points in the two figures on the left represents part of the steady state manifold which should not be marked with a symbol. Only alternate leaves of the folded steady state manifold are marked with symbols.

The corresponding part of the steady state manifold in the $(\mathbf{k}, \mathbf{C})$ parameters can be visualized by folding along the saddle node bifurcation line. The middle picture of Figure 7 shows the folding. At the right is the resulting dynamical phase diagram. A saddle node line along the lower edge has a TB point. The Hopf curve meets the SN curve tangentially at the TB point and becomes a SE curve.

This notation is used in Figure 8, which shows some of the complexity of the Gray-Scott phase diagram.

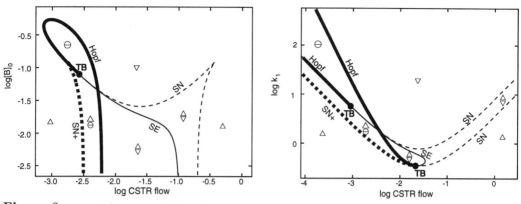

Figure 8   Two 2-dimensional phase diagrams of the Gray-Scott reaction network.

6. CHEMICAL APPLICATIONS OF BIFURCATION THEORY

6.1 The Belousov-Zhabotinskii System

The most widely studied oscillating chemical system is the Belousov-Zhabotinskii system (Field *et al.*, 1972). A simplified model due to Showalter, Noyes and Bar-Eli (1978) has seven species. I will explain how bifurcation theory can be used to show that this model is inadequate to account for the experimental data. The SNB network is given by the diagram in Figure 9.

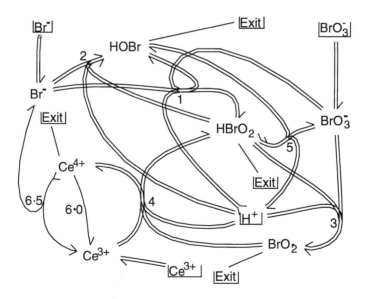

Figure 9   Network diagram of the SNB reaction network. Double curves are reversible reactions.

One of the key experimentally controllable parameters is the input concentration of Br^-. This is called $[Br^-]_0$ and should not be confused with the concentration of the intermediate, $[Br^-]$. This system has an oscillatory region bounded by Hopf bifurcations at low k_0 and low $[Br^-]_0$. At higher k_0 and $[Br^-]_0$, there are two stable steady states. The

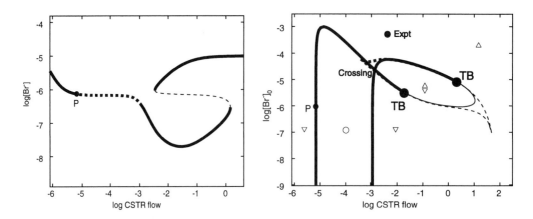

Figure 10   Left: Steady state manifold of the SNB network. Right: Dynamical phase diagram. The curve crossing occurs at the point "Expt" in the experimental system.

steady state manifold showing the Hopf and saddle node regions is at the left in Figure 10, and the dynamical phase diagram is at the right.

The crossing of the Hopf and saddle node bifurcation curves on the phase diagram occurs at the point marked "Expt" in the Belousov-Zhabotinskii system. Is it possible to move the crossing to this point by adjusting parameters? Ringland (1991) made extensive numerical calculations and was unable to move the crossing point. Clarke and Jiang (1993) derived equations for the bifurcation curves and showed analytically that the crossing could not match the experiments. The bifurcation curve equations were derived using only four essential currents. The α-approximation was made and the important terms were extracted using exponent polytopes. The resulting conditions for *instability* which are violated at the bifurcation curves in the figure are

$$j_{12} \geq 3j_1 \qquad \text{(low-flow Hopf bifurcation)}$$
$$j_{12} \geq 3j_{20} \qquad \text{(high-flow Hopf bifurcation)}$$
$$j_{12} \geq j_3, \qquad \text{(alternative high-flow Hopf bifurcation)}$$
$$j_{51} \geq j_{15} \qquad \text{(low-flow saddle node bifurcation)}$$
$$j_{51} \geq j_3. \qquad \text{(high-flow saddle node bifurcation)}$$

The first condition is the vertical line forming the left boundary of the oscillatory region on the right side of Figure 10. The next two conditions are possible equations for the vertical right boundary of the oscillatory region. The last two represent the sloping curves bounding the three steady state region near the TB points.

These can be converted to equations in the experimental parameters using more approximations. The saddle-node condition $j_{15} = j_{51}$ at the low-flow boundary of the three steady state region becomes

$$k_0 = \frac{6k_1k_3[H^+]^2[BrO_3^-]_0^2}{k_2[Br^-]_0 - 2k_3[BrO_3^-]_0}.$$

This and a similar equation for the high-flow boundary are plotted in Figure 11 as dashed lines. The error in the lower dashed line is due to the last set of approximations, and to the omission of a current that becomes more important as k_0 increases.

At the lower SN bifurcation curve, the network consumes Br^- at a critical rate. The bifurcation curve can be raised to the level of the experimental system by adding reactions which consume Br^- to the network. When these are present, a larger input $[Br^-]_0$ is required to compensate, and the critical consumption rate can be made to match the experimental point. This result is not surprising, because the SNB model is known to omit Br^- consuming reactions which are believed to occur in the real system. The 7-species SNB model is too simplified to match the data.

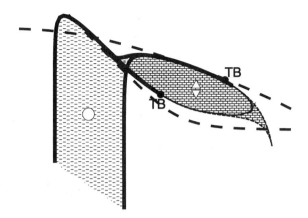

Figure 11 Dashed curves are the approximate equations for the saddle node bifurcation curves. Oscillations occur in the lower left lightly shaded region, and three steady states are present in the darker brickwork region. A single stable steady state exists everywhere else.

Phase diagrams with bifurcation curves in the $(\mathbf{h}, \mathbf{j})$ parameters are similar to Figure 7. Figure 12 shows the $(\mathbf{h}, \mathbf{j})$ phase diagram near the lower Takens-Bogdanov bifurcation point.

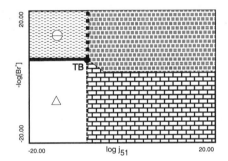

Figure 12 Unstable regions in the $(\mathbf{h}, \mathbf{j})$ parameter domain at the lower Takens-Bogdanov bifurcation point in Figure 11. The shading is the same as in Figure 7. (See text for details.)

The left half of Figure 12 represents the upper steady state and the right half represents the middle steady state. Since the middle state is an unstable intervening state, no symbols are used to mark it on this type of diagram. For the upper steady state, the stable part (bottom) is marked with a triangle and the unstable part (top) is marked with the Hopf region symbol. Whether the Hopf bifurcation is supercritical or subcritical was not determined. If it were supercritical there would be a limit cycle so the Hopf region symbol could be replaced by a circle (limit cycle symbol) where the limit cycle was an attractor. If the Hopf bifurcation were subcritical the Hopf region symbol would mark the unstable region and the presence of a limit cycle attractor associated with the upper steady state would be indicated by a circle surrounding the triangle (fourth symbol from the right in Figure 5) at the lower left of the figure.

6.2 Bromate-Sulfite-Ferrocyanide System

The BSF system is a relatively simple oscillator and is unusual because the chemistry of the important reactions is well understood (Edblom *et al.*, 1989). It should be possible to account for all experimental data using the network given in Figure 13.

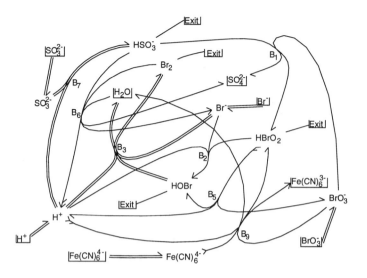

Figure 13   Network diagram of the bromate-sulfite-ferrocyanide reaction network.

The experimental data showed a region of oscillation and a bistable region (two stable steady states) that appeared to lie in opposite quadrants of a typical "cross-shaped phase diagram" (Boissonade *et al.*, 1980). However, when Jiang (1993) calculated the bifurcation curves he found regions that were not cross-shaped, as shown in Figure 14.

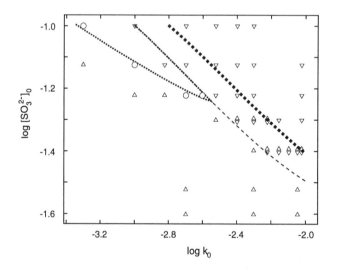

Figure 14   Experiments at 20°C and the observed attractors using the notation of Figure 5. Also shown are Jiang's calculated bifurcation curves. A Hopf curve lies very close to the thick dashed SN+ curve and is not shown. The parameter k_1 was adjusted to fit the SN and SN+ curves to the experimental bistable region.

For typical chemical oscillators a region of oscillation usually coincides with the Hopf region. If the bifurcation is supercritical the oscillations start at the Hopf bifurcation. If the bifurcation is subcritical, the oscillations usually begin in the stable region a short distance before the Hopf bifurcation and roughly coincide with the Hopf region. However,

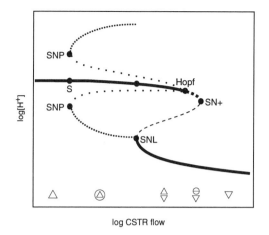

Figure 15 A qualitative bifurcation diagram of the BSF network. The figure shows the steady state manifold and bifurcations branching from it. The sparse-dotted curve is a family of unstable periodic solutions that branch from the subcritical Hopf bifurcation. The dense-dotted curve is a branch of limit cycles. SNP is a saddle-node bifurcation of a periodic solution. SNL is a saddle-node bifurcation on a loop.

the Hopf region on the upper steady state in Figure 14 is so narrow that the Hopf and SN^+ bifurcations coincide at the scale of the plot. What could explain the large oscillation region at the upper left?

The oscillations occur in a region with a single steady state. As shown in Figure 15, the family of periodic solutions which branches from the subcritical Hopf bifurcation extends to much lower k_0.

The oscillation region occurs between a saddle-node of a periodic solution (SNP) bifurcation (low k_0), and a saddle-node loop (SNL) bifurcation (high k_0), which coincides with the fold in the steady state manifold. Based on the qualitative bifurcation diagram and simulations, Jiang produced a corrected phase diagram (Figure 16), which shows the attractors that should be present at each of the parameter values studied experimentally. He also adjusted k_9 to make the SNP curve match the experiments.

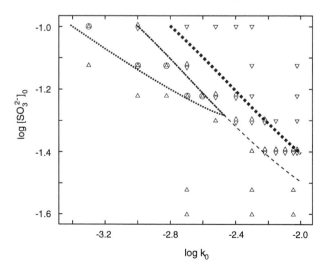

Figure 16 Optimized phase diagram for the BSF network at 20°C.

7. CONCLUSIONS

Bifurcation analysis is just becoming a significant tool in chemistry for the study of chemical mechanisms. So far the successes of the method in chemistry can be summarized as follows.

A mechanism to explain the experimental oscillations in the persulfate-oxalate-silver system has been shown to be capable of matching the experimental bifurcation diagram, however, some clearly identified reactions must be added to the mechanism (Clarke, 1992a).

The SNB model of the Belousov-Zhabotinskii system has been shown to be incapable of matching the experimental data. However, the method also shows that additional reactions which consume Br$^-$ can make the mechanism match experiments and such reactions are known to occur in the real system and were omitted from the SNB model (Clarke and Jiang, 1993).

Finally, the phase diagram of the bromate-sulfite-ferrocyanide systems has been shown to fit extremely well with the proposed mechanism provided that modified rate constants are used and additional experiments confirm attractors that were missed in the original experiments (Jiang, 1993).

Acknowledgements. A number of figures were made by modifying CorelDRAW images created Weimin Jiang for his Ph. D. Thesis, and are used with permission.

REFERENCES

Annabi H., Annabi M. L. and Dumortier F. (1991). Continuous dependence on parameters in the Bogdanov-Takens bifurcation, in *Geometry and analysis in nonlinear dynamics*, edited by H. W. Broer and F. Takens (Longman Scientific and Technical, Harlow, England), pp. 1-21.

Boissonade J. and de Kepper P. (1980). Transitions from bistability to limit cycle oscillations. Theoretical analysis and experimental evidence in an open chemical system, *J. Phys. Chem.* **84**: 501-506.

Clarke B. L. (1974). Graph Theoretic Approach to the Stability Analysis of Steady State Chemical Reaction Networks, *J. Chem. Phys.* **60**: 1481-1492.

Clarke B. L. (1975). Stability of Topologically Similar Chemical Networks, *J. Chem. Phys.* **62**: 3726-3738.

Clarke B. L. (1978). Asymptotes and Intercepts of Real Power Polynomials from the Geometry of the Exponent Polytope, *SIAM J. Appl. Math.* **35**: 755-786.

Clarke B. L. (1980). Stability of Complex Reaction Networks, *Adv. Chem. Phys.* **43**: 1-215.

Clarke B. L. (1981). Complete set of steady states for the general stoichiometric dynamical system, *J. Chem. Phys.* **75**: 970-4979.

Clarke B. L. (1983). Qualitative Dynamics and Stability of Chemical Reaction Networks, in *Chemical Applications of Topology and Graph Theory*, edited by R. B. King (Elsevier, Amsterdam), pp. 322-357.

Clarke B. L. (1988). Stoichiometric Network Analysis, *Cell Biophysics* **12**: 237-253.

Clarke B. L. (1992a). Stoichiometric Network Analysis of the Oxalate-Persulfate-Silver Oscillator, *J. Chem. Phys.* **97**: 2459-2472.

Clarke B. L. (1992b). General method for simplifying chemical networks while preserving overall stoichiometry in reduced mechanisms., *J. Chem. Phys.* **97**: 4066-4071.

Clarke B. L. and Jiang W. (1993). Method for deriving Hopf and saddle-node bifurcation hypersurfaces and application to a model of the Belousov-Zhabotinskii system, *J. Chem. Phys.* **99**: 4464-4478.

Dumortier F., Roussaire R. and Sotomayor J. (1987). Generic 3-parameter families of vector fields on the plane, unfolding a singularity with nilpotent linear part. The cusp case of codimension 3, *Ergod. Th. Dynam. Sys.* **7**: 375-413.

Edblom E. C., Luo Y., Orban M., Kustin K. and Epstein I. R. (1989). Kinetics and Mechanism of the Oscillatory Bromate-Sulfite-Ferrocyanide Reaction, *J. Phys. Chem.* **93**: 2722-2727.

Field R. J., Koros E. and Noyes R. (1972). Oscillations in Chemical Systems II. Thorough Analysis of Temporal Oscillation in the Bromate-Cerium-Malonic Acid System, *J. Am. Chem. Soc.* **94**: 8649-8664.

Gray P. and Scott S. K. (1983). Autocatalytic Reactions in the Isothermal, Continuous Stirred Tank Reactor, *Chem. Engng. Sci.* **38**: 29-43.

Guckenheimer J. (1986). Multiple Bifurcation Problems for Chemical Reactors, *Physica* **20D**: 1-20.

Hale J. and Koçak H. (1991). *"Dynamics and Bifurcations"*, (Springer-Verlag, New York).

Jiang, W. (1993). *"Computer Investigations of the Stability Boundaries of Chemical Reaction Networks"*, (Ph. D. Thesis, University of Alberta, Edmonton).

Newton I. (1673). , in *The Mathematical Papers of Isaac Newton 1670-1673*, edited by D. T. Whiteside (Cambridge University Press, Cambridge, 1969), p. 50.

Orlando L. (1911). Sul problema di Hurwitz relativo alle parti reali delle radici di un'equazione algebrica, *Math. Ann.* **71**: 233.

Porter B. (1968). *"Stability Criteria for Linear Dynamical Systems"*, (Academic Press, London U.K.).

Ringland J. (1991). Rapid reconnaissance of a model of a chemical oscillator by numerical continuation of a bifurcation feature of codimension 2, *J. Chem. Phys.* **95**: 555-562.

Showalter K., Noyes R. and Bar-Eli K. (1978). A modified Oregonator model exhibiting complicated limit cycle behavior in a flow system, *J. Chem. Phys.* **69**: 2514-2524.

Wegscheider (1901). Über simultane Gleichgewichte und die Beziehungen zwischen Thermodynamik und Reaktionskinetik homogener Systeme, *Z. Physik. Chem. Leipzig* **39**: 257-303.

Stability Problems for Volterra Functional Differential Equations

C. CORDUNEANU, Department of Mathematics, University of Texas at Arlington, Arlington, Texas

Let us consider the Volterra functional differential equation

$$(1) \qquad \dot{x}(t) = (Vx)(t), \ t \geq 0,$$

where V stands for a Volterra (causal or nonanticipative) operator on the function space $L_{loc}(R_+, R^n) = L^1_{loc}(R_+, R^n)$.

The characteristic property for Volterra operators can be stated as follows: if $x(s) = y(s)$ a.e. on $[0,t]$, then $(Vx)(s) = (Vy)(s)$ a.e. on $[0,t]$, for any $t > 0$.

The role of Volterra (causal) operators is significant when the equation (1) describes hereditary phenomena and the past must be taken into consideration in order to predict their evolution.

There are two kinds of initial value problem related to a Volterra functional differential equation of the form (1).

The _first_ initial value problem consists in finding the solution of (1), on some interval $[0,T) \subset R_+$, such that

$$(2) \qquad x(0) = x^0 \in R^n.$$

The _second_ initial value problem requires finding a solution of equation (1) on some interval $[t_0, T) \subset R_+$, with $t_0 > 0$, such that

$$(3) \qquad x(t) = \varphi(t), \ t \in [0,t_0), \ x(t_0) = x^0 \in R^n.$$

It is reasonable to assume that

$$(4) \qquad \varphi \in L^1([0,t_0], R^n),$$

in order to assure that $(Vx)(t)$ is well defined for $t \geq 0$.

The solution of the second initial value problem, in case it exists and is unique, will be denoted by

$$(5) \qquad x = x(t; t_0, \varphi, x^0).$$

The existence and uniqueness of solution for the first initial value problem has been discussed, under various assumptions on the data, in [3],[6],[7],[9],[10],[11],[13], [24],[31],[33].

For existence and uniqueness purposes only, the second value initial problem is easily reducible to the first one. If one denotes

(6)
$$x_\varphi(t) = \begin{cases} \varphi(t) & , \ t \in [0, t_0), \\ x^0 & , \ t = t_0, \\ x(t) & , \ t > t_0, \end{cases}$$

for any function $x(t)$ defined only for $t > t_0$, then finding a solution to the second initial value problem (1), (3) is the same as finding $x(t)$, $t \geq t_0$, such that

(7) $\dot{x}_\varphi(t) = (Vx_\varphi)(t), \ t > t_0, \ x_\varphi(t_0) = x^0.$

The existence and uniqueness of solution (even global existence on R_+) for the second initial value problem is secured, for instance, by the following conditions:

a) $V: L_{loc}(R_+, R^n) \to L_{loc}(R_+, R^n)$ satisfies the generalized Lipschitz condition

(8) $\displaystyle\int_0^t |(Vx)(t) - (Vy)(t)| \, ds \leq A(t) \int_0^t |x(s) - y(s)| \, ds, \ \forall t > 0,$

where $A(t)$ is a positive nondecreasing function on R_+;

b) φ satisfies condition (4).

Let us point out that condition (8) implies the fact that the operator V has the Volterra property.

The proof is given in [23] for the linear case, but the same estimates are valid under condition a).

There are two distinct interpretations for the second initial value problem, which is definitely the most adequate for the equations involving Volterra operators.

The first interpretation assumes we are observing the evolution of the system, described by (1), during an initial period $[0, t_0]$, and then use this information to predict the future (past the moment $t_0 > 0$) evolution.

The second interpretation has a control theory flavor, and it looks at the period $[0, t_0]$ as a "controlled" period, during which we impress a certain motion on the system in order to assure a desired behavior or performance in the future. This second interpretation leads to interesting connections between Control Theory and Integral Equations.

Let us concentrate now our attention on the stability concepts related to the initial value problems briefly discussed above. In accordance with the two types of initial value problems, the mathematical literature emphasizes two types of stabilities for the functional differential system (1), in which V stands for an abstract Volterra operator.

The first type of stabilities is predominant in the literature. The corresponding definitions appear in the recent book by Azbelev, Maksimov and Rakhmatullina [3], the first book entirely dedicated to the theory of equations with abstract Volterra operators.

If one assumes that $x = \theta$ is a solution of (1), i.e., $(V\theta)(t) = \theta$ on R_+, then the definitions of various types of stability can be formulated as follows:

Stability: The zero solution of (1) is stable if for every $\epsilon > 0$, there exists $\delta(\epsilon) > 0$, such that

$$(9) \qquad |x^0| < \delta \Rightarrow |x(t;x^0)| < \epsilon \text{ for } t \geq 0$$

Asymptotic stability: The stability property holds, and

$$(10) \qquad \lim |x(t;x^0)| = 0 \text{ as } t \to \infty,$$

for all x^0 such that $|x^0| < \delta_0$.

Exponential asymptotic stability: There exists two positive constraints N and α, such that

$$(11) \qquad |x(t;x^0)| \leq N|x^0|e^{-\alpha t}$$

for $t \geq 0$, and all x^0 with $|x^0| < \delta_0$.

In case of linear equations of the form

$$(12) \qquad \dot{x}(t) = (Lx)(t), \ t > 0,$$

where L stands for a linear abstract Volterra type operator, it is shown that stability (as defined above) is equivalent to the boundedness of all solutions on R_+, asymptotic stability is equivalent to the convergence to zero of all solutions as $t \to \infty$, while exponential asymptotic stability means exactly the fact that all solutions of (12) approach zero exponentially, as $t \to \infty$.

In the book [3], the authors relate these concepts of stability to that of admissibility, and regard the admissibility properties as stability under permanently acting perturbations.

Other authors (see, for instances, L. G. Fedorenko [12]) investigate stability for Volterra functional differential equations in relationship with the second initial value problem. But in his approach t_0 ($= 0$) is kept constant, a feature which should not occur in the definition of stability.

The definitions of stabilities related to the second initial value problem (1), (3) have been formulated by R. Driver [11], under conditions that include also the case of infinite delays. For the solution $x = \theta$, these definitions are:

Stability: For each $\epsilon > 0$ and $t_0 > 0$, there exists $\delta = \delta(\epsilon, t_0) > 0$ such that

(13) $| x(t; t_0, \varphi, x^0) | < \epsilon$ for $t > t_0$,

provided

(14) $| x^0 | < \delta, \displaystyle\int_0^{t_0} | \varphi(s) | \, ds < \delta.$

Uniform stability: As in the above definition of stability, but $\delta(\epsilon, t_0) \equiv \delta(\epsilon)$.

Asymptotic stability: As in the above definition of stability, and for each $t_0 > 0$ there exists $\eta(t_0) > 0$, such that

(15) $| x(t; t_0, \varphi, x^0) | \to 0$ as $t \to \infty$,

whenever

(16) $| x^0 | < \eta, \displaystyle\int_0^{t_0} | \varphi(s) | \, ds < \eta.$

Uniform asymptotic stability: As in the above definition of uniform stability, and there exists $\delta_0 > 0$, and $T(\epsilon) > 0$ for $\epsilon > 0$, such that

(17) $| x(t; t_0, \varphi, x^0) | < \epsilon$ for $t \geq t_0 + T(\epsilon)$,

provided

(18) $| x^0 | < \delta_0, \displaystyle\int_0^{t_0} | \varphi(s) | \, ds < \delta_0.$

Similar definitions can be formulated for other kinds of stability such as exponential asymptotic stability, integral stability or stability under permanent disturbances.

As mentioned above, the predominant type of stability results available in the literature pertains to the stability related to the first initial value problem.

It is certainly useful to develop a stability theory for functional differential equations involving abstract Volterra operators, but related to the second initial value problem (1), (3).

So far, with the notable exceptions of Driver's paper [11], Burton's book [6], the book by Hino, Murakami and Naito [15], which is entirely dedicated to infinite delay equations, as well as a relatively small number of journal papers, there is little progress in this regard. The stability theory of equations with abstract Volterra operators does not seem to have reached the level comparable to that attained for instance, in case of equations with finite delay (see, V. Lakshmikantham [21], A. V. Kim [18], J. K. Hale [13]).

The equation (1) can be also written in the traditional form

(19) $$x(t) = F(t, x_t), \quad t > 0,$$

where x_t stands for the restriction of $x(u)$ at the interval $[0, t]$, $t > 0$. This is due to the fact that V is a Volterra type operator.

Accordingly, the initial conditions (3) can be written in the form

(20) $$x_{t_0} = \varphi, \quad x(t_0) = x^0 \in R^n.$$

Let us notice that, as $t \to \infty$, the delay in (19) also tends to ∞. Fortunately, the phase space is not anymore necessary in this case and the investigation of trajectories becomes possible in R^n.

Lyapunov functional for (19) can be sought in the form

(21) $$W = W(t, x_t),$$

with usual definition for the derivative with respect to the system (19), positive definiteness, "wedges", etc. There are references in this regard, particularly for specific forms of the functional $F(t, x_t)$. See, for instance, [6], [15].

We shall look for Lyapunov functionals in the form suggested by (1), namely

(22) $$W = (Wx)(t), \quad t \geq 0,$$

$W: L_{loc}(R_+, R^n) \to R$, assuming that W is also of Volterra type.

A basic assumption on W will be Fréchet differentiability on $L_{loc}(R_+, R^n)$. Since $L_{loc}(R_+, R^u)$ is not a Banach space, its local convexity (or Fréchet space structure) allows the definition of Fréchet differentiability, as well as use of "chain rule" (see G. Marinescu [25] for this concept).

Therefore, if $W: L_{loc}(R_+, R^u) \to R$ is Fréchet differentiable, and one denotes by W' its Fréchet differential, then along the trajectories of (1)

(23) $$\frac{d}{dt}(Wx)(t) = (W'x)(t)[(Vx)(t)],$$

which means that the derivative of $(Wx)(t)$ is obtained by taking the value of the differential W' at $x(t)$, in the direction $(Vx)(t)$. The relationship (23) may be true only a.e., for those t for which $\dot{x}(t)$ does exist.

If one can find an estimate of the form

(24) $$(W'x)(t)[(Vx)(t)] \leq \omega(t, (Wx)(t)),$$

then the following comparison inequality holds true:

(25) $$\frac{d}{dt}(Wx)(t) \leq \omega(t, (Wx)(t)), \quad t > t_0$$

along the trajectories of (1). The inequality (25) generates the comparison equation

(26) $\dot{y}(t) = \omega(t, y(t)), \; t > t_0$

whose properties can be used to obtain similar properties for $(Wx)(t)$, and, therefore, for $x(t)$. Of course, we need adequate properties for $(Wx)(t)$, in order to convey the results concerning the comparison equation (26) to the system (1).

From (25) and (26) one obtains the inequality

(27) $(Wx)(t) \leq y(t; t_0, y_0), \; t > t_0,$

provided

(28) $(Wx)(t_0) \leq y_0 = y(t_0; t_0, y_0),$

and $y = y(t; t_0, y_0)$ exists on $t \geq t_0$ (which is implicitly admitted in comparison method).

Let us point out that

(29) $(Wx)(t_0) = \lim(Wx)(t)$ as $t \downarrow t_0$,

the solution $x(t)$ being uniquely determined by (1), (3). Actually, based on the continuous dependence of solutions with respect to initial data, one has an estimate of the form

(30) $|(Wx)(t_0)| \leq K(|x^0| + |\varphi|_{L^1}),$

with $K = K(t_0) > 0$.

The following definitions will provide us with the necessary concepts such as positive definiteness, radially decrescent functional, similar to those encountered in classical stability theory.

Definition. The functional $W : L_{loc}(R_+, R^n) \rightarrow R_+, (W\theta)(t) \equiv 0$ on R_+, is said to hold **property A**, if for each $\epsilon > 0$ and each $t_0 > 0$ there exists $\delta = \delta(\epsilon, t_0) > 0$, such that

(31) $(Wx)(t) < \delta$ for $t > t_0$

implies

(32) $|x(t)| < \epsilon$ for $t > t_0$.

We have now all the necessary elements to formulate stability results by comparison method, under the basic assumption that inequality (24) is satisfied for a function $\omega(t, W)$, such that the comparison equation (26) has convenient stability properties.

THEOREM 1.   Assume there exists a Fréchet differentiable functional $W : L_{loc}(R_+, R^n) \rightarrow R_+, (W\theta)(t) \equiv 0$ on R_+, such that the comparison inequality (24) holds. Assume also that the comparison equation (26) enjoys existence and uniqueness of the solution for $y_0 \geq 0$, while $\omega(t, 0) \equiv 0$.

If W satisfies property A, and the zero solution of (26) is stable (asymptotically stable), then the zero solution of (1) is also stable (asymptotically stable).

Proof. Let $\epsilon > 0$, $t_0 > 0$ be arbitrary, and consider the $\delta = \delta(\epsilon, t_0) > 0$ corresponding on behalf of the Definition. We will obviously have (the stability!)

$$(33) \qquad y(t; t_0, y_0) < \delta \text{ for } t \geq t_0,$$

provided $y_0 < \eta(\epsilon, t_0)$, with $\eta > 0$.

From (27), (28) and (30) there results that

$$(34) \qquad (Wx)(t) < \delta \text{ for } t \geq t_0,$$

as soon as x^0 and φ satisfy

$$(35) \qquad K(t_0)(|x^0| + |\varphi|_{L^1(0, t_0)}) < \eta(\epsilon, t_0).$$

The inequality (35) takes place if

$$(36) \qquad |x^0| < \frac{\eta}{2K}, \quad |\varphi|_{L^1(0, t_0)} < \frac{\eta}{2K},$$

where $\eta/2K$ depends on ϵ and t_0 only.

In accordance with the choice of $\delta = \delta(\epsilon, t_0)$, (34) implies on behalf of property A for W:

$$(37) \qquad |x(t)| < \epsilon \text{ for } t \geq t_0,$$

as soon as (36) are satisfied. Hence, the zero solution of (1) is stable.

If the zero solution of the comparison equation (26) is asymptotically stable, then

$$(38) \qquad \lim y(t; t_0, y_0) = 0 \text{ as } t \to \infty,$$

provided $y_0 < \gamma(t_0)$. On behalf of (27), one obtains

$$(39) \qquad \lim (Wx)(t) = 0 \text{ as } t \to \infty,$$

as soon as $|x^0|$ and $|\varphi|_{L^1(0, t_0)}$ are small enough. We shall have

$$(40) \qquad (Wx)(t) < \delta(\epsilon', t_1) \text{ for } t \geq t_1 > t_0,$$

as soon as we choose $|x^0|$ and $|\varphi|_{L^1(0, t_1)}$ sufficiently small. This implies

$$(41) \qquad |x(t)| < \epsilon' \text{ for } t \geq t_1,$$

according to property A, which means that

$$(42) \qquad \lim |x(t; t_0, \varphi, x^0)| = 0 \text{ as } t \to \infty,$$

for sufficiently small $|x^0|$ and $|\varphi|_{L^1(0, t_0)}$ (because $|x^0|$ and $|\varphi|_{L^1(0, t_0)}$ small enough, will imply $|\varphi|_{L^1(0, t_1)}$ also small).

Remark 1. In Theorem 1, no mention has been made with respect to uniform stability or uniform asymptotic stability. It seems that without extra hypotheses, a similar assertion does not necessarily hold.

Nevertheless, the corresponding statement for uniform stability is valid if in property A we can choose $\delta(\epsilon, t_0)$ independent of t_0, while V satisfies the condition (8) with $A(t) = A_0 = $ coast. We leave to the reader the details of the proof.

Remark 2. The assumption concerning the existence of a scalar function $\omega(t, y)$ such that (24) takes place is certainly not the most natural in the context. Indeed, the left hand side of inequality (24) is a Volterra operator in x. Hence, it would be more natural to assume that in the right hand side of (24) we also have a Volterra operator acting on W. In order to pursue such an approach, which leads to a functional differential equation of Volterra type as a comparison equation, it would be necessary to have better results on differential inequalities with Volterra operators. See [27].

In the remaining part of this paper we shall deal only with linear functional differential equations of Volterra type. These equations are described by

$$(43) \qquad\qquad \dot{x}(t) = (Lx)(t), \quad t > 0,$$

where L stands for a linear bounded operator of Volterra type on $L_{loc}(R_+, R^n)$.

The solution of (43), under initial conditions (3) and assumption (4), is represented by the formula

$$(44) \qquad\qquad x(t; t_0, \varphi, x^0) = X(t, t_0)x^0 + \int_0^{t_0} \widetilde{X}(t, t_0, s)\varphi(s)ds, t \geq t_0.$$

By $X(t, t_0)$ one denotes the (unique) matrix solution of (43), such that $X(t_0, t_0) = I = $ the unit matrix, $t_0 \geq 0$.

By $\widetilde{X}(t, t_0, s)$, $0 \leq s \leq t_0$, $t \geq t_0 > 0$ one denotes a matrix whose entries, regarded as function of t, have a weak L^1-derivative such that

$$\int_0^{t_0} \widetilde{X}(t, t_0, s)\varphi(s)ds, \quad t > t_0,$$

represents a solution of (43). It is obvious that $\widetilde{X}(t_0, t_0, s) \equiv 0$, a.e. in s.

The following results characterize various types of stability for the equation (43). These results are similar to those obtained by R. Conti (see [29]) in the case of linear ordinary differential equations. In case

L is a linear Volterra operator on $L^2_{loc}(R_+,R^n)$, similar results are contained in the Ph.D. thesis submitted by Y. Li [23].

THEOREM 2. Consider the linear system (43), with L a linear bounded Volterra operator on $L_{loc}(R_+,R^n)$. Let $x(t;t_0,\varphi,x^0)$ be the solution of (43), under initial conditions (3), represented by (44). The following conditions are necessary and sufficient for:

Stability: For each $t_0>0$, there exists $M(t_0)>0$, such that

$$(45) \qquad |X(t,t_0)| + \int_0^{t_0} |\widetilde{X}(t,t_0,s)|\,ds \le M(t_0),\ t\ge t_0;$$

Uniform stability: There exists a positive number M, such that

$$(46) \qquad |X(t,t_o)| + \int_0^{t_0} |\widetilde{X}(t,t_0,s)|\,ds \le M,\ t\ge t_0;$$

Asymptotic stability: For every $t_o>0$,

$$(47) \qquad \lim_{t\to\infty} \left(|X(t,t_0)| + \int_0^{t_0} |\widetilde{X}(t,t_0,s)|\,ds \right) = 0;$$

Uniform asymptotic stability: The uniform stability, and for each $\epsilon>0$, there exists $T(\epsilon)>0$, such that

$$(48) \qquad |X(t;t_0)| + \int_0^{t_0} |\widetilde{X}(t,t_0,s)|\,ds < \epsilon, t\ge t_0 + T(\epsilon),$$

The proof of Theorem 2 is mostly routine, if one takes into account the corresponding results in [29] for linear ordinary differential equations, and certain conditions of admissibility for integral operators [10]. The case of operators on $L^2_{loc}(R_+,R^n)$ is treated in detail in [23].

The Theorem 2 can be used to prove further results in stability theory related to the perturbed system.

$$(49) \qquad \dot{x}(t) = (Lx)(t) + (fx)(t),$$

under adequate conditions on the nonlinear perturbing term f. See again [23] for similar systems with operators acting on $L^2_{loc}(R_+,R^n)$.

We shall conclude this paper with two examples on the stability of linear integrodifferential systems.

First example appears in [14] and deals with the scalar equation

$$(50) \qquad \dot{x}(t) = a(t)x(t) + \int_0^t b(t-s)x(s)\,ds,\ t\ge 0.$$

By using the Lyapunov functional

$$(51) \qquad (Wx)(t) = x^2(t) - 2 \int_0^t b(t-s)x(s) \left[\int_0^t x(u)du \right] ds,$$

one obtains the <u>asymptotic stability</u> of the zero solution of (50), under the assumptions:

$$a(t) \geq 0 \text{ on } R_+ \text{ and } a \in L^1(R_+);$$

$$b \in C^{(3)}(R_+) \cap L^1(R_+), b(t) \not\equiv b(0);$$

$$(-1)^k b^{(k)}(t) \leq 0 \text{ on } R_+, k = 0,1,2,3.$$

The second example is found in [32] and relates to the system

$$(52) \qquad \dot{x}(t) + A(t)x(t) = \int_{-\infty}^t C(t,s)x(s)ds, \ t \geq t_0,$$

under the initial condition

$$(53) \qquad x(t) = \varphi(t), \ t \leq t_0,$$

φ being continuous and bounded.

The <u>uniform asymptotic stability</u> of the zero solution of (52) is assured by the following conditions on $A(t)$ and $C(t,s)$:

1) There exists a positive definite matrix B (same type as A, say n by n), such that the matrix

$$(54) \qquad R(t) = A^T(t)B + BA(t)$$

is positive definite for almost all $t \geq 0$; let α, β denote the smallest and respectively the largest eigenvalue of B.

2) If one denotes

$$(55) \qquad p(t) = \beta \ inf\{\frac{<x,Rx>}{<x,Bx>}, \ \theta \neq x \in R^n\},$$

then there exist constants $K > 0$ and $\epsilon > 0$, with $\epsilon < (2\beta)^{-1}$, such that

$$(56) \qquad \frac{K}{2} \int_{-\infty}^t |C(t,u)|_B du + \frac{1}{2K} \int_t^\infty |C(v,t)|_B dv \leq \frac{p(t)}{2\beta} - \epsilon[1 + p(t)],$$

for almost all $t \in R$; $|\cdot|_\beta$ stands for the norm of a matrix, when the usual Euclidian scalar product is substituted by $<x,y>_B = <x,By>$.

3) In addition, the functions $p(t)$ and $C(t,s)$ are such that

$$(57) \qquad \sup_{t_0 \in R} \int_{t_0}^\infty [1 + p(t)]^{-1} \{ \int_{-\infty}^{t_0} |C(t,s)|ds \}^2 dt < \infty$$

and

$$(58) \quad \lim_{T \to \infty} \sup_{t_0 \in R} \int_{t_0}^{\infty} [1 + p(t)]^{-1} \Big\{ \int_{-\infty}^{t_0 - T} |C(t,s)| \, ds \Big\}^2 dt = 0.$$

Surprisingly, this result is obtained by using the Lyapunov function $|x|^2$.

REFERENCES

1. N. V. Azbelev and L. M. Berezanskii, Stability of solutions of equations with delay (Russian). *Functional Differential Equations*, 3-15. *Perm*, 1989.

2. N. V. Azbelev, L. M. Berezanskii, P. M. Simonov and A. V. Chistiakov I, II, III. *Differential Equations* (Translation) 23(1987),747-754; 27(1991), 555-562, 1165-1172.

3. N. V. Azbelev, V. P. Maksimov, L. F. Rakhmatullina, *Introduction to the theory of functional-differential equations* (Russian), Moscow, Nauka, 1991, 277 pp. (English translation in preparation)

4. L. M. Berezanskii, Linear functional differential equations on a half-axis; stability of solutions (Russian). *Mat. Fizika i Nel. Mekhanika*, 4(38) (1985), 28-34.

5. L. M. Berezanskii, Stability criterion for differential equations with delay. *Functional Differential Equations, Perm 1986*, pp 21-23.

6. T. A. Burton, *Volterra Integral and Differential Equations*, Academic Press, New York, 1983.

7. C. Corduneanu, Sur certaines équations fonctionnelles de Volterra. *Funkcialaj Ekvacioj*, 9(1966), 119-127.

8. C. Corduneanu, Some global problems for Volterra functional differential equations. *Volterra integro-differential equations in Banach spaces.* Longman, London, 1989, 90-100.

9. C. Corduneanu, Integral representation of solutions of linear abstract Volterra functional differential equations. Libertas Mathematica, IX (1989), 139-146.

10. C. Corduneanu, *Integral Equations and Applications*, Cambridge Univ. Press, 1991.

11. R. Driver, Existence and stability of solutions of a delay-differential system. *Archive for Rational Mechanics*, 10(1962), 401-426

12. L. G. Fedorenko, Stability of functional differential equations. *Differential Equations* (Translation from Russian), 21(1986), 1031-1037.

13. J. K. Hale, *Theory of Functional Differential Equations*, Springer-Verlag, Berlin, New York, 1977.

14. T. Hara, R. Miyazaki, Equivalent conditions for stability of a Volterra integro-differential equation. *J. Math. An. Appl.* 174 (1993), 298-316.

15. Y. Hino, S. Murakami, T. Naito, *Functional Differential Equations with Infinite Delay*. Lecture Notes in Math #1473. Springer-Verlag, Berlin, 1991, 317 pp.

16. S. G. Karnishin, On the stability of solution of the linear functional differential equations with respect to a part of the variables. *Functional Differential Equations, Perm*, 1987, pp. 48-52.

17. J. Kato, On Liapunov-Razumikhin type theorem for functional differential equations. *Funk. Ekvacioj* 16 (1973), 225-239.

18. A. V. Kim, *Lyapunov's direct method in the theory of stability for delay systems* (Russian). Ekaterinburg, Ural University Press, 1992, 144 pp.

19. V. G. Kurbatov, Stability of functional differential equations. *Differential Equations* (Translations) 17 (1981), 963-972.

20. V. G. Kurbatov. Stability of functional differential equations on the whole real line and on a half-line. *Differential Equations* (Translation from Russian), 22(1986), 641-644.

21. V. Lakshmikantham, Recent advances in Lyapunov method for delay differential equations. *Differential Equations* (ed. S. Elaydi), M. Dekker, New York, 1991.

22. V. Lakshmikantham, S. Leela and A. A. Martynyuk, *Stability of Motion by Comparison Method* (Russian), Naukova Dumka, Kiev, 1991.

23. Yizeng Li, Global existence and stability of functional differential equations with abstract Volterra operators. Ph.D. Thesis, University of Texas at Arlington, 1993.

24. Mehran Mahdavi, Contributions to the theory of functional differential equations with abstract Volterra operators, Ph.D. Thesis, University of Texas at Arlington, 1992.

25. G. Marinescu, Différentielles de Gâteaux et Fréchet dans les espaces localement convexes. Bull. Math. Soc. Sci. Math. Phys. Roumanie (N.S.), 1 (49), 1957, 77-86.

26. A. V. Malygina, Some stability criteria for scalar linear delay equations. *Functional Differential Equations*, 27-131, *Perm*, 1992.

27. Alex McNabb, Graham Weir, Comparison theorem for causal functional differential equations. *Proc. AMS*, 104 (1988), 449-452.

28. B. S. Razumikhin, *Stability of Hereditary Systems* (Russian) Moscow, Nauka, 1983, 107 pp.

29. G. Sansone and R. Conti, *Nonlinear Differential Equations.* Pergamon Press, New York, 1964.

30. P. M. Simonov, A. V. Chistiakov, Theorems on uniform exponential stability of delay equations. *Functional Differential Equations, Perm,* 1991, pp. 83-95.

31. Ju N. Smolin, On Cauchy's matrix for functional differential equations (Russian). Izv. Vyssh. Uč. Zav., Mat. 1988, 54-62.

32. O. J. Staffans, A direct Lyapunov approach to Volterra integro-differential equations. *SIAM J. Math. An.* 19 (1988), 879-901.

33. V. A. Tyshkevich, *Some problems in stability theory of functional differential equation* (Russian). Kiev, Naukova Dumka, 1981, 78 pp. 78.

Persistence (Permanence), Compressivity, and Practical Persistence in Some Reaction–Diffusion Models from Ecology

Chris Cosner, Department of Mathematics and Computer, University of Miami, Coral Gables, Florida 33124

1. INTRODUCTION

In recent years mathematical ecology has become a very active field of research. There are both practical and theoretical reasons for all the activity. The practical interest is mostly based on concerns about the impact of human interference and environmental change on natural communities. The theoretical interest is based partly on the practical interest and partly on the fact that the appropriate analytic methods have recently become available. The essential applied problem is to understand how the interactions of biological species with each other and their environment influence their coexistence, extinction, population size, and other vital phenomena. One theoretical approach is to model those interactions using system of differential or difference equations and then try to analyze the models and interpret their results. A fundamental question is whether a given model predicts survival or extinction for a given population. That question raises a "metaquestion": what is meant by survival (for one species) or coexistence (for several species)? Extinction is in a sense simpler; any reasonable definition of extinction requires a prediction that the population becomes or tends toward zero after a sufficient amount of time, at least with high probability if not deterministically. A simple definition for survival in a mathematical model of a population would be the presence of a globally attracting equilibrium with the population or population density positive. There are a few problems with that definition. What if the model has coefficients which vary with time, so that there is never any equilibrium? What if there are several equilibria? How about limit cycles? Two populations whose interactions produce stable oscillations would certainly seem to coexist, but would not be at equilibrium. A number of alternative definitions of survival or coexistence have been proposed in attempts to address such problems. We shall describe some of those and list their strengths and weaknesses. The general ideas are relevant to many sorts of models, but most of the illustrations will be in terms of reaction-diffusion systems.

All of the ecological models we shall discuss will have the property of preserving nonnegativity, so that we can always restrict our attention to the positive cone and its boundary. A typical model system would be

$$
\frac{\partial u_i}{\partial t} = L_i(x,t)u_i + f_i(x,t,\vec{u})u_i \qquad \text{in } \Omega \times (0,\infty),
$$

$$
B_i u_i = 0 \qquad\qquad \text{on } \partial\Omega \times (0,\infty), \quad i = 1,...,n
$$
(1)

where $\Omega \subseteq I\!\!R^m$ is a bounded domain, L_i is a second order uniformly strongly elliptic operator on Ω for each fixed t and B_i is a classical boundary operator. To facilitate the use of maximum principles we shall assume that L_i has no zero order terms, so

$$
L_i u \equiv \sum_{i,j=1}^{m} a_{ij}(x,t)\frac{\partial^2 u}{\partial x_i \partial x_j} + \sum_{i=1}^{m} a_i(x,t)\frac{\partial u}{\partial x_i}.
$$

We also assume that for each i we have $B_i u = \alpha_i(x)u + \beta_i(x)(\partial u/\partial n)$ with $\alpha_i\beta_i \geq 0$ and $\alpha_i + \beta_i = 0$. For technical reasons we shall always assume that L_i is either constant in time or T-periodic, and that everything in the model is smooth. We define positivity in the standard way, so that $\vec{u} > 0$ means $u_i > 0$ on Ω for each i. For Dirichlet boundary conditions we require also $\partial u_i/\partial n < 0$ on $\partial\Omega$; for all others, $u_i > 0$ on $\partial\Omega$. The case which has received the most attention is the autonomous one, where there is no explicit t dependence. In that case (1) can be viewed as a semi-dynamical system, and treated via powerful techniques from the theory of dynamical systems. The first of our alternative definitions for coexistence or survival comes from the theory of dynamical systems; it is called permanence or uniform persistence. A dynamical or semi-dynamical system is said to be uniformly persistent if the flow restricted to the interior of the positive sone has a global attractor which is uniformly bounded away from the boundary of the positive cone. If the system is also dissipative, so that the attractor is compact, it is said to be permanent. These ideas and various methods for establishing uniform persistence/permanence are discussed in (Butler, Freedman, and Waltman 1986), (Hofbauer and Sigmund 1988), (Hale and Waltman 1989), (Hutson and Schmitt 1992), and (Cantrell, Cosner, and Hutson 1993). Another special case is where the system (1) is order preserving and autonomous or periodic. The ordering that is preserved may be different from that defined by the positive cone; for example, a system of mutualists in (1) will have $\partial f_i/\partial u_j \geq 0$ for $i \neq j$ and will preserve the ordering $u_i \geq v_i$, $i = 1,...,n$, if $\vec{u}$ and $\vec{v}$ are solutions, but a system of two competitors will preserve the ordering $u_1 \geq v_1$, $u_2 \leq v_2$. Not all systems of the form (1) are order preserving; for example systems with three competitors or with predator-prey interactions generally are not. Our second alternative definition of coexistence or survival is limited to the order preserving case; it is called compressivity. An order preserving system is said to be compressive if it has steady states which bound (relative to the ordering) a globally attracting order interval. The idea of compressivity was introduced by Hess (1991) and his collaborators. Finally, a system (not necessarily autonomous or order preserving) is said to be practically persistent if it admits an attracting positive set (into which all positive solutions eventually go) and the location of the positive attracting set can be determined, at least approximately. The term practical persistence is introduced in (Cao and Gard 1993). The context of the work in (Cao and Gard 1993) is a model consisting of ordinary differential equations with delays, but the idea makes sense in the context of (1) or in many other settings. A treatment of practical persistence for models such as (1) is given in (Cantrell and Cosner preprint).

2. QUALITATIVE COMPARISON OF DEFINITIONS OF COEXISTENCE

We now consider the pros and cons of these three alternative definitions of survival and coexistence. All have the advantage that they allow a conclusion of coexistence even when there is no globally stable positive equilibrium. The idea of uniform persistence is the most qualitative and thus most natural from the theoretical viewpoint of dynamical or semi-dynamical systems. It does not greatly restrict the sort of dynamics the system can have. On the other hand, most theoretical methods of establishing uniform persistence require a good understanding of the dynamics of subsystems where one component is identically zero. This understanding can be difficult to achieve for large systems. Also, the qualitative nature of the conclusions of uniform persistence which is so attractive from the theoretical viewpoint has some disadvantages in applications. If a system has a globally attracting set which predicts a population of between 1/2 and 1/3 of an individual for one of the species, any predictions of coexistence are dubious at best. In contrast, compressivity is typically established via arguments resembling monotone iteration from sub- and super-solutions. That means there is a good chance of finding some bounds on the location of the steady states bounding the attracting order interval, and it may sometimes be possible to determine the rate at which solutions approach the attracting order interval. However, the requirement that the system be order preserving is quite strong. Both uniform persistence and compressivity impose restrictions on the time dependence of the coefficients of the model. Theoretical results on compressivity assume that the coefficients are constant or periodic in time. Uniform persistence requires that the model be cast as a dynamical system. That can be done for some nonautonomous systems via the method of skew product flows as in (Burton and Hutson 1991) but the conditions are still somewhat restrictive. Practical persistence imposes the weakest requiremnts on time dependence - it can be used for some nonperiodic nonautonomous systems - and typically imposes structure requirements stronger than those needed for uniform persistence but weaker than those needed for the order-preserving properties on which compressivity is based. Like compressivity, practical persistence has the advantage of giving some quantitative information about the asymptotic behavior of solutions. At this time there is little general work on practical persistence, so the arguments for it are typically more *ad hoc* than those used for uniform persistence or compressivity. This last drawback could of course be remedied by further research.

We shall further compare and contrast the various ideas of coexistence described above, but to do so requires a bit more background. Clearly any reasonable definition of coexistence in a system such as (1) must require that any equilibrium with one or more components zero be locally unstable, at least for large t. If that condition is violated, trajectories starting near some such equilibrium would be drawn toward it, implying extinction for those species with zero density at the equilibrium. To establish long term coexistence it is not always sufficient that any equilibrium with some component zero be locally unstable. For example, a Lotka-Volterra system for three competitors where in pairwise competition the first competitor dominates the second, the second dominates the third, and the third dominates the first, may have trajectories that spiral toward the boundary of the positive cone even though all equilibria with some component zero are unstable. (This phenomenon already arises for autonomous systems of ordinary differential equations; see (Hutson and Schmitt 1992) and the references therein. The simplest additional hypotheses to state, and often to verify, are those for uniform persistence/permanence. Essentially if a model can be cast as a dissipative semidynamical system on the positive cone of some Banach space, if the ω-limit set in the boundary of the positive cone consists of equilibria whose stable manifolds do not intersect the interior of the positive cone, and if there are no cycles of connecting orbits linking the boundary equilibria, then by results of Hale and Waltman (1989) the system is uniformly persistent or permanent. (Another approach to permanence/uniform persistence uses a generalization of Lyapunov functions; see (Hofbauer and Sigmund 1988), (Hutson and Schmitt 1992), and (Cantrell, Cosner, and Hutson 1993). For compressivity, we must first know that our system generates a semiflow with suitable order preserving properties and then must construct something like sub- and super-solutions. It has been known for a long time that trajectories for a single reaction-diffusion equation which start at a sub-solution must increase with time; see for example (Aronson and Weinberger 1975). The same is true for systems with suitable order preserving properties, but rather strong hypotheses are needed to insure that a system such as (1) is order preserving. In (1) it would suffice to require $\partial f_i / \partial u_j \geq 0$ for $i \neq j$, (see, e.g. (Fife 1979)) or to require $n = 2$, $\partial f_i / \partial u_j \leq 0$ for $i \neq j$, $i, j = 1, 2$ and to use the ordering $(u_1, u_2) \leq (v_1, v_2)$ if $u_1 \leq v_1$, $u_2 \geq v_2$ (see (Cosner and Lazer 1984)).

The construction of sub- and super-solutions, or related objects, is something more of an art than a science, although it is sometimes natural to build sub- and/or super-solutions from solutions of related problems. We shall return to that point later. Construction of auxilliary functions is also relevant for practical persistence. The idea of practical persistence is more recent and has had less abstract development than uniform persistence/permanence or compressivity, so the conditions needed to apply it are less clear cut. The approach to establishing practical persistence introduced by Cao and Gard (1993) used multiple Lyapunov functions; that used in (Cantrell and Cosner preprint) is similar but treats the system one component at a time. (Multiple Lyapunov functions are generally not vector Lyapunov functions since they need not act simultaneously.) A condition that has typically been imposed for practical persistence is that each component of the model have some type of self-limitation, so that if all other components are held at any fixed value in the ith equation the nonlinearity predicts an upper bound on the ith population or density. The most common sort of self-limitation is logistic growth, but other forms are possible.

3. SOME DETAILS AND EXAMPLES

To check the stability of equilibria and to construct comparison functions in the context of reaction-diffusion systems we must consider some associated linear eigenvalue problems. Suppose that L and B are a second order elliptic operator and boundary operator on Ω as in (1) and that $r(x, t)$ is a smooth function on $\Omega \times (0, \infty)$.

Theorem 1: Suppose that L and r are T-periodic. The eigenvalue problem

$$\frac{\partial \phi}{\partial t} - L\phi - r\phi = \sigma \phi \qquad \text{on } \Omega \times \mathbb{R}$$

$$B\phi = 0 \qquad \text{on } \partial\Omega \times \mathbb{R}, \ \phi \ T - \text{periodic}$$

(2)

has a unique principal eigenvalue σ_1 characterized by having a positive eigenfunction $\phi_1(x,t)$. (If L and r are constant in t, then so is the eigenfunction and (2) becomes an elliptic eigenvalue problem.)

Discussion: The idea of periodic-parabolic eigenvalue problems was introduced by Lazer (1982), (Castro and Lazer 1982) and developed and applied in the work of Hess (1991). The eigenvalue problem (2) provides a characterization of the behavior of solutions to the associated diffusive logistic equation which we shall need for the analysis of systems.

Theorem 2: Suppose that L, B, and r are as in Theorem 1 and $c(x,t)$ is smooth and T-periodic with $c(x,t) \geq c_0 > 0$. The logistic equation

$$\frac{\partial u}{\partial t} = Lu + ru - cu^2 \qquad \text{in } \Omega \times (0,\infty)$$

$$Bu = 0 \qquad \text{on } \partial\Omega \times (0,\infty)$$

(3)

has a unique positive T-periodic steady state $\theta(L,r,c)$ which is globally attracting among nontrivial nonnegative solutions provided the principal eigenvalue σ_1 for (2) is negative. If $\sigma_1 \geq 0$ all positive solutions of (3) tend toward zero as $t \to \infty$. If we have $\sigma_1 < 0$ and normalize the eigenfunction ϕ_1 by $\sup \phi_1 = 1$ then

$$\inf(-\sigma_1/c)\phi_1(x,t) \leq \theta(L,r,c) \leq \sup(r/c). \qquad (4)$$

If L, r, and c are constant in time so is $\theta(L,r,c)$. If θ is not constant in space (e.g. under Dirichlet boundary conditions) the inequality in the upper bound is strict.

Discussion: Theorem 2 is essentially Theorem 28.1 of (Hess 1991). Related results are discussed in (Cantrell and Cosner 1989, 1991).

We now can treat some examples of systems of the form (1) where uniform persistence/permanence, compressivity, and/or practical persistence can be established. We shall give only brief discussions of the first two ideas since they are described in some detail in (Hutson and Schmitt 1992) and (Cantrell, Cosner, and Hutson 1993) and in (Hess 1991) respectively. The three sample systems we shall consider are

$$\frac{\partial u_1}{\partial t} = D_1 \Delta u_1 + (r_1 + c_{12}u_2)u_1$$

$$\frac{\partial u_2}{\partial t} = D_2 \Delta u_2 + (r_2 - c_{21}u_1 - c_{22}u_2)u_2 \qquad \text{in } \Omega \times (0,\infty),$$

(5)

$$u_1 = u_2 = 0 \qquad \text{on } \partial\Omega \times (0,\infty),$$

and

$$\frac{\partial u_1}{\partial t} = D_1 \Delta u_1 + (r_1 - c_{11}u_1 + c_{12}u_2)u_1$$

$$\frac{\partial u_2}{\partial t} = D_2 \Delta u_2 + (r_2 - c_{21}u_1 - c_{22}u_2)u_2 \qquad \text{in } \Omega \times (0, \infty) \tag{6}$$

$$u_1 = u_2 = 0 \qquad \text{on } \partial\Omega \times (0, \infty),$$

and

$$\frac{\partial u_1}{\partial t} = D_1 \Delta u_1 + (r_1 - c_{11}u_1 - c_{12}u_2)u_1$$

$$\frac{\partial u_2}{\partial t} = D_2 \Delta u_2 + (r_2 - c_{21}u_1 - c_{22}u_2)u_2 \qquad \text{in } \Omega \times (0, \infty), \tag{7}$$

$$u_1 = u_2 = 0 \qquad \text{on } \partial\Omega \times (0, \infty).$$

We shall always assume that the coefficients of (5)-(7) are smooth in all arguments and are bounded for all values of those arguments. In general, the coefficients may depend on x, t, u_1, and u_2, although we will need to restrict their dependence for some results. We shall also always assume that $c_{ij} \geq c_0 > 0$ in $\overline{\Omega}$ for some constant c_0, so that (5) and (6) are predator-prey systems with u_1 corresponding to the predator density and (7) is a competition system. The crucial difference between (5) and (6) is that (6) includes the self-limitation term $-c_{11}u_1$ in the predator equation while (5) does not. In (5) we shall assume, as is usual, that $r_1 \leq -r_0 < 0$ for some constant r_0 so the predator population cannot grow in the absence of prey. In (6) and (7) we make no a priori assumption about the sign of r_1; in all the models we shall need $r_2 > 0$ at least for some $x \in \Omega$ to have any sort of coexistence at all, and for (7) we also need $r_1 > 0$ for at least some x. Under the above hypotheses all three systems will be dissipative by results in (Cantrell, Cosner, and Hutson 1993).

Theorem 3: Suppose that none of the coefficients in (5), (6), (7) depend on t. Suppose also that the equation

$$\frac{\partial u_2}{\partial t} = D_2 \Delta u_2 + (r_2(x, 0, u_2) - c_{22}(x, 0, u_2)u_2)u_2 \qquad \text{in } \Omega \times (0, \infty)$$

$$u_2 = 0 \qquad \text{on } \partial\Omega \times (0, \infty) \tag{8}$$

has a unique positive equilibrium $u_2^*(x)$ which is globally attracting among positive solutions. In (6) and (7) suppose further that the equation

$$\frac{\partial u_1}{\partial t} = D_1 \Delta u_1 + (r_1(x, u_1, 0) - c_{11}(x, u_1, 0)u_1)u_1 \qquad \text{in } \Omega \times (0, \infty)$$

$$u_1 = 0 \qquad \text{on } \partial\Omega \times (0, \infty) \tag{9}$$

also has a unique positive equilibrium $u_1^*(x)$ which is globally attracting among positive solutions.

We then have the following conclusions:

i) The system (5) is uniformly persistent/permanent in the space $[C_0^1(\overline{\Omega})]^2$ if the principal eigenvalues σ_1 are both negative in

$$-D_1\Delta\phi - (r_1(x,0,u_2^*) + c_{12}(x,0,u_2^*)u_2^*)\phi = \sigma\phi \quad \text{in } \Omega,$$

$$\phi = 0 \quad \text{on } \partial\Omega, \tag{10}$$

and

$$-D_2\Delta\phi - r_2(x,0,0)\phi = \sigma\phi \quad \text{in } \Omega,$$

$$\phi = 0 \quad \text{on } \partial\Omega. \tag{11}$$

ii) The system (6) is uniformly persistent/permanent in $[C_0^1(\overline{\Omega})]^2$ if the principal eigenvalues σ_1 are negative in (10) and

$$-D_2\Delta\phi + (r_2(x,u_1^*,0) - c_{21}(x,u_1^*,0)u_1^*)\phi = \sigma\phi \quad \text{in } \Omega,$$

$$\phi = 0 \quad \text{on } \partial\Omega. \tag{12}$$

iii) The system (7) is uniformly persistent in $[C_0^1(\overline{\Omega})]^2$ if the principal eigenvalues σ_1 are negative in (12) and

$$-D_2\Delta\phi + (r_1(x,0,u_2^*) - c_{12}(x,0,u_2^*)u_2^*)\phi = \sigma\phi \quad \text{in } \Omega,$$

$$\phi = 0 \quad \text{on } \partial\Omega. \tag{13}$$

Discussion: Theorem 3 follows immediately from results in (Cantrell, Cosner, and Hutson 1993), specifically Theorems 5.3 and 5.7. An alternative derivation could be based on the methods of (Hale and Waltman 1989). The hypotheses that u_2^* and (for (6) and (7)) u_1^* exist and have the required properties will be met if the coefficients r_i and c_{ii} depend only on x and $r_1(x)$, $r_2(x)$ are sufficiently large (at least on parts of Ω) that $\sigma_1 < 0$ for $i = 1, 2$ in

$$-D_i\Delta\phi - r_i(x,0,0)\phi = \sigma\phi \quad \text{in } \Omega$$

$$\phi = 0 \quad \text{on } \partial\Omega. \tag{14}$$

This follows from Theorem 2. More general forms of dependence are possible; in particular c_{ii} may depend on u_i and u_j as long as $\partial c_{ii}/\partial u_i < 0$ when $u_j \equiv 0$, $j \neq i$; see (Cantrell and Cosner 1989), Theorem 2.3. The meaning of the hypotheses that $\sigma_1 < 0$ in (10)-(13) is simply that the equilibria with one or both species absent are unstable. It is probably possible to extend Theorem 3 to cases where the coefficients are T-periodic in t by using the methods of (Burton and Hutson 1991). (The eigenvalue conditions would still make sense by virtue of Theorem 1.)

It is clear from the statement of Theorem 3 that uniform persistence/permanence imposes few restrictions on the form of the nonlinear terms in (5)-(7). On the other hand it provides no information about the location of the positive attracting set or the way that trajectories approach that set. In contrast, compressivity places rather strict demands on the nonlinearities but may give more information about the attracting set or at least about the way that trajectories approach it.

Theorem 4: Suppose that the coefficients in (7) are T-periodic in t, and that the equations

$$\frac{\partial u_1}{\partial t} = D_1 \Delta u_1 + (r_1(x,t,u_1,0) - c_{11}(x,t,u_1,0)u_1)u_1 \quad \text{in } \Omega \times (0,\infty)$$

$$u_1 = 0 \qquad\qquad\qquad\qquad\qquad\qquad\qquad \text{on } \partial\Omega \times (0,\infty)$$

(15)

and

$$\frac{\partial u_2}{\partial t} = D_2 \Delta u_2 + (cr_2(x,t,0,u_2) - c_{22}(x,t,0,u_2))u_2 \quad \text{in } \Omega \times (0,\infty)$$

$$u_2 = 0 \qquad\qquad\qquad\qquad\qquad\qquad\qquad \text{on } \partial\Omega \times (0,\infty)$$

(16)

have unique positive T-periodic steady states $u_1^*(x,t)$, $u_2^*(x,t)$ which are globally attracting among positive solutions of (15) and (16) respectively. Suppose further that for $u_1, u_2 \geq 0$

$$\frac{\partial[(r_i - c_{ii}u_i - c_{ij}u_j)u_i]}{\partial u_j} \leq 0 \quad \text{for } j \neq i, \ i,j = 1,2.$$

(17)

If the principal eigenvalues σ_1 are negative in the problems

$$\frac{\partial\phi}{\partial t} - D_1\Delta\phi - (r_1(x,t,0,u_2^*) - c_{12}(x,t,0,u_2^*)u_2^*)\phi = \sigma\phi \quad \text{in } \Omega \times \mathbb{R},$$

(18)

$$\phi = 0 \quad \text{on } \partial\Omega \times \mathbb{R}, \ \phi \ T - \text{periodic}$$

and

$$\frac{\partial\phi}{\partial t} - D_2\Delta\phi - (r_2(x,t,u_1^*,0) - c_{21}(x,t,u_1^*,0)u_1^*)\phi = \sigma\phi \quad \text{in } \Omega \times \mathbb{R},$$

(19)

$$\phi = 0 \quad \text{on } \partial\Omega \times \mathbb{R}, \ \phi \ T - \text{periodic}$$

then the system (7) is compressive, relative to the ordering where $(v_1, v_2) \leq (w_1, w_2)$ means $v_1 \geq w_1$, $v_2 \leq w_2$. That is, there exist pairs of positive T-periodic functions $(\overline{u}_1, \underline{u}_2)$ and $(\underline{u}_1, \overline{u}_2)$ satisfying (7) with $\overline{u}_1 \geq \underline{u}_1$ and $\underline{u}_2 \leq \overline{u}_2$ such that the order interval defined by

$$(\overline{u}_1, \underline{u}_2) \leq (u_1, u_2) \leq (\underline{u}_1, \overline{u}_2)$$

(20)

is globally attracting among positive solutions.

Remark: It follows from (17) that

$$(r_1(x,t,u_1,0) \ -c_{11}(x,t,u_1,0)u_1)u_1$$

$$\geq (r_1(x,t,u_1,u_2) - c_{11}(x,t,u_1,u_2) - c_{12}(x,t,u_1,u_2))u_1$$

for $u_2 \geq 0$, so that for any solution (u_1, u_2) of (7), u_1 is a subsolution of (15) and hence for any $\varepsilon > 0$, $u_1 \leq (1 + \varepsilon)u_1^*$ for t sufficiently large. Thus we will have $\overline{u}_1 \leq u_1^*$ and for similar reasons $\bar{u}_2 \leq u_2^*$. In the case where r_i and c_{ii} are T-periodic in t and do not depend on u_1 or u_2 then (15) and (16) are diffusive logistic equations, so criteria for the existence of u_1^* and u_2^* and estimates on their size are given in Theorem 2. In any case, $u_i^* \leq \sup(r_i/c_{ii})$ by the maximum principle as in the logistic case. We shall see that in simple cases estimates of $\underline{u}_1$ and $\underline{u}_2$ can also be obtained.

Discussion: Theorem 4 is a variation on Theorem 33.3 of (Hess 1991). The condition (17) is needed to insure that the system (7) is genuinely competitive and thus has the necessary order preserving properties. In the Lotka-Volterra case, the left side of (17) is simply $-c_{ij}$ so the condition always holds.

The essential idea in proving results such as Theorem 4 for suitable order preserving systems is that in such systems those solutions which have subsolutions as initial data will increase monotonically (with respect to the ordering) toward steady state solutions and starting at supersolutions will decrease toward steady states. This idea was used (Aronson and Weinberger 1975); related ideas were used in (Cosner and Lazer 1984) and are discussed systematically in (Hess 1991) for Lotka-Volterra competition systems. As is noted in the Remark above, the system (7) will have $u_i \leq (1 + \varepsilon)u_i^*$ for t sufficiently large for any $\varepsilon > 0$. By continuity the principal eigenvalues in (18) and (19) will remain negative if u_i^* is replaced by $(1 + \varepsilon)u_i^*$ for ε sufficiently small. If we denote the eigenfunctions by ϕ_1^1 and ϕ_1^2, it is straightforward to check that if $(p_1, p_2) = ((1+\varepsilon)u_1^*, \delta\phi_1^2)$ then for $\delta > 0$ small

$$\frac{\partial p_1}{\partial t} \geq D_1 \Delta p_1 + (r_1 - c_{11}p_1 - c_{12}p_2)p_1$$

$$\frac{\partial p_2}{\partial t} \leq D_2 \Delta p_2 + (r_2 - c_{21}p_1 - c_{22}p_2)p_2$$

so (p_1, p_2) is a subsolution to (7); similarly $(q_1, q_2) = (\delta\phi_1^1, (1 + \varepsilon)u_2^*)$ is a supersolution. But by the strong maximum principle any solution (u_1, u_2) of (7) which is nonnegative and nontrivial in both components will have u_i strictly positive in Ω with negative normal derivative on $\partial\Omega$ at any time $t = t_0 > 0$, and hence $u_i \geq \delta\phi_1^i$ at $t = t_0$ for δ small enough. It follows that any solution (u_1, u_2) can be "trapped" so that $(p_1, p_2) \leq (u_1, u_2) \leq (q_1, q_2)$ for some $t = t_0 > O_0$. If (v_1, v_2) and (w_1, w_2) are solutions to (7) with initial data (p_1, p_2) and (q_1, q_2) respectively, then since (p_1, p_2) is a subsolution, (q_1, q_2) is a supersolution, and $(p_1, p_2) \leq (q_1, q_2)$ we must have $(v_1, v_2) \uparrow (\bar{u}_1, \underline{u}_2)$ and $(w_1, w_2) \downarrow (\underline{u}_1, \bar{u}_2)$ with $(\bar{u}_1, \underline{u}_2) \leq (\underline{u}_1, \bar{u}_2)$. Thus, not only is (u_1, u_2) attracted to the order interval defined in (20) but the dynamics are bounded by those of solutions which approach the interval monotonically. In many cases the steady states $(\bar{u}_1, \underline{u}_2)$ and $(\underline{u}_1, \bar{u}_2)$ can be estimated in terms of solutions to diffusive logistic equations. We have already seen that $\bar{u}_i \leq u_i^* \leq \sup(r_i/c_{ii})$. Lower bounds on $\underline{u}_i$ can also be obtained; the arguments involved are similar to those used in establish practical persistence, so we shall briefly defer their discussion.

Both uniform persistence/permanence and compressivity are obtained by working directly with the original system. We shall establish criteria for practical persistence via comparison methods. The comparisons will be between solutions of systems such as (6) or (7) with solutions of single diffusive logistic equations. Since making comparisons necessarily involves making choices of what to compare, there may not be any optimal results. Choosing simple comparisons may lead to simple but rough results; choosing complicated comparisons might produce results that are sharper but more complicated and thus harder to interpret. The results in the next theorem are representative but many others are possible. A theory covering a reasonable range of models for n interacting species is developed in (Cantrell and Cosner preprint), but the derivation is somewhat involved and we shall not describe it in detail here.

Theorem 5: Let $\bar{c}_{ij} = \sup c_{ij}$, $\underline{c}_{ij} = \inf c_{ij}$, $\bar{r}_i = \sup r_i$, $\underline{r}_i = \inf r_i$ and suppose that $\rho_i(t)$ is a T-periodic function with $\rho_i(t) \leq r_i$.

i) Suppose (u_1, u_2) satisfies (6). Let $M_1 = (\bar{r}_1 + \bar{c}_{12}(\bar{r}_2/\underline{c}_{22}))/\underline{c}_{11}$. If the principal eigenvalues in the problems

$$-D_2\Delta\phi - (\underline{r}_2 - \bar{c}_{21}M_1)\phi = \sigma\phi \quad \text{in } \Omega$$

$$\phi = 0 \quad \text{on } \partial\Omega \tag{21}$$

and

$$-D_1\Delta\phi - (\underline{r}_1 + \underline{c}_{12}\theta(D_2\Delta, \underline{r}_2 - \bar{c}_{21}M_1, \bar{c}_{22}))\phi = \sigma\phi \quad \text{in } \Omega,$$

$$\phi = 0 \quad \text{on } \partial\Omega \tag{22}$$

are negative then all solutions of (6) with initial data nonnegative and nonzero in both component must approach or enter the set of (u_1, u_2) satisfying

$$\theta(D_1\Delta, \underline{r}_1 + \underline{c}_{12}\theta(D_2\Delta, \underline{r}_2 - \bar{c}_{21}M_1, \bar{c}_{22}), \bar{c}_{11}) \leq u_1 \leq M_1, \tag{23}$$

and

$$\theta(D_2\Delta, \underline{r}_2 - \bar{c}_{21}M_1, \bar{c}_{22}) \leq u_2 \leq \bar{r}_2/\underline{c}_{22} \quad \text{as } t \to \infty. \tag{24}$$

(Negative principal eigenvalues in (21), (22) imply the existence of the θ's in (23), (24)).

ii) Suppose that (u_1, u_2) satisfies (7). Suppose also that the principal eigenvalues in

$$-D_i\Delta\phi - \bar{r}_i\phi = \sigma\phi \quad \text{in } \Omega,$$

$$\phi = 0 \quad \text{on } \partial\Omega, \ i = 1, 2, \tag{25}$$

and

$$\frac{\partial\phi}{\partial t} - D_i\Delta\phi - (\rho_i(t) - \bar{c}_{ij}\theta(D_j\Delta, \bar{r}_j, \underline{c}_{jj}))\phi = \sigma\phi$$

$$\text{in } \Omega \times \mathbb{R}, \ \text{for } j \neq i, \ i = 1, 2 \ \text{with} \tag{26}$$

$$\phi = 0 \ \text{on } \partial\Omega \times \mathbb{R}, \ \phi \ T - \text{periodic},$$

are negative. As $t \to \infty$, (u_1, u_2) must approach or enter the set of (u_1, u_2) satisfying

$$\theta(D_i\Delta, \rho_i - \bar{c}_{ij}\theta(D_j\Delta, \bar{r}_j, \underline{c}_{jj}), \bar{c}_{ii}) \leq u_i \leq \theta(D_i\Delta, \bar{r}_i, \underline{c}_{ii}) \ \text{for } j \neq i, \ i, j = 1, 2. \tag{27}$$

Remark: In the case of a true Lotka-Volterra system in (7) the coefficients will be constants and $\theta(D_i\Delta, \bar{r}_i, \underline{c}_{ii})$ in (27) will coincide with u_i^* in Theorem 4, so that the upper bound in (27) would imply $\bar{u}_i \leq u_i^*$ as in the remarks following Theorem 4.

Discussion: We shall sketch the proof for (i); the analysis in (ii) is similar.

We first observe that

$$\frac{\partial u_2}{\partial t} = D_2\Delta u_2 + (r_2 - c_{21}u_1 - c_{22}u_2)u_2$$

$$\leq D_2\Delta u_2 + (\bar{r}_2 - \underline{c}_{22}u_2)u_2$$

so that u_2 is a subsolution to

$$\frac{\partial u}{\partial t} = D_2\Delta u + (\bar{r}_2 - \underline{c}_{22}u)u \qquad \text{on } \Omega \times (0,\infty)$$

$$u = 0 \qquad\qquad \text{on } \partial\Omega \times (0,\infty). \tag{28}$$

By (21) and the monotonicity of eigenvalues the principal eigenvalue of

$$-D_2\Delta\phi - \bar{r}_2\phi = \sigma\phi \qquad \text{in } \Omega$$

$$\phi = 0 \qquad \text{on } \partial\Omega$$

is negative, so by Theorem 2, there exists a unique attracting steady state (in this case an equilibrium) $\theta(D_2\Delta, \bar{r}_2, \underline{c}_{22}) < \bar{r}_2/\underline{c}_{22}$ for (28). If u satisfies (28) with the same initial data as u_2, then $u_2 \leq u \to \theta(D_2\Delta, \bar{r}_2, \underline{c}_{22}) < \bar{r}_2/\underline{c}_{22}$ as $t \to \infty$, so $u_2 \leq \bar{r}_2/\underline{c}_{22}$ for $t \geq t_0$ if t_0 is large enough. For $t \geq t_0$, u_1 is a subsolution to

$$\frac{\partial u}{\partial t} = D_1\Delta u + (\bar{r}_1 + \bar{c}_{12}(\bar{r}_2/\underline{c}_{22}) - \underline{c}_{11}u)u \qquad \text{in } \Omega \times (0,\infty)$$

$$u = 0 \qquad\qquad \text{on } \partial\Omega \times (0,\infty). \tag{29}$$

Since $\theta(D_2\Delta, \underline{r}_2 - \bar{c}_{21}M_1, \bar{c}_{22}) < \bar{r}_2/\underline{c}_{22}$, it follows from Theorem 2 that (29) has a positive equilibrium $\theta(D_1\Delta, \bar{r}_1 + \bar{c}_{12}(\bar{r}_2/\underline{c}_{22}), \underline{c}_{11}) < M_1$ which attracts all positive solutions. Thus, if u satisfies (29) with $u(x, t_0) = u_1(x, t_0)$ then $u_1 \leq u \to \theta < M_1$ as $t \to \infty$ so that $u_1 \leq M_1$ for $t \geq t_1$ with t_1 sufficiently large. We have now obtained the upper bounds in (23), (24). The upper bound on u_1 will next yield a lower bound on u_2 (as $t \to \infty$) and that will in turn yield a lower bound on u_1. Since $u_1 \leq M_1$ for t large we see that for t large u_2 is a supersolution of

$$\frac{\partial u}{\partial t} = D_2\Delta u + (\underline{r}_2 - \bar{c}_{21}M_1 - \bar{c}_{22}u)u \qquad \text{in } \Omega \times (0,\infty)$$

$$u = 0 \qquad\qquad \text{on } \partial\Omega \times (0,\infty). \tag{30}$$

By (21), (30) has the unique positive attracting equilibrium $\theta(D_2\Delta, \underline{r}_2 - \bar{c}_{21}M_1, \bar{c}_{22})$; thus, if we take $t \geq t_2$ for t_2 large enough and set $u(x, t_2) = u_2(x, t_2)$ then as $t \geq t_2$, $t \to \infty$, we have $u_2(x, t) \geq u(x, t) \to \theta(D_2\Delta, \underline{r}_2 - \bar{c}_{21}M_1, \bar{c}_{22})$ so that for any $\varepsilon > 0$, $u_2 \geq (1-\varepsilon)\theta(D_2\Delta, \underline{r}_2 - \bar{c}_{21}M_1, \bar{c}_{22})$ for t large. In this sense u_2 must at least be attracted to the interval shown in (24). Finally, since $\sigma_1 < 0$ in (22) we will have $\sigma_1 < 0$ if the function θ in (22) is replaced with $(1 - \varepsilon)\theta$ for $\varepsilon > 0$ small. For any sufficiently small $\varepsilon > 0$ we have $u_2 \geq (1 - \varepsilon)\theta(D_2\Delta, \underline{r}_2 - \overline{c}_{21}M_1, \overline{c}_{22})$ for large t, so that u_1 will be a supersolution to

$$\frac{\partial u}{\partial t} = D_1\Delta u(\underline{r}_1 + \underline{c}_{12}(1 - \varepsilon)\theta(D_2\Delta, \underline{r}_2 - \overline{c}_{21}M_1, \overline{c}_{22}) - \overline{c}_{11}u)u \quad \text{in } \Omega \times (0, \infty),$$
$$(31)$$
$$u = 0 \quad \text{on } \partial\Omega \times (0, \infty).$$

Since (31) will have the positive attracting equilibrium $\theta(D_1\Delta, \underline{r}_1 + \underline{c}_{12}(1 - \varepsilon)\theta(D_2\Delta, \underline{r}_2 - \overline{c}_{21}M_1, \overline{c}_{22}), \overline{c}_{11})$ and $\varepsilon > 0$ is arbitrarily small, we can conclude that the interval in (23) is attracting for u_1.

The proof of (ii) is similar in spirit to that of (i), but exploits fully the periodic-parabolic aspects of Theorems 1 and 2 to give a result which is more complicated but in some cases sharper than those which could be obtained using constant bounds on the coefficients.

Remarks: It is natural to ask how "practical" the conclusions of Theorem 5 really are. By using estimates such as those in (4), asymptotic bounds on u_1, u_2 in terms of solutions θ to diffusive logistic equations can be translated into terms of the principal eigenfunctions of the Laplacian or related operators. Such eigenfunctions are reasonably well known for simple geometries. If Ω is the unit interval the principal eigenvalue of

$$-D\Delta\phi - r\phi = \sigma\phi \text{ on } (0, 1), \quad \phi(0) = 0, \quad \phi(1) = 0$$

is $D\pi^2 - r$, with eigenfunction $\phi = \sin(\pi x)$ having $\sup \phi = 1$. Thus, (21) will hold if $D_2\pi^2 - \underline{r}_2 + \overline{c}_{21}M_1 = \sigma_1 < 0$; in that case the asymptotic lower bound in (24) would imply $\liminf_{t\to\infty} u_2 \geq (\underline{r}_2 - D_2\pi^2 - \overline{c}_{21}M_1)\sin(\pi x)/\overline{c}_{22}$ by (4). Other than π, all the coefficients in this last estimate are algebraic combinations of coefficients or bounds on coefficients in (6), so they are readily computable. To obtain an explicit lower bound in (23) is somewhat more delicate, especially if $\underline{r}_1 < 0$, but still is possible. The asymptotic lower bound on u_2 is such that for any $\varepsilon > 0$ and t sufficiently large, $u_2(x, t) \geq [(1-\varepsilon)/2](\underline{r}_2 - D_2\pi^2 - \overline{c}_{21}M_1)/\overline{c}_{22}$ for $1/6 \leq x \leq 5/6$. Thus, the principal eigenvalue in (22) is smaller than that for

$$-D_1\Delta\phi - (\underline{r}_1 + \underline{c}_{12}[(1 - \varepsilon)/2]\chi_{(1/6,5/6)}(x)(\underline{r}_2 - D_2\pi^2 - \overline{c}_{21}M_1)/\overline{c}_{22}))\phi = \sigma\phi$$

$$\text{on } (0, 1), \quad \phi = 0 \quad \text{for } x = 0, 1.$$

That eigenvalue in turn is (by variational properties of eigenvalues) smaller than that for

$$-D_1\phi - (\underline{r}_1 + \underline{c}_{12}[(1 - \varepsilon)/2](\underline{r}_2 - D_2\pi^2 - \overline{c}_{21}M_1)/\overline{c}_{22})\phi = \sigma\phi \quad \text{on } (1/6, 5/6),$$

$$\phi = 0 \quad \text{for } x = 1/6, 5/6.$$
$$(32)$$

The principal eigenvalue for (32) is

$$\sigma_1 = 9\pi^2 D_1/4 - (\underline{r}_1 + \underline{c}_{12}[(1-\varepsilon)/2](\underline{r}_2 - D_2\pi^2 - \overline{c}_{21}M_1)/\overline{c}_{22}); \qquad (33)$$

so if $\sigma_1 < 0$ in (33) then $\sigma_1 < 0$ in (22). The expression in (32) again depends only on coefficients bounds. If $\sigma_1 < 0$ in (33) then there will be a positive equilibrium $\underline{\theta}$ for

$$\frac{\partial u}{\partial t} = D_1\Delta u + [(\underline{r}_1 + \underline{c}_{12}[(1-\varepsilon)/2](\underline{r}_2 - D_2\pi^2 - \overline{c}_{21}M_1)/\overline{c}_{22}) - \overline{c}_{11}u]u \qquad (34)$$

$$\text{on } (1/6, 5/6), \quad u = 0 \quad \text{for } x = 1/6, 5/6,$$

which can be bounded below by (4):

$$\underline{\theta} \geq \{[\underline{r}_1 + \underline{c}_{12}[(1-\varepsilon)/2](\underline{r}_2 - D_2\pi^2 - \overline{c}_{12}M_1)/\overline{c}_{22}]/\overline{c}_{11}\} \sin\left[\frac{3\pi}{2}\left(x - \frac{1}{6}\right)\right] \qquad (35)$$

If we extend $\underline{\theta}$ to be zero on $[0,1]\backslash(1/6, 5/6)$ then we obtain a subsolution to the problem (31) for which u_1 is ultimately a supersolution; since (31) will have a unique equilibrium θ we can argue that u_1 must eventually be larger than θ while $\underline{\theta} < \theta$ on $(1/6, 5/6)$ and $0 < \theta$ on $(0,1)\backslash(1/6, 5/6)$ so that for large t u_1 will satisfy $u_1 > \underline{\theta}$ with $\underline{\theta}$ bounded below by (35). Thus, in a simple geometry, we can obtain bounds which are not necessarily sharp but are quite explicit and computable.

4. CONCLUSIONS

Theorems 3-5 are representative of the results which can be obtained via the ideas of uniform persistence/permanence, compressivity, and practical persistence. The first two methods work directly with the original system, and hence impose rather strong conditions on the time dependence of coefficients and require information about the equilibria with one of the components zero. Practical persistence uses comparison methods and thus evades those issues. Uniform persistence/permanence imposes the weakest conditions on the nonlinearity but yields the least detailed conclusions. Compressivity imposes rather strong conditions on the nonlinearity but gives the most information; the strong hypotheses are needed so that the system will have suitable monotonicity properties and the strong conclusions follow from those properties. Practical persistence gives conclusions which are in practice similar to those of compressivity (since in the compressive case the attracting order interval often can be computed only approximately via Theorems 1 and 2) but requires much weaker conditions on the nonlinearity. The main limitation for practical persistence is that it seems to require some type of of self-regulation in all equations, at least in the reaction-diffusion setting. (In ODE problems it is sometimes possible to use Lyapunov methods to get similar sorts of results without self-regulation.) The implications of self-regulation may be more than mere technicalities. Butler, Schmid and Waltman (1990) have shown that self-regulation restricts the dynamics of finite dimensional systems in a manner similar to that of a competitive structure.

Many other applications of practical persistence are possible. A fairly general theory is presented and a number of examples are analyzed in (Cantrell and Cosner, preprint.) Related ideas have been used in various contexts; (Cantrell and Cosner, preprint) also has references. There are also many open questions. In particular it would be interesting if some sort of practical persistence could be shown for systems such as (5). The problem is in finding suitable upper bounds on the predator density. Usually for such systems some type of L^p bound is obtained and then strengthened via regularity theory and/or Sobolev embedding theorems; see (Cantrell, Cosner, and Hutson 1993). Such bounds are adequate for global existence, dissipativity, and other qualitative properties but are too complicated or too weak for practical persistence.

REFERENCES

Aronson, D.G. and Weinberger, H.F. (1975). *Nonlinear diffusion in population genetics, combustion, and nerve pulse propagation*, **Partial Differential Equations and Related Topics, Lecture Notes in Mathematics 446**, Springer-Verlag, Berlin, p.5-49.

Burton, T.A. and Hutson, V. (1991). *Permanence for nonautonomous predator-prey systems*, **Differential and Integral Equations 4**: 1269-1280.

Butler, G., Freedman, H., and Waltman, P. (1986). *Uniformly persistent dynamical systems*, **Proc. American Math. Soc. 96**: 425-430.

Butler, G., Schmid, R., and Waltman, P. (1990). *Limiting the complexity of limit sets in self-regulating systems*, **J. Math. Anal. Appl. 147**: 63-68.

Cantrell, R.S. and Cosner, C. (1989). *Diffusive logistic equations with indefinite weights: population models in disrupted environments*, **Proc. Royal Soc. Edinburgh 112A**: 293-318.

Cantrell, R.S. and Cosner, C. (1991). *Diffusive logistic equations with indefinite weights: population models in disrupted environments II*, **S.I.A.M. J. Math. Anal. 22**: 1043-1064.

Cantrell, R.S. and Cosner, C. (1993). *Should a park be an island?*, **S.I.A.M. J. Appl. Math 53**: 219-252.

Cantrell, R.S. and Cosner, C. (preprint). *Practical persistence in ecological models via comparison methods*.

Cantrell, R.S., Cosner, C., and Hutson, V. (1993). *Permanence in ecological systems with spatial heterogeneity*, **Proc. Royal Soc. Edinburgh 123A**: 533-559.

Cao, Y. and Gard, T.C. (1993). *Uniform persistence for population models with time delay using multiple Lyapunov functions*, **Differential and Integral Equations 6**: 883-898.

Castro, A. and Lazer, A.C. (1982). *Results on periodic solutions of parabolic equations suggested by elliptic theory*, **Bull. Un. Mat. Ital. 6 I-B**: 1089-1104.

Cosner, C. and Lazer, A.C. (1984). *Stable coexistence states in the Volterra-Lotka competition model with diffusion*, **S.I.A.M. J. Appl. Math. 44**: 1112-1132.

Fife, P.C. (1979). **Mathematical Aspects of Reacting and Diffusing Systems, Lecture Notes in Biomathematics 28**, Springer-Verlag, Berlin.

Hale, J.K. and Waltman, P. (1989). *Persistence in infinite dimensional systems*, **S.I.A.M. J. Math. Anal. 20**: 388-395.

Hess, P. (1991). **Periodic-Parabolic Boundary Value Problems and Positivity**, Pitman Lecture Notes in Mathematics 247, Longman Scientific and Technical, Harlow, Essex, U.K.

Hofbauer, J. and Sigmund, K. (1988). **Dynamical Systems and the Theory of Evolution.** Cambridge University Press, Cambridge U.K.

Hutson, V. and Schmitt, K. (1992). *Permanence in dynamical systems*, **Math. Biosci. 111**: 1-71.

Lazer, A.C. (1982). *Some remarks on periodic solutions of parabolic equations*, **Dynamical Systems II**, ed. A. Bednarek and L. Cesari, Academic Press, New York, p.227-246.

Perturbing Vector Lyapunov Functions and Applications to Large-Scale Dynamic Systems

ZAHIA DRICI, Florida Institute of Technology, Department of Applied Mathematics, Melbourne, Florida

1. Introduction

It is well known that in the application of the Lyapunov theory to practical problems it is often difficult to find a Lyapunov function that meets all the criteria of Lyapunov theorems. The method of vector Lyapunov functions [1] which employs several Lyapunov functions instead of one was developed in an attempt to enlarge the class of Lyapunov functions that can be used. Another concept which has served the same purpose is that of perturbing the Lyapunov functions [2] whereby it is shown that a Lyapunov function which satisfies some but not all the required criteria can still be used to establish a desired stability property if a perturbed version of it which satisfies less stringent conditions is found.

Recently, a new direction in the method of vector Lyapunov functions was introduced [3]. It was shown that combining the two concepts described above and performing the perturbation in an expanded version of the original state space enlarges the applicability of this method by further weakening the requirements on the Lyapunov functions. This new approach can be directly applied to large-scale dynamic systems with overlapping decompositions. In such systems an expansion of the original state space is performed in order to apply standard methods [4]. The stability laws established in the expanded system are then contracted for implementation in the original system. This expansion process provides a natural set-up for the application of the new method mentioned above. Following this approach we present in this paper, some nonuniform stability results for large-scale dynamic systems.

2. Preliminaries

Consider the dynamic systems

$$S_x: \quad x' = f(t,x), \ x(t_0) = x_0, \ t_0 \geq 0, \tag{2.1}$$

and

$$S_y: \ y' = \tilde{f}(t,y), \ y(t_0) = y_0, \tag{2.2}$$

where $f \in C[R_+ \times R^n, R^n]$ and $\tilde{f} \in C[R_+ \times R^m, R^m]$, $n < m$. We suppose that the functions f and $\tilde{f}$ are sufficiently smooth to guarantee existence and uniqueness of solutions of (2.1) and (2.2) for $t \geq t_0$. Also assume that $f(t,0) \equiv 0$, $\tilde{f}(t,0) \equiv 0$ so that the systems (2.1) and (2.2) possess the trivial solutions.

Our aim is to enlarge the given system (2.1) into (2.2) so as to bring out hidden good properties, if any, of (2.1) and then investigate its stability properties in the best possible manner. For this purpose, we need the following known [4] results. Let us begin with the following definition.

Definition 2.1: A system S_x is said to be included in a system S_y (written $S_x \subset S_y$) if there exists an ordered pair of matrices (U,V) such that $UV = I$, and if for any initial state (t_0, x_0) of S_x, $y_0 = V x_0$ implies

$$x(t; t_0, x_0) = U y(t; t_0, V x_0) \quad \forall t \geq t_0. \tag{2.3}$$

Conditions for inclusion of S_x in S_y are given in the following theorem.

Theorem 2.1: *Assume that*

(i) *there exists an ordered pair of matrices (U,V) such that $UV = I$, where V and U are $m \times n$ full row rank and $n \times m$ full column rank, respectively,*

(ii) *$\tilde{f}(t,y) = V f(t, Uy) + m(t,y)$, where $m \in C[R_+ \times R^m, R^m]$ is a complementary function satisfying either*

(a) *$m(t, Vx) = 0$ for $(t,x) \in R_+ \times R^n$*

or

(b) *$U m(t,y) = 0$ for $(t,y) \in R_+ \times R^m$*

then $S_x \subset S_y$.

For a proof see [4].

Next, we shall state a known [5] comparison result which is also required. For any function $V \in C[R_+ \times R^n, R_+^N]$ we define, for $(t,x) \in R_+ \times R^n$,

$$D^+ V(t,x)_{(2.1)} = \lim_{h \to 0} \sup \frac{1}{h} \left[V(t+h, x + hf(t,x)) - V(t,x) \right].$$

When we write $D^+ V(t,x)_{(2.1)}$, it means that $D^+ V(t,x)$ is computed relative to system (2.1). We need to specify this since we shall use later generalized derivatives of V relative to both system (2.1) and system (2.2).

Theorem 2.2: Let $V \in C[R_+ \times R^n, R_+^N]$ be locally Lipschitzian in x. Assume that

$$D^+V(t,x)_{(2.1)} \leq g(t,V(t,x)), \ (t,x) \in R_+ \times R^n$$

where $g \in C[R_+ \times R_+^N, R^N]$ and $g(t,u)$ is quasimonotone in u, that is if $u \leq v$ and $u_i = v_i$ for some $1 \leq i \leq N$, then $g_i(t,u) \leq g_i(t,v)$. Let $r(t) = r(t,t_0,u_0)$ be the maximal solution of

$$u' = g(t,u), \ u(t_0) = u_0 \geq 0 \tag{2.4}$$

existing for $t \geq t_0$. Then, if $V(t_0,x_0) \leq u_0$, we have

$$V(t,x(t)) \leq r(t,t_0,u_0), \quad t \geq t_0,$$

where $x(t) = x(t;t_0,x_0)$ is any solution of (2.1) existing on $[t_0,\infty)$. The inequalities between vectors are understood to be componentwise.

Corollary 2.1: Suppose that in Theorem 2.2 we have $g(t,u) \equiv 0$. Then, $V(t,x)$ is nonincreasing in t and $V(t,x(t)) \leq V(t_0,x_0)$ for $t \geq t_0$.

Corresponding to the standard stability notions of (2.1) and (2.2) we require a similar notion relative to the comparison system (2.4).

Definition 2.2: The trivial solution of (2.4) is said to be equistable if given $\epsilon > 0$ and $t_0 \in R_+$, there exists a $\delta = \delta(t_0,\epsilon) > 0$ such that

$$\sum_{i=1}^{N} u_{0_i} < \delta \text{ implies that } \sum_{i=1}^{N} u_i(t,t_0,u_0) < \epsilon, \quad t \geq t_0,$$

where $u(t,t_0,u_0)$ is any solution of (2.4). It is said to be uniformly stable if δ does not depend on t_0.

Next, we state some known nonuniform stability results under weaker conditions employing a set of vector Lyapunov functions. Proofs of these results can be found in [3]. Let $S_n(\rho) = [x \in R^n : |x| < \rho]$ for some $\rho > 0$.

Theorem 2.3: Assume that
(i) there exists $V_1 \in C[R_+ \times S_n(\rho), R_+^N]$, $V_1(t,0) \equiv 0$, $\sum_{i=1}^{N} V_{1i}(t,x) \geq 0$, and for $(t,x) \in R_+ \times S_n(\rho)$

$$D^+V_1(t,x)_{(2.1)} \leq g_1(t,V_1(t,x)),$$

where $g_1 \in C[R_+^2, R^N]$ and $g_1(t,u)$ is quasimonotone nondecreasing in u;
(ii) the conditions of Theorem 2.1 hold;
(iii) there exists $V_2 \in C[R_+ \times S_m(\rho_0) \cap S_m^c(\eta), R_+^M]$, for every $0 < \eta < \rho_0$ where

$\rho_0 = \frac{\rho}{|U|}$, *such that*

$$D^+V_2(t,y)_{(2.2)} \le g_2(t, V_2(t,y)), \quad (t,y) \in S_m(\rho_0) \cap S_m^c(\eta),$$

where $g_2 \in C[R_+ \times R_+^M, R^M]$, $g_2(t,u)$ *is quasimontone nondecreasing in* u, *and for* $(t,y) \in R_+ \times S_m(\rho_0) \cap S_m^c(\eta)$

$$b(|y|) \le \sum_{i=1}^{M} V_{2i}(t,y) \le a(|y|) + \sum_{i=1}^{N} V_{1i}(t, U_y)),$$

where $a, b \in K = \{\phi \in C[[0, \rho_0), R_+] : \phi(u) \text{ strictly increasing and } \phi(0) = 0\}$;

(iv)   *the trivial solution of* $u' = g_1(t,u)$, $u(t_0) = u_0$ *is equistable*;

(v)   *the trivial solution of* $v' = g_2(t,v)$, $v(t_0) = v_0$ *is uniformly stable*,

then the trivial solution of (2.1) *is equistable*.

Corollary 2.2: *The conclusion of Theorem 2.3 holds true if* $g_1(t,u) \equiv 0$ *and* $g_2(t,u) \equiv 0$.

The following result gives a set of conditions for asymptotic stability under weaker conditions.

Theorem 2.4: *Assume that conditions* (ii) − (v) *of Theorem 2.3 hold true and that condition* (i) *is replaced by*

(i)   *there exists* $V_1 \in C[R_+ \times S_n(\rho), R_+^N]$, $V_1(t,x)$ *locally Lipschitzian in* x, $V_1(t,0) \equiv 0$, *and there exists* $w \in C[R_+ \times S_n(\rho), R_+]$ *such that* $w(t,x)$ *is locally Lipschitzian in* x, *positive definite, and* $D^+w(t,x)$ *is bounded from above or below, and for* $(t,x) \in R_+ \times S_n(\rho)$

$$D^+V_{1p}(t,x)_{(2.1)} \le -w(t,x) \quad 1 \le p \le N$$

$$D^+V_{1i}(t,x)_{(2.1)} \le g_{1i}(t, V_1(t,x)) \quad 1 \le i \le N, \, i \ne p \qquad (2.5)$$

where $g_1 \in C[R_+ \times R_+^N, R^N]$ *and* $g_1(t,u)$ *is quasimonotone in* u. *Then, the trivial solution of* (2.1) *is asymptotically stable*.

The next result offers another set of conditions for asymptotic stability.

Theorem 2.5: *Assume that the conditions of Theorem 2.3 hold except for condition* (iii) *which is replaced by*

(iii)   *there exists* $V_2 \in C[R_+ \times S_m(\rho_0) \cap S_m^c(\eta), R_+^M]$, $V_2(t,y)$ *locally Lipschitzian in* y, $V_2(t,0) \equiv 0$, *and for* $(t,y) \in R_+ \times S_m(\rho_0) \cap S_m^c(\eta)$,
$$b(|y|) \le V_2(t,y) \le a(|y|) + V_1(t, Uy),$$

and that there exists $w \in C[R_+ \times S_m(\rho_0), R_+]$, *such that* $w(t,y)$ *is locally*

Lipschitzian in y, *and* $D^+w(t,y)$ *is bounded from above or below, and, for* $(t,y) \in R_+ \times S_m(\rho_0) \cap S_m^c(\eta)$,

$$c_1(|y|) \le w(t,y) \le c_2(|y|), \quad c_1, c_2 \in K \tag{2.5}$$

$$D^+V_{2p}(t,y) \le -w(t,y) \quad 1 \le p \le M \tag{2.6}$$

$$D^+V_{2i}(t,y) \le g_{2i}(t, V_2(t,y)) \quad 1 \le i \le M, \, i \ne p,$$

where $g_2 \in C[R_+ \times R_+^M, R^M]$ *and* $g_2(t,u)$ *is quasimonotone nondecreasing in* u. *Then, the trivial solution of* (2.1) *is asymptotically stable.*

3. Main results

Assume that (2.1) admits a decomposition of the form:

$$x_i' = F_i(t, x_i) + R_i(t, x) \quad x_i(t_0) = x_{i0} \tag{3.1}$$

where

$$x = (x_1, x_2, \ldots, x_N)$$

$$x_i \in R^{n_i}, \quad i = 1, 2, \ldots, N$$

$$n = \sum_{i=1}^{N} n_i.$$

The functions $F_i(t, x_i)$ represent isolated decoupled subsystems

$$x_i' = F_i(t, x_i), \quad x_i(t_0) = x_{i0} \tag{3.2}$$

and $R_i(t, x)$ are the interconnections among subsystems (3.2).

Similarly, assume

$$y_i' = \tilde{F}_i(t, y_i) + \tilde{R}_i(t, y) \quad y_i(t_0) = y_{i0} \tag{3.3}$$

where

$$y = (y_1, y_2, \ldots, y_N)$$

$$y_i \in R^{m_i}, i = 1, 2, \ldots, N$$

$$m = \sum_{i=1}^{N} m_i$$

The functions $\tilde{F}_i(t, y_i)$ represents isolated decoupled subsystems

$$y_i' = \tilde{F}_i(t, y_i), y_i(t_0) = y_{i0} \tag{3.4}$$

and $\tilde{R}_i(t, x)$ are the interconnections among subsystems (3.4).

Theorem 3.1: *Assume that for each $i = 1, 2, \ldots, N$*

(i)   *there exists $V_{1i} \in C[R_+ \times S_n(\rho), R_+^{r_i}]$ such that for $1 \leq j \leq r_i$ $V_{1ij}(t, x_i)$ is locally Lipschitzian in x_i with Lipschitz constant L_{1ij}, $V_{1i}(t, 0) \equiv 0$, and for $(t, x) \in R_+ \times S_n(\rho)$*

$$D^+_{(3.2)} V_{1i}(t, x) \leq g_{1i}(t, V_1(t, x))$$

where $g_{1i} \in C[R_+ \times R_+^{r_i}, R_+^{r_i}]$, $g_{1i}(t, 0) = 0$, and $g_{1i}(t, u)$ is quasimonotone nondecreasing in u,

(ii)   *for $(t, x) \in R_+ \times S_n(\rho)$ and $1 \leq j \leq r_i$,*

$$L_{1ij}(t, x_i) \, |R_i(t, x)| \leq \lambda_{ij}(t) V_{1ij}(t, x), \quad \lambda_{ij} \in L^1,$$

(iii)   *the conditions of theorem 2.2 hold,*

(iv)   *for every $0 < \eta < \rho_0$ where $\rho_0 = \frac{\rho}{|u|}$ there exists $V_{2i} \in C[R_+ \times S_m(\rho_0) \cap S_m^c(\eta)$, $R_+^{\tilde{r}_i}]$ such that for $1 \leq j \leq \tilde{r}_i$, $V_{2ij}(t, y_i)$ is locally Lipschitzian in y_i with Lipschitz constant L_{2ij}, and for $(t, y_i) \in R_+ \times S_m(\rho_0) \cap S_m^c(\eta)$*

$$D^+_{(3.4)} V_{2i}(t, y_i) \leq g_{2i}(t, V_{2i}(t, y_i))$$

where $g_{2i} \in C[R_+ \times R_+^{\tilde{r}_i}, R^{\tilde{r}_i}]$, $g_{2i}(t, u)$ quasimonotone nondecreasing in u, and

$$b(|y|) \leq \sum_{i=1}^{N} \sum_{j=1}^{\tilde{r}_i} V_{2ij}(t, y_i) \leq \sum_{i=1}^{N} \sum_{j=i}^{r_i} V_{1ij}(t, Uy_i) + a(|y|)$$

where $a, b \in \kappa = \{\phi \in C[[0, \rho_0), R_+] : \phi(u) \text{ is strictly increasing and } \phi(0) = 0\}$,

(v)   *for $(t, y) \in R_+ \times S_m(\rho_0) \cap S_m^c(\eta)$*

$$L_{2ij}(t, y_i) \, |\tilde{R}_i(t, y)| \leq W_{ij}(t, |y|)$$

with $W_{ij}(t, u)$ nondecreasing in u,

(vi)   *the trivial solution of $u' = g_3(t, u)$, $u(t_0) = u_0$ where $g_3(t, u) = \lambda(t)u + g_1(t, u)$ is equistable,*

(vii)   *the trivial solution of $v' = g_4(t, v)$, $v(t_0) = v_0$ where $g_4(t, v) = W(t, b^{-1}(v)) + g_2(t, v)$ is uniformly stable,*

then the trivial solution of (2.1) is equistable.

    Proof: For $(t, x) \in R_+ \times S_n(\rho)$ and $1 \leq j \leq r_i$, we compute $D^+ V_{1ij}(t, x_i)_{(3.1)}$. We have

$$V_{1ij}(t + h, x_i + hf_i(t, x)) - V_{1ij}(t, x_i) = V_{1ij}(t + h, x_i + hF_i(t, x_i) + hR_i(t, x))$$

$$- V_{1ij}(t + h, x_i + hF_i(t, x_i)) + V_{1ij}(t + h, x_i + hF_i(t, x_i)) - V_{1ij}(t, x_i)$$

$$\leq L_{1ij}(t,x_i)h \mid R_i(t,x) \mid + V_{1ij}(t+h,x_i+hF_i(t,x_i)) - V_{1ij}(t,x_i).$$

Dividing by $h > 0$ and taking the limit as $h \to 0$ we have

$$D^+V_{1ij}(t,x_i)_{(3.1)} \leq L_{1ij}(t,x_i) \mid R_i(t,x) \mid + D^+V_{1ij}(t,x_i)_{(3.2)}$$

which, in view of assumptions (i) and (ii), yields

$$D^+V_{1ij}(t,x_i)_{(3.1)} \leq \lambda_{ij}(t)V_{1ij}(t,x_i) + g_{1ij}(t,V_{1i}(t,x_i)) \equiv g_{3ij}(t,V_{1i}(t,x_i)).$$

Hence,

$$D^+V_1(t,x)_{(3.1)} \leq g_3(t,V_1(t,x))$$

where $g_3 \in C[R_+ \times R_+^r, R^r]$, $r = \sum_{i=1}^{N} r_i$, and $g_3(t,u)$ is quasimonotone nondecreasing in u.

Similarly, for $(t,y) \in R_+ \times S_m(\rho_0) \cap S_m^c(\eta)$ and $1 \leq j \leq \tilde{r}_i$ we compute $D^+V_2(t,y)_{(3.3)}$ so that in view of assumptions (iv) and (v) we arrive at

$$D^+V_{2ij}(t,y_i)_{(3.3)} \leq L_{2ij}(t,y_i) \mid \tilde{R}_i(t,y) \mid + D^+V_{2ij}(t,y_i)_{(3.4)}$$

$$\leq W_{ij}(t,\mid y \mid) + g_{2ij}(t,V_{2i}(t,y_i))$$

$$\equiv g_{4ij}(t,V_{2i}(t,y_i)).$$

Hence, $D^+V_2(t,y) \leq g_4(t,V_2(t,y))$ where $g_4 \in C[R_+ \times R_+^{\tilde{r}}, R^{\tilde{r}}]$, $\tilde{r} = \sum_{i=1}^{N} \tilde{r}_i$, $g_4(t,u)$ is quasimonotone nondecreasing in u.

Then the conclusion of theorem 3.1 follows directly from theorem 2.3.

Next, we prove a result on nonuniform asymptotic stability.

Theorem 3.2: _Assume that conditions $(ii) - (vii)$ of theorem 3.1 hold true and condition (i) is replaced by_

   (i) _there exists $V_{1i} \in C[R_+ \times S_n(\rho), R_+^{r_i}]$, and for $1 \leq j \leq r_i$ $V_{1ij}(t,x_i)$ is locally Lipschitzian in x_i with Lipschitz constant L_{1ij}, $V_{1i}(t,0) \equiv 0$, and that there exists $w \in C[R_+ \times S_n(\rho), R_+]$ such that $w(t,x)$ is locally Lipschitzian in x, positive definite, and $D^+w(t,x)$ is bounded from above or below, and for $(t,x) \in R_+ \times S_n(\rho)$_

$$D^+V_{1ip}(t,x_i)_{(3.1)} \leq -w(t,x_i) \qquad 1 \leq i \leq N, \quad 1 \leq p \leq \tilde{r}_i$$

$$D^+V_{1i}(t,x_i)_{(3.1)} \leq g_{1i}(t,V_{1i}(t,x_i)) \qquad i \neq p$$

where $g_1 \in C[R_+ \times R_+^r, R_+^r]$, $r = \sum_{i=1}^{N} r_i$ $g_1(t,u)$ is quasimonotone non-decreasing in u.

Then the trivial solution of (2.1) *is asymptotically stable.*

Proof: By theorem 3.1 with $g_{1ip}(t, u) = 0$ it follows that system (2.1) is equistable. Hence, it is enough to prove that given $t_0 \in R_+$, there exists a $\delta = \delta(t_0) > 0$ such that $|x_0| < \delta$ implies that $|x(t, t_0, x_0)| \to 0$ as $t \to \infty$. For $\epsilon = \lambda \le \rho_0$ let $\delta = \delta(t_0, \lambda)$ be the δ of equistability and let $|x_0| < \delta$. Since $w(t, x)$ is positive definite, it is enough to prove that $\lim_{t\to\infty} w(t, x(t)) = 0$ for any solution $x(t) = x(t, t_0, x_0)$ of (2.1) with $|x_0| < \delta$. First, we prove that $\lim_{t\to\infty} \inf w(t, x(t)) = 0$. If not, there would exist $\beta > 0$ and $T = T(\beta) > 0$ such that

$$w(t, x(t)) \ge \beta \text{ for } t \ge T. \tag{3.5}$$

Since $\lambda_{ij} \in L^1[t_0, +\infty)$ and $\lambda_{ij}(t) \ge 0$ for $t \in [t_0, +\infty) \exists N > 0$:

$$\int_{t_0}^{t} \lambda_{ij}(s)ds \le \int_{t_0}^{\infty} \lambda_{ij}(s)ds < N_{ij}. \tag{3.6}$$

Let $m(t) = V_{1ip}(t, x(t)) + \int_{t_0}^{t} w(s, x(s))ds - \int_{t_0}^{t} \lambda_{ip}(s)ds$. Then,

$$D^+ m(t)_{(3.1)} \le D^+ V_{1ip}(t, x(t)) + w(t, x(t)) - \lambda_{ip}(t) \le 0$$

where the last inequality results from (i) and (ii). Using Corollary 2.1 we conclude that $m(t) \le m(t_0)$ or

$$V_{1ip}(t, x(t)) \le V_{1ip}(t_0, x_0) - \int_{t_0}^{t}(w(s, x(s)) + \int_{t_0}^{t} \lambda_{ip}(s)ds$$

$$\le V_{1ip}(t_0, x_0) - \int_{T}^{t} w(s, x(s))ds + N_{ip}$$

$$\le V_{1ip}(t_0, x_0) - \beta(t - T) + N_{ip}$$

where the last inequality results from (3.5). For t sufficiently large, $V_{1ip}(t_0, x_0) - \beta(t_1 - T) + N_{ip} < 0$ which contradicts the assumption that $V_1(t, x(t)) \ge 0$. Consequently,

$$\lim_{t\to\infty} \inf w(t, x(t)) = 0. \tag{3.7}$$

Next, we prove that $\lim_{t\to\infty} \sup w(t, x(t)) = 0$. Suppose that $\lim_{t\to\infty} \sup w(t, x(t)) \ne 0$. Then, there exist divergent sequences $\{t_n\}$ and $\{t_n^*\}$ and $\epsilon > 0$ such that

$$\begin{cases} t_i < t_i^* < t_{i+1} & i = 1, 2, 3, \ldots \\ w(t_i, x(t_i)) = \frac{\epsilon}{2} \\ w(t_i^*, x(t_i^*)) = \epsilon \\ \frac{\epsilon}{2} < w(t, x(t)) < \epsilon & \text{for } t_i < t < t_i^*. \end{cases} \qquad (3.8)$$

Suppose that $D^+ w(t, x(t)) \leq M$, a positive constant. Then it is easy to show that $t_i^* - t_i \geq \frac{\epsilon}{2M}$. Furthermore,

$$V_{1ip}(t_n^*, x(t_n^*)) \leq V_{1ip}(t_0, x_0) - \sum_{i=1}^{w} \int_{t_i}^{t_i^*} w(s, x(s)) ds + N_{ip}$$

$$\leq V_{1ip}(t_0, x_0) - \left(\frac{\epsilon}{2}\right) \left(\frac{\epsilon}{2M}\right) n + N_{ip}$$

where the last inequality is obtained using (3.5), (3.6), and (3.8). The foregoing estimate yields a contradiction for sufficiently large n. Hence, $\lim_{t \to \infty} \sup w(t, x(t)) = 0$. The argument is similar if $D^+ w(t, x(t))$ is bounded from below. This result together with (3.7) implies that $\lim_{t \to \infty} w(t, x(t)) = 0$. Therefore, $x(t, t_0, x_0) \to 0$ as $t \to \infty$.

The next result offers another set of conditions for asymptotic stability whose proof can be constructed with suitable modifications following the proof of theorem 3.2. The details will be omitted.

Theorem 3.3: _Assume that conditions of theorem 3.2 hold except for condition (iii) which is replaced by_

   (iii)   _there exists_ $V_{2i} \in C[R_+ \times S_n(\rho), R_+^{\tilde{r}_i}]$ _such that for_ $1 \leq j \leq \tilde{r}_i$, $V_{2ij}(t, y)$ _is locally Lipschitz in_ y _and for_ $(t, y) \in R_+ \times S_m(\rho_0) \cap S_m^c(\eta)$

$$b(|y|) \leq \sum_{i=1}^{N} \sum_{j=1}^{\tilde{r}_i} \leq a(|y|) + \sum_{i=1}^{N} \sum_{j=1}^{r_i} V_{1ij}(t, U y_i)$$

and there exists $w \in C[R_+ \times S_m(\rho_0), R_+]$ _such that_ $w(t, y)$ _is locally Lipschitzian in_ y, $D^+ w(t, x)$ _is bounded from above or below, and for_ $(t, y) \in R_+ S_m(\rho_0) \cap S_m^c(\eta)$

$$c_1(|y|) \leq w(t, y) \leq c_2(|y|), \ c_1, c_2 \in \kappa.$$

$$D^+ V_{2ip}(t, y) \leq -w(t, y), \ 1 \leq q \leq N, \ 1 \leq p \leq \tilde{r}_q$$

$$D^+ V_{2i}(t, y) \leq g_{2i}(t, y) \quad i \neq q$$

where $g_2 \in C[R_+ \times R_+^{\tilde{r}}, R^{\tilde{r}}]$, $\tilde{r} = \sum_{i=1}^{N} \tilde{r}_i$, _and_ $g(t, u)$ _is quasimonotone nondecreasing in_ u.

Then the trivial solution of (2.1) *is asymptotically stable.*

References

[1] V. Lakshmikantham, V.M. Matrosov, and S.Sivasundharam, *Vector Lyapunov Functions and Stability Analysis of Nonlinear Systems*, Kluwer Academic Publishers, 1991.

[2] V. Lakshmikantham, S. Leela, A.A. Martynyuk, *Stability Analysis of Nonlinear Systems*, Marcel Dekker, Inc. 1989.

[3] Z. Drici, New directions in the method of vector Lyapunov functions, to appear in *JMAA*.

[4] D.D. Šiljak, *Decentralized Control of Complex Systems*, Academic Press, 1991.

[5] V. Lakshmikantham and S. Leela, *Differential and Integral Inequalities*, Vol. I, Academic Press, 1969.

Nonlinear Boundary Value Problems

L.H. Erbe   Department of Mathematics, University of Alberta, Edmonton, Alberta, Canada T6G 2G1

Shouchuan Hu   Department of Mathematics, Southwest Missouri State University, Springfield, Missouri 65804

ABSTRACT

We consider the nonlinear boundary value problem $-u'' = f(t, u)$ with linear boundary conditions and establish criteria for the existence of several positive solutions via fixed point index theory.

§1. INTRODUCTION

Consider the following second order boundary value problem

$$-u'' = f(t, u), \quad 0 < t < 1 \tag{1}$$

$$\begin{cases} \alpha u(0) - \beta u'(0) & = 0 \\ \gamma u(1) + \delta u'(1) & = 0 \end{cases} \tag{2}$$

127

where f is continuous and $f(t, u) \geq 0$ for $t \in [0, 1]$ and $u \geq 0,\quad \alpha, \beta, \gamma, \delta \geq 0$ and

$$\rho := \gamma\beta + \alpha\gamma + \alpha\delta > 0. \tag{3}$$

The BVP (1), (2) was considered in Erbe et al. [7], Erbe and Wang [8], Bandle et al. [2], Bandle and Kwong [3], Coffman and Marcus [4], and Garaizar [9]. And the work here is a continuation of [7]. Often one is interested in establishing criteria for the existence of one (or several) positive solutions. In particular, in [7], [8] criteria were given for the case when f is superlinear or sublinear (in the u variable) at one or both endpoints (zero or infinity). The techniques involved fixed point theorems of cone expansion or compression type. In this paper we shall establish the existence of more than two positive solutions by combining some of the earlier techniques of [7] and [8], along with certain additional assumptions. To be precise, we introduce the following notation.

For $u > 0$, we define

$$m(u) := \min_{t \in [0,1]} \frac{f(t, u)}{u} \tag{4}$$

$$M(u) := \max_{t \in [0,1]} \frac{f(t, u)}{u}. \tag{5}$$

We shall write $m(0), M(0), m(\infty), M(\infty)$ to denote the limits of $m(u), M(u)$ as $u \to 0+$ or as $u \to \infty$, provided these limits exist.

The conditions $m(0) = \infty = m(\infty)$ correspond to the assumption that $f(t, u)$ is sublinear at $u = 0$ and superlinear at $u = \infty$ (f is sub-superlinear). Similarly, $M(0) = 0 = M(\infty)$ corresponds to the assumption that f is superlinear at $u = 0$ and sublinear at $u = \infty$ (f is super-sub-linear). If $M(0) = 0$, $m(\infty) = \infty$ then f is superlinear at $u = 0$ *and* $u = \infty$; if $m(0) = \infty$ and $M(\infty) = 0$ then f is sublinear at $u = 0$ *and* $u = \infty$. We shall investigate these four cases in §2 below and in §3, with the help of Leggett and Williams [13], shall extend some of these further. Finally, in §4 we shall show that if f is sublinear at both zero and infinity, then one may still obtain the existence of a positive solution of (1), (2) even if f is not continuous, provided f is now increasing in u.

The function $G(t, s)$ denotes the Green's function for the problem $-u'' = 0$ subject to the boundary conditions (2) and is given explicitly by

$$G(t, s) = \frac{1}{\rho} \begin{cases} (\gamma + \delta - \gamma t)(\beta + \alpha s), & 0 \leq s \leq t \leq 1 \\ (\beta + \alpha t)(\gamma + \delta - \gamma s), & 0 \leq t \leq s \leq 1. \end{cases} \tag{6}$$

It is well-known that the BVP (1), (2) has a solution u iff u solves the integral equation

$$u(t) = \int_0^1 G(t,s)f(s,u(s))ds := (Fu)(t). \tag{7}$$

Thus, (1), (2) is equivalent to the fixed point equation $u = Fu$ in $X = C[0,1]$. The operator $F : X \to X$ is completely continuous and if we define the cone $K \subset X$ by

$$K = \{u \in X : u(t) \geq 0, \quad \min_{\frac{1}{4} \leq t \leq \frac{3}{4}} u(t) \geq \sigma \|u\|\}, \tag{8}$$

where $\sigma = \min \{\frac{\gamma+4\delta}{4(\gamma+\delta)}, \frac{\alpha+4\beta}{4(\alpha+\beta)}\}$ and $\|u\| = \sup_{t \in [0,1]} |u(t)|$, then one can verify that $G(t,s)/G(s,s) \geq \sigma$ for $\frac{1}{4} \leq t \leq \frac{3}{4}$ and $0 \leq s \leq 1$. Consequently, it follows (cf. [7]) that $F(K) \subset K$.

For convenience we introduce the notation

$$\eta = \left(\int_0^1 G(s,s)ds \right)^{-1} = \frac{6\rho}{6\delta_\beta + 3\gamma_\beta + \alpha\gamma + 3\alpha\delta} \tag{9}$$

$$\lambda = \left(\int_{1/4}^{3/4} G(\tfrac{1}{2},s)ds \right)^{-1}. \tag{10}$$

In addition to various assumptions on superlinearity or sublinearity, we introduce the conditions

(C_1) There is a $p > 0$ such that $0 \leq u \leq p$ and $0 \leq t \leq 1$ implies $f(t,u) \leq \eta p$.

(C_2) There is a $p > 0$ such that $\sigma p \leq u \leq p$ and $\frac{1}{4} \leq t \leq \frac{3}{4}$ implies $f(t,u) \geq \lambda p$.

We conclude this section with the following lemma, see [5] for a proof and a discussion of the fixed point index.

LEMMA 1. *Let X be a Banach space, $K \subset X$ a cone in X. For $p > 0$ define $K_p = \{x \in K : \|x\| \leq p\}$. Assume that $F : K_p \to X$ is a completely continuous map such that $Fx \neq x$ for $x \in \partial K_p = \{x \in K : \|x\| = p\}$. Then*

(i) if $\|x\| \leq \|Fx\|$ for $x \in \partial K_p \Longrightarrow i(F, K_p, K) = 0$

(ii) if $\|x\| \geq \|Fx\|$ for $x \in \partial K_p \implies i(F, K_p, K) = 1$.

§2. MULTIPLE POSITIVE SOLUTIONS

We state our first two results for the cases when f is sub-superlinear or super-sublinear.

THEOREM 1. (*f sub-superlinear*). *Assume that* $m(0) = \infty = m(\infty)$ *and there exist* $0 < p_1 < q < p_2$ *such that* (C_1) *holds with* $p = p_1$ *and* p_2 *and* (C_2) *holds with* $p = q$. *Then the BVP* (1), (2) *has at least 4 positive solutions satisfying*

$$0 < \|u_1\| < p_1 < \|u_2\| < q < \|u_3\| < p_2 < \|u_4\|. \tag{11}$$

THEOREM 2. (*f super-sublinear*). *Assume that* $M(0) = 0 = M(\infty)$ *and there exist* $0 < p_1 < q < p_2$ *such that* (C_2) *holds for* $p = p_1$ *and* $p = p_2$ *and* (C_1) *holds for* $p = q$. *Then the conclusion of Theorem 1 holds.*

PROOF OF THEOREM 1: As in [7], $m(0) = \infty$ implies that there exists $r > 0$, $0 < r < p_1$ such that $f(t, u) \geq Bu$ for $0 \leq u \leq r$ where $B > \frac{\lambda}{\sigma}$. It follows that for $u \in \partial K_r$

$$(Fu)(\tfrac{1}{2}) = \int_0^1 G(\tfrac{1}{2}, s) f(s, u(s)) ds \geq B \int_0^1 G(\tfrac{1}{2}, s) u(s) ds$$

$$\geq B\sigma \int_{1/4}^{3/4} G(\tfrac{1}{2}, s) \|u\| ds > \|u\| \quad \text{by (10)}.$$

Hence $\|Fu\| > \|u\|$ for $u \in \partial K_r$ so $i(F, K_r, K) = 0$. Similarly $m(\infty) = \infty$ implies there exists $R_1 > 0$ such that $f(t, u) \geq Bu$ for $u \geq R_1$. Choosing $R > \max\{p_2, \frac{R_1}{\sigma}\}$ it follows that for $u \in \partial K_R$ we have $(Fu)(\tfrac{1}{2}) > \|u\|$ so $\|Fu\| > \|u\|$ and thus $i(F, K_R, K) = 0$. Now for $u \in \partial K_{p_1}$ condition (C_1) implies

$$\|Fu\| = \max_{0 \leq t \leq 1} \int_0^1 G(t, s) f(s, u(s)) ds \leq \int_0^1 G(s, s) f(s, u(s)) ds$$

$$\leq \int_0^1 G(s, s) \eta \|u\| ds = \|u\|,$$

where η is given in (9). Thus, Lemma 1 implies $i(F, K_{p_1}, K) = 1$. Similarly, $i(F, K_{p_2}, K) = 1$.

Now for $u \in \partial K_q$ we have

$$(Fu)(\tfrac{1}{2}) = \int_0^1 G(\tfrac{1}{2}, s) f(s, u(s)) ds \geq \lambda q \int_{1/4}^{3/4} G(\tfrac{1}{2}, s) ds$$

$$= q = \|u\|$$

so that by Lemma 1 we have $i(F, K_q, K) = 0$. By the additivity and excision properties of the fixed point index we conclude

$$i(F, K_{p_1} \backslash \overset{\circ}{K}_r, K) = 1, \quad i(F, K_q \backslash \overset{\circ}{K}_{p_1}, K) = -1, \tag{12}$$

$$i(F, K_{p_2} \backslash \overset{\circ}{K}_q, K) = 1, \quad i(F, K_R \backslash \overset{\circ}{K}_{p_2}, K) = -1. \tag{13}$$

Therefore, the existence of 4 positive solutions satisfying (11) follows from (12) and (13). This proves Theorem 1.

Similarly, one may establish Theorem 2 by analogous arguments. We shall omit the proof.

The following two theorems are stated without proof and give the existence of at least three solutions in the superlinear or sublinear case.

THEOREM 3. (*f superlinear*). *Assume* $M(0) = 0$ *and* $m(\infty) = \infty$. *Let there exist* $0 < q_1 < p_1$ *such that* (C_2) *holds for* $p = q_1$ *and* (C_1) *holds for* $p = p_1$. *Then the BVP (1), (2) has at least three positive solutions satisfying*

$$0 < \|u_1\| < q_1 < \|u_2\| < p_1 < \|u_3\|.$$

THEOREM 4. (*f sublinear*). *Assume* $m(0) = \infty$ *and* $M(\infty) = 0$. *Let there exist* $0 < p_1 < q_1$ *such that* (C_1) *holds for* $p = p_1$ *and* (C_2) *holds for* $p = q_1$. *Then the BVP (1), (2) has at least three positive solutions satisfying* $0 < \|u_1\| < p_1 < \|u_2\| < q_1 < \|u_3\|$.

Although it is clear how one can obtain even more positive solutions by imposing condition (C_1) and (C_2) on a finite sequence of points $0 < p_1 < \cdots < p_m$, assumptions would become more restrictive on $f(t, u)$ as m gets larger. Therefore it is also useful to investigate other approaches for obtaining multiple positive solutions.

§3. ON THE CONDITION (C_2)

From the proof of Theorems 1-4, it is clear that the role played by condition (C_2) is to create a region $K_q = \{x \in K : \|x\| \leq q\}$ such that $i(F, K_q, K) = 0$. It is possible to generate this sort of "zero-index" region via concave positive functionals, a technique introduced by Leggett and Williams [13]. We illustrate this for the case when f is super-sublinear.

THEOREM 5. *Assume that $M(0) = 0 = M(\infty)$ and let there exist an $a > 0$ such that*

$$f(t, u) \geq \tfrac{a}{\xi} \quad for \quad a \leq u \leq \tfrac{a}{\sigma} \tag{14}$$

where

$$\xi = \min_{\frac{1}{4} \leq t \leq \frac{3}{4}} \int_{1/4}^{3/4} G(t, s)\,ds. \tag{15}$$

Then the BVP (1), (2) has at least two positive solutions.

Before giving the proof, we shall need the following definition and lemma. A continuous map $\alpha : K \to [0, \infty)$ is a concave positive functional if

$$\alpha\big(\lambda x + (1 - \lambda)y\big) \geq \lambda\alpha(x) + (1 - \lambda)\alpha(y),$$

for $0 \leq \lambda \leq 1$ and $x, y \in K$.

LEMMA 2. *([13]) Assume that $F : K_c \to K_c$ is completely continuous and there is a concave positive functional $\alpha(u)$ such that $\alpha(u) \leq \|u\|$ for $u \in K_c$. Assume also that there are $0 < d < a < b \leq c$ such that*

(i) $a \leq \alpha(u)$, $|u| \leq b \Longrightarrow \alpha(Fu) > a$ where we assume

$$\{x \in K : a < \alpha(x), \ |x| \leq b\} \neq \emptyset$$

(ii) $a \leq \alpha(u)$, $|u| \leq c$ and $\|Fu\| > b \Longrightarrow \alpha(Fu) > a$
(iii) $u \in K_d \Longrightarrow \|Fu\| < d$.

Then F has at least two positive fixed points.

PROOF OF THEOREM 5: The hypotheses imply there exists $0 < r < a < \frac{a}{\sigma} < R$ such that $\|Fu\| < \|u\|$ for $u \in \partial K_r$ and also for $u \in \partial K_R$. Let $d = r$, $b = \frac{a}{\sigma}$ and $c = R$ in Lemma 2. Define $\alpha(u) = \min_{\frac{1}{4} \leq t \leq \frac{3}{4}} u(t)$. We need to check that all the conditions in Lemma 2 are satisfied. First, if $a \leq \alpha(u)$ and $\|u\| \leq b$, then since $a \leq u(t) \leq b$ for $\frac{1}{4} \leq t \leq \frac{3}{4}$ we have by (14)

$$\alpha(Fu) = \min_{\frac{1}{4} \leq t \leq \frac{3}{4}} \int_0^1 G(t,s)f(s,u(s))ds$$

$$> \min_{\frac{1}{4} \leq t \leq \frac{3}{4}} \int_{1/4}^{3/4} G(t,s)a\xi^{-1}ds = a.$$

Second, if $a \leq \alpha(u)$ and $\|Fu\| > b$ then since $G(t,s) \geq \sigma G(\varphi,s)$ for $\frac{1}{4} \leq t \leq \frac{3}{4}$ and any φ, $s \in [0,1]$, we obtain $\alpha(Fu) \geq \sigma\|Fu\| > \sigma b = a$. Thus, an application of Lemma 2 yields the theorem.

REMARKS: (1) In some cases, $\xi > \sigma\lambda$ so that (14) improves (C_2) in those situations. (2) If we set $\Omega = \{x \in K : a < \alpha(x), \|x\| < R\}$, then we have $i(F, \Omega, K) = 1$ and thus $i(F, K_R \backslash \Omega, K) = 0$. Here the set $K_R \backslash \Omega$ plays the role of K_q as explained at the beginning of this section.

§4. THE DISCONTINUOUS CASE

In this section we do not assume f is continuous in u but shall assume f is monotone. We shall use the following result due to Amann [1].

LEMMA 3. *Let (X, K) be an ordered Banach space with a normal cone K. Assume that for u, $v \in X$ with $u < v$, $F : [u, v] \rightarrow X$ is compact and increasing and satisfies*

$$u \leq Fu, \quad Fv \leq v \tag{16}$$

Then F has maximal and minimal fixed points in $[u, v]$.

As an easy consequence of this lemma, if f is increasing in u, one can obtain the existence of solutions of the BVP (1), (2) provided one can obtain lower and upper solutions of (1), (2). In this last section we shall show that the sublinearity of $f(t, u)$ at zero and infinity does indeed imply the existence of lower and upper solutions and hence the existence of a positive solution of (1), (2).

THEOREM 4. *Assume $m(0) = \infty$ and $M(\infty) = 0$ and that $f(t, u)$ is increasing in u. Then the BVP (1), (2) has at least one positive solution.*

PROOF: Again, BVP (1), (2) is equivalent to (7). It suffices to find $u \leq v$ which satisfy (16). By the sublinearity of f, for any $\varepsilon > 0$ there is an

$M > 0$ such that $f(t, u) \leq M + \varepsilon u$ for $u \geq 0$, $t \in [0, 1]$. Hence, for $u \in K$,

$$(Fu)(t) \leq \int_0^1 G(t, s)(M + \varepsilon u(s))ds. \tag{17}$$

From (17), $v \equiv R$, for some sufficiently large R, will satisfy $Fv \leq v$. Now let $\alpha_\varepsilon(t) = \varepsilon \int_0^1 G(t, s)ds$. By the sublinearity, for any $M > 0$ and sufficiently small $\varepsilon > 0$, we have

$$f(t, \alpha_\varepsilon(t)) \geq M\alpha_\varepsilon(t). \tag{18}$$

Let $\delta_0 = \min_{\frac{1}{4} \leq t \leq \frac{3}{4}} \int_0^1 G(t, s)ds > 0$. Since $e(t) := \int_0^1 G(t, s)ds > 0$ for $t \in (0, 1)$, and $e(t)$ satisfies $e'(0) > 0$ if $e(0) = 0$ and $e'(1) < 0$ if $e(1) = 0$. (see Hu [11]). Therefore, one can show that there is $\delta_1 > 0$ such that

$$\int_{1/4}^{3/4} G(t, s)ds \geq \delta_1 \int_0^1 G(t, s)ds, \quad t \in [0, 1]. \tag{19}$$

Thus, by (18) and (19) we have

$$(F(\alpha_\varepsilon))(t) \geq M \int_0^1 G(t, s)\alpha_\varepsilon(s)ds \geq M\delta_0\varepsilon \int_{1/4}^{3/4} G(t, s)ds$$

$$\geq M\delta_0\delta_1\varepsilon \int_0^1 G(t, s)ds = M\delta_0\delta_1\alpha_\varepsilon(t).$$

Noting that δ_0, δ_1 are independent of M so one can choose M so large that $M\delta_0\delta_1 > 1$ and let $u = \alpha_\varepsilon(t)$. Thus we obtain $u \leq Fu$ and $u \leq v$ so that Lemma 3 implies that BVP (1), (2) has maximal and minimal solutions in $[u, v]$. This completes the proof.

REFERENCES

1. H. Amann, Order structures and fixed points, *Atti* 2^0 *Sem. Anal. Funz. Appl.*, Univ. Cosenga (Italy), 1-50 (1977).
2. C. Bandle, C.V. Coffman and M. Marcus, Nonlinear elliptic problems in annular domains, *J. Diff. Eqns.*, **69**: 322-345 (1987).
3. C. Bandle and M.K. Kwong, Semilinear elliptic problems in annular domains, *J. Appl. Math. Phys. (Zamp)*, **40**: 245-257 (1989).

4. C.V. Coffman and M. Marcus, Existence and uniqueness results for semilinear Dirichlet problems in annuli, *Arch. Rat. Mech. Anal.,* **108**: (1989).

5. K. Deimling, "Nonlinear Functional Analysis," Springer 1985.

6. L.H. Erbe, Boundary value problems for ordinary differential equations, *Rocky Mountain Math. J.,* **1**: 709-729 (1970).

7. L.H. Erbe, S. Hu and H. Wang, Multiple positive solutions of some boundary value problems, *J. Math. Anal. Appl.,* (to appear).

8. L.H. Erbe and H. Wang, On the existence of positive solutions of ordinary differential equations, *Proc. Amer. Math. Soc.,* (to appear).

9. X. Garaizar, Existence of positive radial solutions for semilinear elliptic problems in the annulus, *J. Diff. Eqns.,* **70**: 69-72 (1987).

10. G.B. Gustafson and K. Schmitt, Nonzero solutions of boundary value problems for second order ordinary and delay-differential equations, *J. Diff. Eqns.,* **12**: 127-147 (1972).

11. S. Hu, Nonlinear elliptic boundary value problems via cone analysis, *Nonlinear Analysis,* **18**: 713-730 (1992).

12. G. Iffland, Positive solution of a problem of Emden-Fowler type with a free boundary, *SIAM J. Math. Anal.,* **18**: 283-292 (1987).

13. R.W. Leggett and L.R. Williams, Multiple positive fixed points of nonlinear operators on ordered Banach spaces, *Indiana University Math. J.,* **28**: 673-688 (1979).

Gradient and Gauss Curvature Bounds for H–Graphs

ROBERT FINN   Department of Mathematics, Stanford University, Stanford, CA 94305

We survey in this paper various a priori gradient and curvature bounds that are imposed on the solutions of certain nonlinear elliptic equations arising in geometry and in mechanics by the particular nonlinearity. These bounds differ in significant ways from the bounds that are characteristic for linear equations.

1. One of the basic results in the theory of linear second order elliptic equations in the plane is the a priori gradient bound of the solutions. Under suitable hypotheses on the equation one obtains, for a solution $u(x)$ satisfying $|u(x)| < M$ in a disk $B_R(p)$, a bound of the form

$$|\nabla u(p)| < \mathcal{F}\left(\frac{M}{R};\frac{m}{M}\right) \tag{1}$$

where $m = |u(p)|$; the structure of $\mathcal{F}$ depends only on the coefficients in the equation, and not on the particular solution considered. In the particular case of the Laplace equation

$$\Delta u = 0 \tag{2}$$

we are able to obtain explicitly a best possible result. We assume for simplicity that p is the origin of coordinates, and write $x = (x,y)$.

__Theorem 1:__ *Let $u(x)$ satisfy (2) interior to a disk $B_R(0)$, with $|u(x)| < M$ and $|u(0)| = m$. Then*

$$|\nabla u(0)| \le \frac{4}{\pi}\frac{M}{R}\cos\frac{\pi}{2}\frac{m}{M} \;. \tag{3}$$

The bound is achieved by the function

$$w(z) = \mathcal{J}m\left\{\frac{2M}{\pi}\ln\frac{(1+i\,\lambda)R + i\,(1-i\,\lambda)z}{(1-i\,\lambda)R - i\,(1+i\,\lambda)z}\right\}, \tag{4a}$$

with $z = x+iy$, and

$$\lambda = \tan\frac{\pi m}{4M} \;. \tag{4b}$$

A proof can be given readily using conformal mapping and the Schwarz Lemma. We indicate here another procedure which will form the basis for our results to follow.

We may assume without loss of generality that ∇u is directed along the positive x-axis and that $u(0) = m$. Let $w(x)$ be the (unique) solution of (2) that is bounded in $B_R(0)$, with $w(x) = M$ and $w(x) = -M$ on the interiors of two arcs Σ^+ and Σ^- symmetric with respect to the x-axis, as indicated in Figure 1; the two arcs are chosen so that $w(0) = m$. The function $w(x)$ can be written explicitly in the form (4), and formal calculation yields the bound (3) with equality, for $w(x)$.

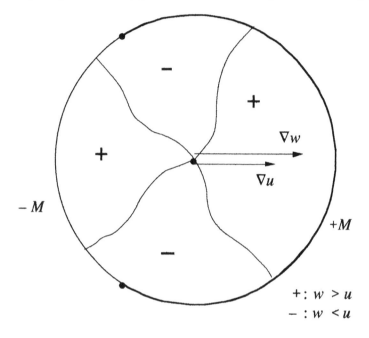

Figure 1: Comparison regions for u, w.

We assert that $|\nabla u(0)| \leq |\nabla w(0)|$, equality holding if and only if $u \equiv w$. If $|\nabla u(0)| > |\nabla w(0)|$ then by decreasing the radius of the disk in which w is defined, we could increase $|\nabla w(0)|$ without changing $w(0)$, until $\nabla w(0) = \nabla u(0)$. We would then have u and w defined in a common disk (which we again denote by $B_R(0)$), with u harmonic up to the boundary $\partial B_R(0) = \Sigma^+ \cup \Sigma^- \cup \Sigma_0$, and (in view of the strong maximum principle) $u - w > 0$ on Σ^+, $u - w < 0$ on Σ^-; Σ_0 consists of the two points joining Σ^+ to Σ^-. On the other hand (cf. [1], Lemma III.2) there must be at least four level curves $u - w = 0$ emanating from the origin, creating at least four regions in which (alternatively) $u - w > 0$ and $u - w < 0$, see Figure 1. Such a configuration is readily seen to conflict with the maximum principle; the technical difficulties arising from the singularities of w at Σ_0 are easily overcome using the asymptotic constancy of w on radial segments emanating from these points. A different technical reasoning obviates the case $\nabla w(0) = \nabla u(0)$ (in the original configuration) unless $u \equiv w$.

2. It is remarkable that an inequality of the general form (1) holds also for the minimal surface equation.

Theorem 2 [1] . *Let u(x) be a solution of the minimal surface equation*

$$\operatorname{div} Tu = 0 , \qquad Tu = \frac{1}{W}\, Du , \qquad W = \sqrt{1+|Du|^2} \tag{5}$$

in $B_R(0)$, *with* $|u| < M$ *and* $u(0) = m$. *Then*

$$|\nabla u(0)| < \frac{1}{1 + e^{\frac{-\pi m}{R}}}\, e^{\frac{\pi(M-m)}{2R}} + Ce^{\frac{-\pi(M-m)}{2R}} \tag{6}$$

for an (explicitly known) constant C, independent of the particular solution considered.

This result is asymptotically, for large (*M*–*m*), best possible.

Outline of proof: We consider a solution $\varphi(x)$ of (5) with boundary values M and $-M$ on segments of ∂B_R as before. A reasoning analogous to the one just above shows that $|\nabla\varphi|$ majorizes the gradient magnitude of any other solution with the same bound and the same value at the origin. In the present case φ is not known explicitly, however its gradient at the origin can be estimated. We observe that the equation (5) is an integrability condition for a function $\psi(x)$ such that

$$\begin{aligned}\psi_x &= \frac{1}{W}\, \varphi_y \\ \psi_y &= \frac{-1}{W}\, \varphi_x\end{aligned} \tag{7}$$

with $\psi(0) = 0$. It turns out that $\varphi + i\psi$ yields a 1-1 conformal map of the solution surface S determined by φ onto a rectangle of width $2M$ and height $h < 2R$, and that $h \to 2R$ as $M - m \to \infty$. The function

$$\chi(\zeta) = \frac{\varphi_x - i\varphi_y}{1+W} \tag{8}$$

(stereographic projection of the spherical image) is analytic in the rectangle, $|\chi(\zeta)| = 1$ on the upper and lower boundary segments, and $\arg \chi$ is known asymptotically on the left and right segments. From this information, $|\chi|$ can be estimated at the point $(0,m)$, yielding the stated result.

    3. There are several natural generalizations of Theorems 1 and 2. We indicate some of them.

    3.1. Uniformly elliptic equations: These include equations of the form

$$au_{xx} + 2bu_{xy} + cu_{yy} = 0 \tag{9}$$

with the property that the metric

$$ds^2 = ad x^2 + 2bdxdy + cd y^2 \tag{10}$$

is quasi-conformal over the (x,y) plane, uniformly in all arguments of a, b, c. Such equations have been extensively studied and it is known that an estimate of the type (1) holds in great generality; for references see, e.g. [2].

3.2. **Equations of minimal surface type:** These are equations of the form (9) for which the metric (10) is quasi-conformal on the surface represented by any solution, with eccentricity independent of the particular solution considered. The condition is equivalent to quasi-conformality of the spherical image mapping. The estimate (6) was extended in [1] to an estimate of the same general form that holds for any such equation.

3.3. **Prescribed mean curvature equations:** An a priori interior gradient estimate of the form (1) for equations of the form

$$\operatorname{div} Tu = 2H(\mathrm{x}; u), \quad H'(u) \geq 0 \tag{11}$$

was first given by Bombieri and Giusti [3]. Improvements were provided by a number of authors, most recently by Korevaar [4] and by Lieberman [5]. All these papers provide results conceptually very similar to the preceding ones, indicating some overlap in the geometric information provided by the equation.

3.4. **Equations of mean curvature type:** These are equations of the form

$$au_{xx} + 2bu_{xy} + cu_{yy} = 2H(x; u; Du) \tag{12}$$

with conditions on the coefficients as in 3.2 above, and with conditions on H that have the effect of requiring the spherical image mapping to be elliptic in the sense of [6]. Simon [7] showed that results analogous to those above continue to hold for such equations.

In the remainder of this paper we intend to show that in fact *there are very significant distinctions in the geometric structure that arises in* (11) *and in* (12), *relative to the* (*homogeneous*) *cases previously considered, and that these distinctions can impose on the solutions a priori estimates of basically different character than those described in the preceding cases.*

4. The motivation in what follows derives from the physical problem of liquid rise (or fall) in a capillary tube; for this reason we provide here a brief outline of some of the relevant features in that theory. We will be concerned with a semi-infinite cylindrical tube Z of general section, partly filled with a volume V of liquid over its base domain Ω, and possibly subject to an external (gravity) field directed along the generators of Z downward into the fluid (Figure 2).

We suppose that the liquid covers Ω and that the free surface interface S can be described by a graph $z = u(x)$. The equation for S then takes the form (see, e.g. [8], Chapter 1)

$$\text{div } Tu = \kappa u + \lambda \tag{13}$$

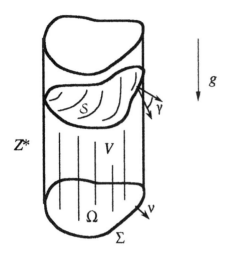

Figure 2: Free surface S in capillary tube.

where $\lambda = \text{const}$ is a Lagrange parameter arising from the volume constraint, and $\kappa = \rho g/\sigma$ is the "capillarity constant"; here ρ is density change across S, g the gravitational acceleration and σ the surface tension. When $\kappa \geq 0$ (liquid not less dense than air) we recognize (13) as a special form of (11). On $\Sigma = \partial\Omega$, S must meet Z in a prescribed angle γ, depending only on the materials; thus

$$\nu \cdot Tu = \cos \gamma \tag{14}$$

on Σ.

We consider at first the case in which gravity is absent; the equation (13) can then be written in the form

$$\text{div } Tu = 2H \equiv \text{const}; \tag{15}$$

here H is the *mean curvature* of S and is determined by the relation

$$2H|\Omega| = \oint_\Sigma \cos \gamma \, ds \tag{16}$$

obtained by integrating (15) over Ω and using the divergence theorem together with the boundary condition (14). We note in particular that if $\gamma \equiv \text{const} \neq \pi/2$ then $H \neq 0$.

5. *In general, the problem* (15), (16), (14) *cannot be expected to admit a solution.*
Consider for example the case in which Ω contains a corner bounded by a straight segment with opening 2α, as in Figure 3 . We introduce a segment Γ cutting off a subdomain Ω^* as shown, and

a segment Λ cutting off the vertex P, since data cannot be prescribed at P. Integration of (15) yields, in the case $\gamma \equiv$ const,

$$2H|\Omega^*| = |\Sigma^*|\cos \gamma + \int_\Gamma v \cdot Tu \ ds + \lim_{\Lambda \to P} \int_\Lambda v \cdot Tu \ ds \quad . \tag{17}$$

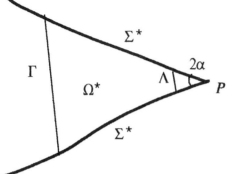

Figure 3: Corner configuration.

Since $|v \cdot Tu| < 1$ for any function $u(x,y)$, it follows that the limit exists and is zero. Placing the same estimate into the other integral, we find immediately

$$|\Sigma^*|(\sin \alpha - \cos \gamma) + 2H|\Omega^*| > 0. \tag{18}$$

Letting Γ move by translation to the vertex, *we obtain* $\sin \alpha \geq \cos \gamma$, *that is*

$$\alpha \geq \left|\frac{\pi}{2} - \gamma\right| \tag{19}$$

as necessary condition for existence of a solution. (This condition is known to be sharp.)
In particular, *if* $\gamma = 0$ *or* π *then no configuration with a protruding corner can admit a solution.*

We note that in the above proof, *no growth condition has been imposed at the vertex P*; the result applies to the most general solution of (15) that assumes the prescribed data on the smooth parts of the boundary. An analysis of the proof shows that the difficulty does not arise from the boundary singularity at the vertex; the vertex can be rounded and a similar result is obtained. This seemingly paradoxical behavior is clarified physically by the observation that if a boundary continuum of locally large curvature appears, the fluid may be strongly attracted to that part of the cylinder; in the absence of gravity the restoring forces can be insufficient to prevent the fluid from uncovering the base and flowing out along the generators to infinity.

A gravity field directed downward into the fluid can greatly ameliorate the singular behavior, and it can be shown [9] that when $\kappa > 0$ solutions of (13) will exist in very general circumstances under capillary boundary conditions. Nevertheless, a discontinuous dependence on data at corners continues to occur, the solution transforming discontinuously from a bounded one with finite (minimizing) mechanical energy into one with infinite surface energy, see [10] or [8] Chapter 5.

6. It is thus clear that solutions of (15) can respond to very different rules than do solutions of (2), and it has to be expected that these distinctions will be reflected in the kinds of a priori estimates that hold. A clue as to what should be expected can be found in the observation that in contrast to Dirichlet type boundary conditions, for which an "extremal" character relates to a bound in magnitude, extremal data for the capillary problem have a character that is universal among all solutions. That is, extremal data are simply the data $v \cdot Tu = \pm 1$, corresponding physically to the contact angles $\gamma = 0$ or π. If solutions corresponding to such data can be found, they should presumably majorize (in appropriate senses) among all possible solutions that can occur, without regard to bounds on the solutions. In analogy with the results found just above, the estimates obtained can be expected to hold without any need to impose growth conditions at "small" singular sets on the boundary. A further clue can be found in the results of [11] and of [12] Theorems 2 and 3, which suggest that extremal surfaces be sought in domains bounded by circular arcs of particular radius, related to *H*.

All these considerations depend ultimately on the validity of comparison principles related to capillary as opposed to Dirichlet conditions. In fact, such principles hold under remarkable generality. The essential features of what is needed are contained in the following statement ([8], Chapter 5).

Comparison principle: *Define* $Nu \equiv divTu - 2H(u)$ *We suppose* $\partial\Omega = \Sigma = \Sigma_\alpha \cup \Sigma_\beta \cup \Sigma_0$, *with* $\Sigma_\beta \in C^{(1)}$ *and* Σ_0 *of (linear) Hausdorff measure zero. We suppose* $H'(u) \geq 0$ *and that*

$$\text{i)} \quad Nu \geq Nw \text{ in } \Omega$$

$$\text{ii)} \quad \lim_{p \to \Sigma_u} \inf \left(w(p) - u(p) \right) \geq 0$$

$$\text{iii)} \quad v \cdot Tw \geq v \cdot Tu \quad \text{on } \Sigma_\beta .$$

Then either $w > u$ *in* Ω *or else* $w \equiv u + const.$

For some particular purposes we need a refinement of the principle, see [13], §2. We note that no growth conditions are imposed at Σ_0. In this respect the principle as stated does not hold for harmonic functions (i.e., *Nu* replaced by Δu) even when $\Sigma_\beta = \emptyset$. An example is provided by the Poisson kernel, which is positive in the unit disk and vanishes on the boundary except at a single point.

7. With the above considerations in mind, we choose as initial configuration, corresponding to a constant $H \neq 0$, that of Figure 4, bounded by two circular arcs (moon domain). We may assume without loss of generality that $H > 0$.

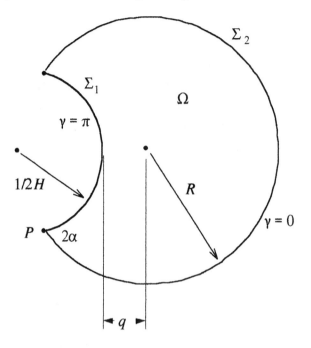

Figure 4: Moon domain.

Theorem 3 [12]: *Suppose $1/2H < R < 1/H$. Then there is a unique value of q for which a solution of $\text{div} Tw = 2H$ will exist in Ω, with $v \cdot Tw = 1$ on Σ_2 and $v \cdot Tw = -1$ on Σ_1. There exists $R_0 = (0.5654062332...)/H$ such that $q > 0$ if and only if $R > R_0$. The solution w is bounded above, is negative infinite on Σ_1, and is unique up to an additive constant. It can be obtained as limit, uniform in compact subsets of Ω, of solutions over the full disk B_R .*

Here the boundary data must be interpreted in a limiting sense, as they impose infinite gradients on the solution at all boundary points.

In this result, the range indicated for R is the largest that could be encountered. If $R = 1/H$ then Σ_1 degenerates to a point and the unique solution in Ω becomes a lower hemisphere; if $R > 1/H$ then there is no solution in Ω [14]. If R decreases to $1/2H$ then Σ_1 and Σ_2 coincide and Ω degenerates.

Note that Ω has a protruding vertex at P, although the data are extremal. In the remark following (19) above, we pointed out that a protruding vertex is not possible when $\gamma = 0$ or π. In the present case, we have $\gamma = 0$ on one side of the vertex, $\gamma = \pi$ on the other side. In accordance with a general result of [15], the unit surface normal is discontinuous at P. In fact, the solution is shown in [12] to have an infinite jump discontinuity at P.

From Theorem 3 we obtain the result [16] :

Theorem 4: *There exists $\mathcal{F}(R\,;\,H)$ decreasing in R, with $\mathcal{F}(R_0\,;\,H) = \infty$ and $\mathcal{F}(1/H\,;\,H) = 0$, such that if u(x) satisfies* div $Tu = 2H$ *interior to the disk $B_R(0)$ with $R > R_0$, then*

$$|\nabla u(0)| < \mathcal{F}(R\,;\,H) \tag{20}$$

The value R_0 cannot be improved, in the sense that if $R \leq R_0$ then solutions can be found for which $|\nabla u(0)|$ *is as large as desired.*

Thus, for any solution defined in a disk of radius $R > R_0$, the gradient at the center is bounded depending only on R, without any regard to a bound on the solution height.

Outline of proof: If $R > R_0$ we can position a moon domain interior to the disk as indicated in Figure 5, such that ∇w coincides in direction with ∇u at the origin. If $|\nabla w| < |\nabla u|$ then we decrease the radius of the outer disk in the moon domain until equality is attained. The remainder of the proof follows conceptually that of Theorem 1; the usual maximum principle is replaced by the above comparison principle. The singularities at the two vertex points cause no difficulty, as these two points form a set of Hausdorff measure zero.

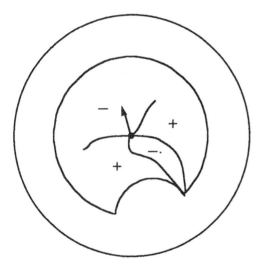

Figure 5: Proof of gradient bound.

The function $\mathcal{F}(R\,;\,H)$ can be estimated explicitly, see Chua [17]. The proof just outlined clearly extends to a gradient estimate in a disk of (small enough) radius about the origin. Using more sophisticated methods, Liang [18] showed that *if $\hat{R} < R$ determines the largest disk $B_{\hat{R}}(0)$ concentric to $B_R(0)$ and contained in the corresponding moon domain for which Σ_2 has radius R, an analogous bound holds throughout the interior of $B_{\hat{R}}(0)$*.

Additionally, the following result is true:

Theorem 5: *If R > 1/2H, and if to every point p of distance r to the origin we associate the function*

$$\mathcal{P}(r \, ; \, R) = \frac{1 + 2H^2(r^2 - R^2)}{2Hr} \tag{21}$$

then there corresponds to each such p an outward radial sector **S** *of opening*

$$\psi(R) = \begin{cases} 2\cos^{-1}\mathcal{P} & if \; |\mathcal{P}| < 1 \\ 0 & if \; \mathcal{P} > 1 \\ 2\pi & if \; \mathcal{P} \leq -1 \end{cases} \tag{22}$$

and a finite $\mathcal{M}(r \, ; \, R)$, *such that for any solution u(x) in* $B_R(0)$ *for which* $\nabla u(\,p)$ *is directed into* **S**, *there holds* $|\nabla u(\,p)| < \mathcal{M}$. *That is, if the gradient has large magnitude at interior points, it cannot be directed outward. For details, see* [18] *or* [19].

One can obtain also a form of Harnack inequality, that holds for any *R*.

Theorem 6: *There exists* $\rho(m; R) > 0$ *and a function* $\varphi(m \, ; \, R \, ; \, r) < \infty$ *in* $r < \rho$, *such that if u(x) is a non vanishing solution of* div $Tu = 2H$ *in* $B_R(0)$ *and u(0) = m, then* $|u| < \varphi(m \, ; \, R \, ; \, r)$ *in* $B_r(0)$. *There holds* $\lim_\rho \varphi(m \, ; \, R \, ; \, r) = \infty$, *while* $\rho(m; R) \to 0$ *as* $|m| \to \infty$.

Up to this point, the theorem is essentially that of Serrin [20]. The above methods yield a conceptually simpler proof, and provide also the additional information [19]:

I) *If* $R > R_0$ *then* $\rho(m; R) \geq \hat{R}$, *independent of m. There exist* $A_0^-(R \, ; \, r)$, $A_0^+(R \, ; \, r)$, *with* $\lim_{R \to 1/H} A_0^-(R \, ; \, r) = \lim_{R \to 1/H} A_0^+(R \, ; \, r) = 0$, *all* $r < 1/H$, *and* $\lim_{r \to R - \epsilon} A_0^+(R \, ; \, r) < \infty$, *all* $\epsilon > 0$, *and such that*

$$A_0^-(R \, ; \, r) < u - m - H^{-1} + H\sqrt{H^{-2} - r^2} < A_0^+(R \, ; \, r) \tag{23}$$

II) *If* $R > R_0$, $m > 0$, *then* $\rho(m) = \infty$, *independent of m. Specifically, there holds*

$$u - m - H^{-1} + H\sqrt{H^{-2} - r^2} < A_1(R) \tag{24}$$

throughout $B_R(0)$, *with* $A_1(R) \to 0$ *as* $R \to 1/H$.

8. We turn our attention to situations in which $H(u)$ need not be constant. It is not difficult to see that if $0 < H_1 = \lim_{t \to -\infty} H(t) < H(u) < \lim_{t \to \infty} H(t) = H_2 < \infty$ and if $H'(u) \geq 0$ then a result analogous to Theorem 4 continues to hold; specifically, *there exists* $R_0 > 0$ *and* $\mathcal{F}(R \, ; \, H_1 \, ; \, H_2)$ *such that if u(x) is a solution in a disk strictly containing* $B_{R_0}(0)$ *then* $|\nabla u(0)| < \mathcal{F}(R \, ; \, H_1 \, ; \, H_2)$. However the only estimate presently known for R_0 is $R_0 < 1/H_2$,

which is clearly much too large in the limit as $H_1 \to H_2$. There is on the other hand a different kind of extension, which could less easily have been anticipated [21].

Theorem 7. *Suppose* $-\infty < H_1 = \lim\limits_{t \to -\infty} H(t) < H(u) < \lim\limits_{t \to \infty} H(t) = H_2 < \infty$, $H'(u) \geq 0$. *Then there is a function* $\mathcal{F}(R;u_0)$ *such that if* $u(x)$ *satisfies* div $Tu = 2H(u)$ *in* $B_R(0)$ *with* $u(0) = u_0$ *then*

$$|\nabla u(0)| < \mathcal{F}(R \, ; u_0). \tag{25}$$

Thus, the bound depends only on *R* and on the surface height at the single point of evaluation. In this theorem there is no restriction on *R*. The theorem is clearly false if $H \equiv$ constant; it is essential that $H_1 < H_2$. The dependence on u_0 in (25) is essential; a bound depending only on *R* would be false.

The proof of Theorem 7 depends on an existence theorem, which follows, essentially, from Theorem 7.10 of [8].

Theorem 8: *Suppose* $-\infty < H_1 = \lim\limits_{t \to -\infty} H(t) < H(u) < \lim\limits_{t \to \infty} H(t) = H_2 < \infty$, $H'(u) \geq 0$. *Then in the configuration of Figure 6, for all* q^* *sufficiently small there exists a solution* $w(x)$ *of* div $Tu = 2H(u)$ *in* Ω, *with* $v \cdot Tw = -1$ *on* Σ_1, $v \cdot Tw = +1$ *on* Σ_2. *There holds* $w \to -\infty$ *on* Σ_1, $w \to +\infty$ *on* Σ_2, *and w is monotone increasing on the horizontal symmetry line joining* Σ_1 *to* Σ_2.

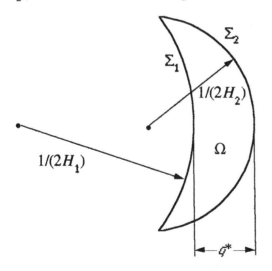

Figure 6: Moon crescent.

If $u(x)$ is a solution in $B_R(0)$, we choose q^* small enough that the crescent domain will lie interior to $B_R(0)$ with any point of the symmetry line at the origin. We choose that point of the line at which $w = u_0$ to be at the origin, and orient the crescent so that the gradient directions coincide. The remainder of the proof follows as in Theorem 4.

9. We consider finally the case $\lim_{u \to \pm\infty} |H(u)| = \infty$, $H'(u) \geq 0$. We then have

Theorem 9: *Under the hypothesis just made, there exists* $\mathcal{F}(R)$ *such that if* $u(x)$ *satisfies* div $Tu = 2H(u)$ *in* $B_R(0)$, *then*

$$|\nabla u(0)| < \mathcal{F}(R) .\tag{26}$$

This result follows from Theorem 8 and from the known result [13]

Theorem 10: *Suppose* $\lim_{u \to -\infty} H(u) = -\infty$, $\lim_{u \to \infty} H(u) = \infty$. *Set* $N_R = \min\{u : H(u) \geq -1/R\}$, $M_R = \max\{u : H(u) \leq 1/R\}$. *If* $u(x)$ *satisfies* div $Tu = 2H(u)$ *in* $B_R(0)$, *then*

$$N(R) - R \leq u(x) \leq M(R) + R \tag{27}$$

throughout $B_R(0)$.

The hypothesis that $H'(u) \geq 0$ is not needed for this result. The proof of the right side of (27) follows easily by observing that if $R' < R$ and a lower hemisphere over $B_{R'}$ is raised above the solution surface and then lowered until an initial point of contact occurs, then at that point the hemisphere has mean curvature not less than that of the surface. The left side follows similarly.

10. For equations of the kind considered in this survey, it should be expected that particular a priori restrictions will also be imposed on the Gauss curvature K. This question has been less intensively investigated, but striking results have been obtained in particular situations. In the case $H \equiv$ const., Spruck [22] showed that $|K|$ is bounded at the center of B_R depending only on R. A more precise form of his result appears in [23] and in [24], and the result was extended to more general equations of mean curvature type in [7]. This bound is quite remarkable, especially in view of the fact that the gradient (and also second derivatives) can become unbounded at the center, cf. the discussion in [23]. Caffarelli, Nirenberg and Spruck [25], and later Ecker and Huisken [26] extended the bound to non-constant H in higher dimensions, but their results depend on gradient bounds. The methods described above can be used to extend both the curvature estimate and the Harnack inequality of §7 to situations of variable mean curvature, analogous to the above extensions of the gradient estimates; this work is currently in progress, and a detailed form for the results is not yet available.

In each of the individual cases we have considered, our method has yielded the best of the known results, and it has the appeal of conceptual simplicity (although the required existence proofs are in general not elementary). The most significant deficiency for the method is that the procedure does not extend directly to more than two dimensions. A counterexample due to H. Lewy [27] for the particular case of harmonic functions strongly suggests that the comparison procedure will fail in three or more dimensions unless auxiliary hypotheses are introduced.

The research described above was supported in part by the National Aeronautics and Space Administration and in part by the National Science Foundation. I wish to thank the Augustus Universität in Leipzig for its hospitality during the preparation of the manuscript.

REFERENCES

1. R. FINN, *New estimates for equations of minimal surface type*; Arch. Rat. Mech. Anal. **14** (1963) 337-375.

2. O. A. LADYZHENSKAIA & N. N. URAL'TSEVA, *Linear and Quasilinear Elliptic Equations*; Academic Press, New York, 1968.

3. E. BOMBIERI & E. GIUSTI, *Local estimates for the gradient of non-parametric surfaces of prescribed mean curvature*; Comm. Pure Appl. Math. **26** (1973) 381-394.

4. N. J. KOREVAAR, *An easy proof of the prescribed gradient bound for solutions to the prescribed mean curvature equation;* Proc. Symp. Pure Math. **45** (1986) Part 2, 81-89.

5. G. M. LIEBERMAN, *Gradient estimates for capillary-type problems via the maximum principle*; Comm. in P.D.E. **13** (1988) 33-59.

6. R. FINN & J. B. SERRIN, *On the Hölder continuity of quasi-conformal and elliptic mappings*; Trans. Amer. Math. Soc. **89** (1958) 1-16.

7. L. SIMON, *Equations of mean curvature type in 2 independent variables*; Pac. J. Math. **69** (1977) 245-268.

8. R. FINN, *Equilibrium Capillary Surfaces*; Springer-Verlag, New York, 1986.

9. R. FINN & C. GERHARDT, *The internal sphere condition and the capillary problem*; Ann. Mat. Pura Appl. **112** (1977) 13-31.

10. P. CONCUS & R. FINN, *On capillary free surfaces in a gravitational field*; Acta Math. **132** (1974) 207-223.

11. R. FINN, *A note on the capillary problem*; Acta Math. **132** (1974) 199-206.

12. R. FINN, *Moon surfaces, and boundary behaviour of capillary surfaces for perfect wetting and nonwetting*; Proc. London Math. Soc. **57** (1988) 542-576.

13. R. FINN, *Comparison methods in capillarity,* Springer-Verlag Lecture Notes **1357** (1988) 156-197.

14. R. FINN, *Remarks relevant to minimal surfaces and to surfaces of prescribed mean curvature*; J. Analyse Math. **14** (1965) 139-160.

15. P. CONCUS & R. FINN, *Capillary wedges revisited*; LBL Report 33553 (1993), Univ. of Calif., Berkeley.

16. R. FINN & E. GIUSTI, *On nonparametric surfaces of constant mean curvature*; Ann. Scuola Normale Sup. Pisa **4** (1977) 13-31

17. K.-S. CHUA, *Absolute gradient bounds for surfaces of constant mean curvature*; Ann. Scuola Normale Sup. Pisa, in press.

18. F.-Ts. LIANG, *An absolute gradient bound for nonparametric surfaces of constant mean curvature*; Ind. Univ. Math. J. **41** (1992) 569-604.

19. R. FINN, *The inclination of an H-graph,* Springer-Verlag Lecture Notes **1340** (1988) 40-60.

20. J. B. SERRIN, *The Dirichlet problem for surfaces of constant mean curvature*; Proc. Lon. Math. Soc. **21** (1970) 361-384.

21. R. FINN & J. LU, *A new a-priori gradient bound for surfaces of prescribed mean curvature*; in preparation.

22. J. SPRUCK, *Gauss curvature estimates for surfaces of constant mean curvature*; Comm. Pure Appl. Math. **27** (1974) 547-557.

23. R. FINN, *The Gauss curvature of an H-graph*; Göttingen Nachrichten, Nr. 2, (1987).

24. Y. SHIH, *An estimate on Gaussian curvature of constant mean curvature surfaces*; Chinese J. Math. **16** (1988) 111-122.

25. L. CAFFARELLI, L. NIRENBERG & J. SPRUCK, *On a form of Bernstein's Theorem*; Analyse Mathématique et Applications (ded. J. L. Lions). Gauthier-Villars, Paris (1988) 55-66.

26. K. ECKER & G. HUISKEN, *Interior curvature estimates for hypersurfaces of prescribed mean curvature*; Ann. Inst. H. Poincaré. Anal. Non Linéaire **6** (1989) 251-260.

27. H. LEWY, *On the minimum number of domains in which the nodal lines of spherical harmonics divide the sphere;* Comm. Partial Diff. Eqns. 2 (1977) 1233-1245.

Some Applications of Geometric Methods in Mechanics

ZHONG GE and W.F. SHADWICK , The Fields Institute for Research in Mathematical Sciences, 185 Columbia St. West, Waterloo, Ontario N2L 5Z5

Introduction.

Many differential equations arising in mechanics and physics can be described as Hamiltonian systems. These equations enjoy many special properties which are geometric in nature, e.g. they possess some integral invariants (Poincare, Cartan [4]). Moreover, many problems in mechanics, being non-linear, are best formulated in the language of geometry and consequently solved by geometric methods, so it is not surprising that geometric methods play a very important role in the study of Hamiltonian systems. In this note we will discuss some of applications of geometric methods in mechanics.

1 Geometric Algorithms

In the classical Hamiltonian formalism, one usually considers a system of n particles on $\mathbb{R}^3$. Let H be the Hamiltonian (i.e. the total energy, say, the kinetic energy plus the potential energy), then the motion is described by

$$
\begin{aligned}
\frac{dp_i}{dt} &= H_{q_i}, & i = 1, 2, \cdots, 3n \\
\frac{dq_i}{dt} &= -H_{p_i}, & i = 1, 2, \cdots, 3n
\end{aligned}
\tag{1}
$$

where q_i's denote the coordinates for the particles, and p_i's the conjugate momentums. The coordinates (p_i, q_i) are called the canonical coordinates. Let $z = (p_i, q_i)$, then (1) can be written as

$$
\frac{dz}{dt} = J \; \bigtriangledown H
$$

Supported by the Ministry of Colleges and Universities of Ontario and the Natural Sciences and Engineering Research Council of Canada

where

$$J = \begin{pmatrix} 0 & I \\ -I & 0 \end{pmatrix},$$

and I is the identity matrix. Poincare and Cartan showed that the following differential 2-form (**the symplectic 2-form**)

$$\sum dp_i \wedge dq_i \qquad (2)$$

is an invariant of the dynamical system.

A transformation which preserves the 2-form (2) is called a **symplectic transformation.** Thus every motion in a conservative mechanical system is a symplectic transformation.

Though this 2-form does not have any physical meaning, it plays a very important role. A part of the reason is that if a Hamiltonian system has symmetries, we can simplify the equations by reducing the number of equations (the Marsden-Weinstein reduction), usually at the expense of introducing curved spaces. To properly understand the Hamiltonian structure in curved spaces, one has to generalize the Hamiltonian formalism in two ways: generalize from linear vector spaces to manifolds and from the case where there are global canonical coordinates (p, q) to the case where there no such coordinates. Since the concept of a symplectic 2-form is easy to generalize to manifolds, it plays a most important role in the generalization. These generalizations are the starting point for modern geometrical mechanics, as developed by Arnold, Marsden, Weinstein, et al. see Abraham-Marsden [1].

Recently people are interested in developing numerical methods for Hamiltonian systems which preserve the symplectic structure. These methods are called **symplectic algorithms.** More precisely, an algorithm is symplectic if its iteration map

$$z^n \to z^{n+1}, \quad z^{n+1} = \phi(z^n)$$

is a symplectic transformation. For example, an Euler method of the form

$$\frac{z^{n+1} - z^n}{\triangle t} = J^{-1} \bigtriangledown H((1 - \lambda)z^n + \lambda z^{n+1})$$

is symplectic if and only if $\lambda = 1/2$, in which case the iteration map $z^n \to z^{n+1}$ is symplectic.

However, more complicated Hamiltonian systems will appear if we study rigid body dynamics, or, continuum mechanics. This leads to the study of Hamiltonian systems of the form

$$\frac{dz}{dt} = J(z) \bigtriangledown H \qquad (3)$$

where $J(z)$ is an anti-symmetric matrix which may depend on z and can be *degenerate*. In this case we define a bilinear map

$$(f, g) \to \{f, g\} = (\bigtriangledown f)^{\mathsf{T}} J (\bigtriangledown g), \quad f, g \in C^{\infty}(M).$$

If this bracket satisfies the following property

$$\{f,\{g,h\}\} + \{g,\{h,f\}\} + \{h,\{f,g\}\} = 0$$

then $\{,\}$ is called a **Poisson bracket,** and the system (3) is called a **Poisson Hamiltonian system.**

Note that if J is non-degenerate, then after a change of variables, we can write J in the form (1). But in many examples J is necessarily degenerate, and hence cannot be reduced to a simple form. In this case the notion of symplectic transformations should be replaced by that of **Poisson transformations**, i.e. maps T which satisfy

$$\{f \circ T, g \circ T\} = \{f,g\} \circ T,$$

and hence preserve the bracket structure. The simplest example is the motion of a free rigid body.

Example.

Consider the motion of a free rigid body

$$\begin{aligned}
\dot{\Pi}_1 &= \frac{I_2 - I_3}{I_2 I_3} \Pi_2 \Pi_3 \\
\dot{\Pi}_2 &= \frac{I_3 - I_1}{I_3 I_1} \Pi_3 \Pi_1 \\
\dot{\Pi}_3 &= \frac{I_1 - I_2}{I_1 I_2} \Pi_1 \Pi_2
\end{aligned} \qquad (4)$$

where (Π_1, Π_2, Π_3) is the angular momentum, I_1, I_2, I_3 positive constants. After introducing $z = (\Pi_1, \Pi_2, \Pi_3)$, (4) can be rewritten in the form (1), with

$$J = \begin{pmatrix} 0 & \Pi_3 & -\Pi_2 \\ -\Pi_3 & 0 & \Pi_1 \\ \Pi_2 & -\Pi_1 & 0 \end{pmatrix},$$

and

$$H = \frac{1}{2}\left(\frac{\Pi_1^2}{I_1} + \frac{\Pi_2^2}{I_2} + \frac{\Pi_3^2}{I_3}\right).$$

Though the phase flow of (4) does not preserve any two-form, it does preserve the area-forms when restricted to a family of spheres. In fact, as there is no gravity, the total angular momentum is conserved:

$$\|\Pi\|^2 = \|\Pi_1\|^2 + \|\Pi_2\|^2 + \|\Pi_3\|^2 = constant$$

So eq. (4) can be considered as a dynamical system on the sphere, that is, we can replace any one of the original equations by the constraint equation $\|\Pi\| = const$. So we have reduced the number of equations from 3 to 2, at the expense of introducing curved spaces (in this case, the sphere). Moreover, the phase flow of (4), when restricted to the spheres, preserves the area-form.

What is important here is that the rigid body is best understood as a Hamiltonian system on the sphere. This would imply, by the general result that any Hamiltonian system on a surface is integrable by quadrature, that the rigid body is integrable, without any computations.

The free rigid body is an example of Lie-Poisson systems. By definition, a **Lie-Poisson system** is a Hamiltonian system of the form (1) in which $J(z)$ is linear in z. Needless to say, among Poisson Hamiltonian systems, Lie-Poisson systems are the simplest next to the non-degenerate ones.

The Hamiltonian structure for the rigid body was not understood until 1965, when Arnold showed that it is best described in terms of Lie theory. It was later found out that this structure was actually known to Lie and has become known as the Lie-Poisson structure. Moreover, Arnold, Marsden, Ratiu, Weinstein, et al. showed that the underlying Hamiltonian structures in fluid mechanics, plasma physics, etc. are Lie-Poisson.

For Lie-Poisson Hamiltonian systems, it is a much more difficult problem to develop symplectic algorithms. Ge-Marsden [8] solved this problem and developed the so-called Lie-Poisson integrator, which has been applied by Channell-Scovel [5] to numerical analysis. These numerical methods have the following advantages

1. The method preserves the Hamiltonian structure and there is no artificial dissipation.

2. They preserve the angular momentum.

3. They have very good long time stability.

In the case of a free rigid body, an example of such difference schemes is $\Pi^k \to \Pi^{k+1}$:

$$\Pi^k = \frac{1}{2}\{\frac{1}{4}[A\bar{I}(A - A^\top) + \bar{I}(A - A^\top)A^\top]\triangle t + (A - A^\top)\} \tag{5}$$

$$\Pi^{k+1} = \frac{1}{2}\{\frac{1}{4}[\bar{I}(A - A^\top)A + A^\top\bar{I}(A - A^\top)]\triangle t + (A - A^\top)\}. \tag{6}$$

where eq. (5) is to be solved for the rotation matrix A and the result substituted into (6), $\bar{I} : so(3) \to so(3)$ the transformation corresponding to the map $\mathbb{R}^3 \to \mathbb{R}^3$, $(\Pi_1, \Pi_2, \Pi_3) \to (I_1\Pi_1, I_2\Pi_2, I_3\Pi_3)$, $so(3)$ being identified with $\mathbb{R}^3$ in the usual way. The iteration map $\Pi^k \to \Pi^{k+1}$ preserves the spheres $|\Pi| = c$ and the area-forms on the spheres (i.e. the symplectic 2-forms).

It is easy to verify that method (5)-(6) is a 1st-order approximation of the free rigid body eq. (4). One can obtain higher-order methods by using the method (5)-(6).

Let $\Theta(\triangle t) : \Pi^k \to \Pi^{k+1}$ be the map determined by eq. (5), (6). Note that Θ is not time-symmetric, i.e. $\Theta(\triangle t)^{-1} \neq \Theta(-\triangle t)$. To make up for this let $\theta(2\triangle t) = \Theta(-\triangle t)^{-1}\Theta(\triangle t)$, which is time-symmetric. Then the multiple-step method

$$\theta(\alpha\triangle t) \circ \theta((1-2\alpha)\triangle t) \circ \theta(\alpha\triangle t), \qquad \alpha = \frac{1}{2-2^{1/3}}$$

is fourth order, as easily verified by a Taylor expansion in $\triangle t$.

Note that, however, these methods are implicit in nature. Inspired by an idea of McLachlan[11], Reich [15] , we will develop explicit methods below.

First note that the Hamiltonian system for $H = \Pi_1^2/I_1$ with the same J as in the rigid body is given by

$$\begin{aligned} \dot{\Pi}_1 &= 0 \\ \dot{\Pi}_2 &= -\frac{\Pi_1\Pi_3}{I_1} \\ \dot{\Pi}_3 &= \frac{\Pi_1\Pi_2}{I_1} \end{aligned}$$

so the solution is given by

$$\begin{pmatrix} \Pi_1 \\ \Pi_2 \\ \Pi_3 \end{pmatrix} = A_1(\triangle t) \begin{pmatrix} \Pi_1(0) \\ \Pi_2(0) \\ \Pi_3(0) \end{pmatrix}, \quad A_1(\triangle t) = \begin{pmatrix} 1 & 0 & 0 \\ 0 & cos(\frac{\Pi_1\triangle t}{I_1}) & sin(\frac{\Pi_1\triangle t}{I_1}) \\ 0 & -sin(\frac{\Pi_1\triangle t}{I_1}) & cos(\frac{\Pi_1\triangle t}{I_1}) \end{pmatrix}$$

Similarly, the solutions for the Hamiltonian system with Π_2^2/I_2 and Π_3^2/I_3 are given by the matrices

$$A_2 = \begin{pmatrix} cos(\frac{\Pi_2\triangle t}{I_2}) & 0 & sin(\frac{\Pi_2\triangle t}{I_2}) \\ 0 & 1 & 0 \\ -sin(\frac{\Pi_2\triangle t}{I_2}) & 0 & cos(\frac{\Pi_2\triangle t}{I_2}) \end{pmatrix}, \quad A_3 = \begin{pmatrix} cos(\frac{\Pi_3\triangle t}{I_3}) & sin(\frac{\Pi_3\triangle t}{I_3})) & 0 \\ -sin(\frac{\Pi_3\triangle t}{I_3}) & cos(\frac{\Pi_3\triangle t}{I_3}) & 0 \\ 0 & 0 & 1 \end{pmatrix}.$$

Then, as a first order method, we may take

$$z^{k+1} = A_3(\triangle t)A_2(\triangle t)A_1(\triangle t)z^k$$

where $z^k = (\Pi_1^k, \Pi_2^k, \Pi_3^k)$. This is a method which preserves the Hamiltonian structure of the rigid body. As before, we can use this first-order method as a building block to develop higher-order methods.

There is an on going program in Los Alamos to apply these ideas to climate prediction.

2   Second-order equations

It is well-known that the Newton's equation $F = ma$ is a 2-nd order equation, but not all 2-nd order equations describe a *conservative* mechanical system. So one may ask the following question: When does a system of 2-nd order O. D. E.

$$\frac{d^2x^i}{dt^2} = F_i(x, \frac{dx}{dt}, t) \tag{7}$$

describe a mechanical system? That is, when there is a Lagrangian

$$L = \frac{1}{2}a_{ij}(x,t)\dot{x}^i\dot{x}^j - V(x,t) = kinetical\ \ energy - potential\ \ energy$$

such that (7) is the Euler-Lagrange equation of the variational problem

$$\delta \int L dt = 0?$$

And if L exists, we would like to find the Lagrangian.

We first consider the case of a single equation

$$\frac{d^2x}{dt^2} = F(x, \frac{dx}{dt}, t). \tag{8}$$

If two such equations can be transformed into one another after a change of variables, $x \to X(x,t), t \to T(t)$, then we consider them to be the same, or, in the same equivalence class.

Kamran-Lamb-Shadwick [12] classified all such equivalence classes. They have computed that there are three basic local invariants of eq. (8), the first two of which are

$$I_1 = -\frac{A}{2B^2}F_{ppp}$$
$$I_2 = \frac{1}{2AB}(\frac{D}{Dt}F_{pp} - F_{py})$$

where $D/Dt = \partial/\partial t + p\partial/\partial x + F\partial/\partial p$, $p = dx/dt$.

Obviously, a necessary condition for eq. (8) to be the Euler-Lagrange equation of some functional of the form $\frac{1}{2}m(x,t)(\dot{x})^2 - V(x,t)$ is $I_1 = 0$. Furthermore, a simple computation shows that eq. (8) is the Euler-Lagrange equation of some functional of mechanical type if and only if it takes the form of

$$\frac{d^2x}{dt^2} = \frac{1}{2}(\frac{dx}{dt})^2 M_x - M_t\frac{dx}{dt} + N(t,x),$$

and the Lagrangian takes the form of

$$exp(M)(\dot{x})^2 - \int exp(M)N(t,x)dx.$$

In particular, this is true for all the six Painleve transcendents in Ince [10], p. 345, and so all come from mechanics (non-autonomous in general). These six classes are

$(i) \quad \dfrac{d^2x}{dt^2} = 6x^2 + t$

$(ii) \quad \dfrac{d^2x}{dt^2} = 2x^3 + xt + a$

$(iii) \quad \dfrac{d^2x}{dt^2} = \dfrac{1}{x}(\dfrac{dx}{dt})^2 - \dfrac{1}{t}\dfrac{dx}{dt} + \dfrac{1}{t}(\alpha x^2 + \beta) + \gamma x^3 + \dfrac{\delta}{x}$

(iv) $\quad \dfrac{d^2x}{dt^2} = \dfrac{1}{2x}(\dfrac{dx}{dt})^2 + \dfrac{3x^3}{2} + 4tx^2 + 2(x^2 - \alpha)x + \dfrac{\beta}{x}$

(v) $\quad \dfrac{d^2x}{dt^2} = (\dfrac{1}{2x} + \dfrac{1}{x-1})(\dfrac{dx}{dt})^2 - \dfrac{1}{t}\dfrac{dx}{dt} + \dfrac{(x-1)^2}{t^2}\{\alpha x + \dfrac{\beta}{x}\} + \dfrac{\gamma x}{t} + \dfrac{\delta x(x+1)}{x-1}$

(vi) $\quad \dfrac{d^2x}{dt^2} = \dfrac{1}{2}\{\dfrac{1}{x} + \dfrac{1}{x-1} + \dfrac{1}{x-t}\}(\dfrac{dx}{dt})^2 - \{\dfrac{1}{t} + \dfrac{1}{t-1} + \dfrac{1}{t-x}\}\dfrac{dx}{dt}$

$\qquad + \dfrac{x(x-1)(x-2)}{t^2(t-1)^2}\{\alpha + \dfrac{\beta t}{x^2} + \dfrac{\gamma(t-1)}{(x-1)^2} + \dfrac{\delta t(t-1)}{(x-t)^2}\}$

and the kinetical energy respectively are (i) p^2; (ii) p^2; (iii) $p^2 ln(xt)$; (iv) $ln\sqrt{x}p^2$; (v) $ln(t(x - 1)\sqrt{x})p^2$; (vi) $ln(\sqrt{x(x-1)(x-t)}t(t-1))p^2$.

Next we consider a system of two 2-nd order differential equations

$$\frac{d^2x^1}{dt^2} = f^1(x, \frac{dx}{dt})$$

$$\frac{d^2x^2}{dt^2} = f^2(x, \frac{dx}{dt}) \tag{9}$$

It's easy to see that for (9) to describe a mechanical system, it is necessary that f^1, f^2 have the following special form

$$f^k = \sum \Gamma^k_{ij}\dot{x}^i\dot{x}^j + U^k(x)$$

We may think of Γ^k_{ij} as the Christoffel symbols in Riemannian geometry. One can define curvatures using these Christoffel symbols. For example, we define the Gaussian curvature

$$K = \partial_2\Gamma^1_{21} - \partial_1\Gamma^1_{22} + \Gamma^1_{p2}\Gamma^p_{21} - \Gamma^1_{p1}\Gamma^p_{22}.$$

Using these curvatures, Atkins-Ge [2] give a complete characterization of those ODEs which describe the motion of a mechanical system. A partial solution of this problem was known before, but the complete solution seems to be new.

References

[1] R. Abraham and J. Marsden, Foundations of Mechanics, 2nd ed., Benjamin/Cummings, Reading, 1978.

[2] R. Atkins, and Zhong Ge, *An inverse problem in the calculus of variations and the characteristic curves on SO(3)-bundles,* preprint.

[3] S. Benzel, Z. Ge and C. Scovel, *Elementary construction of higher order Lie-Poisson integrators,* Phys. Letter A 174(1993) 229-232.

[4] E. Cartan, *Lecons sur les invariants integrax,* Hermann, Paris, 1972.

[5] P. J. Channell and J. C. Scovel, *Integrators for Lie-Poisson systems,* Physica D, 50(1991) 80.

[6] Feng Kang, *Difference schemes for Hamiltonian formalism and symplectic geometry*, J. Computational Math., 4(1986), 101-107.

[7] Feng Kang and Qin Meng-zhao, *The symplectic methods for the computation of Hamiltonian equations*, in Lect. Notes in Math., vol. 1297, Numerical Methods for Partial Differential Equations, Springer, 1988.

[8] Z. Ge and J. Marsden, Lie-Poisson Hamilton-Jacobi theory and Lie-Poisson integrators, Phys. Lett. A 133(1988) 135-139.

[9] Z. Ge, Equivariant symplectic difference schemes and generating functions, Physica D 49 (1991), 376-386.

[10] E. L. Ince, *Ordinary differential equations*, Longmans Green, London, 1927.

[11] R. McLachlan, *Explicit Lie-Poisson integration and the Euler equations*, preprint.

[12] N. Kamran, W. F. Shadwick, *A differential geometriccharacterization of the 1st Painleve's transcendent*, J. of Differ. Geom. 22(1985), 139-150.

[13] N. Kamran, K. G. Lamb and W. F. Shadwick, *Local equivalence problem for $d^2x/dt^2 = F(x, t, dy/dt)$* , J. of Differ. Geom. 22(1985), 139-150.

[14] G. Patrick, Two axially axially symmetric coupled rigid bodies: Relative equilibria, stability, bifurcations, and a momentum preserving symplectic integrator, thesis, Berkeley (1990).

[15] S. Reich, *Numerical integration of the generalized Euler equations*, preprint.

Comparison of Even-Order Elliptic Equations

VELMER B. HEADLEY, Department of Mathematics, Brock University, St. Catharines, Ontario, Canada

1. INTRODUCTION

Let Ω be an unbounded open subset of R^n, and let L be a uniformly strongly elliptic differential operator defined by

$$(1) \qquad L u := \sum_{|\alpha|,|\beta|=0}^{m} (-1)^{|\alpha|} D^\alpha \left[A_{\alpha\beta}(x) D^\beta u \right] \qquad \left(x \in \Omega \subseteq R^n \right),$$

where the coefficients $A_{\alpha\beta}(x)$ $\left(|\alpha| \le m, |\beta| \le m, x \in \Omega \right)$ are real-valued, satisfy the symmetry conditions $A_{\alpha\beta}(x) = A_{\beta\alpha}(x)$ $\left(|\alpha| \le m, |\beta| \le m, x \in \Omega \right)$, and are sufficiently smooth on Ω.

The *ellipticity constant* E_0 is defined by

$$(2) \qquad 0 < E_0 := \inf \left\{ \sum_{|\alpha|=|\beta|=m} A_{\alpha,\beta}(x) \xi^{\alpha+\beta} : x \in \Omega, \ \xi \in R^n, |\xi| = 1 \right\}.$$

If k is any positive integer, and if G is any nonempty open subset of Ω, we define the seminorm $|\cdot|_{k,G}$ and the norm $\|\cdot\|_{k,G}$ as in [2], and we let $H_0^k(G)$ denote the completion of $C_0^\infty(G)$ with respect to the norm $\|\cdot\|_{k,G}$.

DEFINITION 1.1. If G is a bounded nonempty open subset of Ω that satisfies the hypotheses of [2, Lemma 9.1], and if the differential equation

$$(3) \qquad\qquad L u = 0$$

has a nontrivial solution u in $H_0^m(G) \cap C^{2m}(G)$ such that each $D^\alpha u$ $\left(|\alpha| \le m-1 \right)$ is continuous in some neighbourhood of each point of the boundary of G, then G is called a *nodal domain* for L. We will say that (3) is *nodally oscillatory* in Ω iff, for every $r > 0$, the region $\left\{ x \in \Omega : |x| > r \right\}$ contains at least one nodal domain for L.

DEFINITION 1.2. Following [10], we will say that a nontrivial C^{2m} solution of (3) is *oscillatory* in Ω iff the set $\left\{ x \in \Omega : u(x) \ne 0 \right\}$ is unbounded and is expressible (see [5, Theorem 4.44]) as the union of a countably infinite collection, $\left\{ G_k : k \ge 1 \right\}$, of mutually disjoint, connected, bounded, open sets such that:

(i) $\|u\|_{m,G_k} < \infty$ ($k \geq 1$);

(ii)   given any $r > 0$,   there exists at least one G_k contained in the region $\{x \in \Omega : |x| > r\}$.

Note that if u is oscillatory in the sense of Definition 1.2, then its zero-set $\{x \in \Omega : u(x) = 0\}$ is unbounded, and therefore u is also oscillatory in the traditional sense. Observe also that in the case where $n = 1$, each G_k in Definition 1.2 is a bounded open interval whose endpoints are zeros of u.

REMARK 1.3.   It is known [3] that if $m = 1$ and $n \geq 1$, then the equation (3) is nodally oscillatory in Ω if, and only if, every C^2 solution has the property that its zero-set $\{x \in \Omega : u(x) = 0\}$ is unbounded.   In other words, if L is of order two, then nodal oscillation of (3) is equivalent to ordinary oscillation of every C^2 solution. It is also known (see [11] and [9]) that if $m \geq 2$ and $n \geq 1$ (i.e., if L is of order four or higher), then nodal oscillation is not equivalent to ordinary oscillation. In an earlier paper [10, Theorem 2.4] we obtained sufficient conditions under which the existence of at least one nontrivial oscillatory solution of the equation

(4) $$M u := (-1)^{|m|} \sum_{|\alpha|=|\beta|=m} D^\alpha\left[A_{\alpha\beta}(x)D^\beta u\right] + a_0(x)u = 0 \qquad \left(x \in \Omega \subseteq R^n\right)$$

implies nodal oscillation of the equation (4). In the present paper (see Theorem 2.8) we will extend that result to the case where M is replaced by L. (Note that M is a special case of L.) In Theorem 2.8 and [10, Theorem 2.4], the comparison method employed is a variational method for comparing the smallest eigenvalues of Dirichlet problems of the form $Lu = \lambda u$, $u \in H_0^m(\Omega_t) \cap C^{2m}(\Omega_t)$, where the family $\{\Omega_t : t > 0\}$ is a nested family of bounded nonempty open subsets of Ω having thickness t in the sense of Definition 2.3.

We will also obtain (in Theorem 2.1) a converse to Theorem 2.8.   The order in which these two theorems are presented is a consequence of the fact that the proof of Theorem 2.1 has fewer prerequisites than the proof of Theorem 2.8.

2. MAIN RESULTS

THEOREM 2.1. If (3) is nodally oscillatory in Ω, then it also has at least one nontrivial

oscillatory solution in Ω.

PROOF: If the differential equation (3) is nodally oscillatory in Ω, then it follows from

Definition 1.1 that, given any $r_1 > 0$, we can find a bounded, nonempty, open set G_1

$\subset \{x \in \Omega : |x| > r_1\}$ and a nontrivial function u_1 in $H_0^m(G_1) \cap C^{2m}(G_1)$ such that

$L u_1 = 0$ throughout G_1 and each $D^\alpha u_1$ $(|\alpha| \leq m - 1)$ is continuous in some

neighbourhood of each point of the boundary of G_1.

Since G_1 is bounded, there exists $r_2 > r_1$ such that $G_1 \subset \{x \in \Omega : r_1 < |x| < r_2\}$.

Using once more the hypothesis that (3) is nodally oscillatory in Ω, we can find a

bounded, nonempty open set $G_2 \subset \{x \in \Omega : |x| > r_2\}$ and a nontrivial function u_2 in

$H_0^m(G_2) \cap C^{2m}(G_2)$ such that $L u_2 = 0$ throughout G_2 and each $D^\alpha u_2 (|\alpha| \leq m - 1)$ is

continuous in some neighbourhood of each point of the boundary of G_2.

Thus, given any $r_1 > 0$, we can construct countable families $\{r_k : k \geq 2\}$,

$\{G_k : k \geq 1\}$ and $\{u_k : k \geq 1\}$ such that for every positive integer k we have: $r_{k+1} > r_k$;

$G_k \subseteq \{x \in \Omega : r_k < |x| < r_{k+1}\}$; $L u_k = 0$ throughout G_k; $u_k \in H_0^m(G_k) \cap C^{2m}(G_k)$; each

$D^\alpha u_k (|\alpha| \leq m - 1)$ is continuous in some neighbourhood of each point of the boundary

of G_k; and (by [2, Lemma 9.1]) $D^\alpha u_k = 0$ $(|\alpha| \leq m - 1)$ at each point of the boundary

of G_k. Let $u := \sum_{k \geq 1} u_k \chi_{G_k}$, where χ_{G_k} denotes the characteristic function of the set

G_k. Then it is clear that the set $\{x \in \Omega : u(x) \neq 0\}$ is unbounded and is contained in

$\bigcup_{k=1}^\infty G_k$. Moreover, u is in $H_0^m(G_k) \cap C^{2m}(\Omega)$ ($k \geq 1$) and is a nontrivial solution of (3).

Furthermore, since the function $u : \Omega \rightarrow R^1$ is continuous, therefore

$G_k \cap \{x \in \Omega : u(x) \neq 0\}$ is an open subset of R^n and is expressible [5, Th. 4.44] as the

union of a unique countable collection, $\{G'_{k_j} : j \geq 1\}$, of mutually disjoint open connected

sets. Since each G_k is bounded, therefore each G'_{k_j} is also bounded. Summarizing the

results obtained so far in this proof, we conclude that the set $\{x \in \Omega : u(x) \neq 0\}$ is unbounded and is expressible as the union of a countably infinite collection, $\left\{G'_{k_j} : k \geq 1, j \geq 1\right\}$, of mutually disjoint, connected, bounded, open sets such that :

(i) $\|u\|_{m,G'_{k_j}} < \infty$ $(k \geq 1, j \geq 1)$;

(ii)   given any $r_1 > 0$, we can find at least one G'_{k_j} contained in the set

$$\left\{x \in \Omega : |x| > r_1\right\}.$$

Thus, we have shown that (3) has a nontrivial oscillatory solution in the sense of Definition 1.2 (and, therefore, in the traditional sense).

REMARK 2.2. Our proof of Theorem 2.8 will depend on a comparison method that uses a version of Poincaré's inequality (a generalization of [2, Lemma 7.3] proved in [7] and restated below in Lemma 2.4 (for completeness)), and a version of Courant's minimum principle (stated below in Lemma 2.5). These auxiliary results make use of the following ideas.

DEFINITION 2.3. Let G be a nonempty open subset of Ω. Reformulating a definition given in [10], we will say that G has *bounded thickness* iff we can find a line Γ such that every line Γ' parallel to Γ has the property that every maximal connected subset of $\Gamma' \cap G$ has bounded diameter; the supremum of all such diameters is called the *thickness of G in the direction* Γ. Note that if G is bounded, then it has bounded thickness in every direction; the supremum of all such thicknesses will be called the *thickness* of G.

As examples, we note that the bounded spherical shell $\left\{x \in R^n : r_1 < |x| < r_2\right\}$, where $-\infty < r_1 < r_2 < \infty$, has thickness $2\left(r_2^2 - r_1^2\right)^{1/2}$ in every direction, and that the unbounded cylindrical shell $\left\{(y_1, y_2, ..., y_{n+1}) \in R^{n+1} : r_1 < \left(\sum_{k=1}^{n} y_k^2\right)^{1/2} < r_2\right\}$ has thickness $2\left(r_2^2 - r_1^2\right)^{1/2}$ in any direction orthogonal to the unit vector $(0, ..., 0, 1)$.

LEMMA 2.4. (Poincaré's Inequality) Suppose that G is a nonempty open subset of R^n which has bounded thickness t in the direction Γ, and that there exists a positive integer k_0 such that every set of the form $\Gamma' \cap G$ in Definition 2.3 has at most k_0

maximal connected subsets. Then there exists a positive constant b_0 (depending on m and n) such that for every real-valued function φ in $C_0^\infty(G)$ and every nonnegative integer j in $\{0, 1, ..., m-1\}$ we have

(5) $$|\varphi|_{j,G} \leq b_0(k_0 t)^{m-j} |\varphi|_{m,G}.$$

PROOF. See [7, Lemma 1].

LEMMA 2.5. (Courant's Minimum Principle) Suppose that G is bounded and satisfies the cone condition [1, p. 66] and the hypotheses of Lemma 2.4, and that each of the functions $A_{\alpha,\beta}$ $(|\alpha| \leq m, |\beta| \leq m)$ satisfies the regularity hypotheses of the global version of Gårding's inequality [2, Theorem 7.6] and is in $C^q(G)$, where

$$q > (n/2) + 2m - 1 + |\alpha| + |\beta|.$$

Let F_G be the quadratic functional defined by

(6) $$F_G[L; \varphi] := \sum_{|\alpha|, |\beta| \leq m} \int_G A_{\alpha\beta}(x) D^\alpha \varphi \, D^\beta \varphi \, dx,$$

and let

(7) $$\lambda_0(L; G) := \inf \left\{ |\varphi|_{0,G}^{-2} F_G[L; \varphi] : 0 \neq \varphi \in C_0^\infty(G) \right\}.$$

Then there exists a nontrivial function u_0 in $H_0^m(G) \cap C^{2m}(G)$ such that

(8) $$L u_0 = \lambda_0(L; G) u_0.$$

PROOF. First, we will show that $\lambda_0(L; G)$ is well-defined. Putting $j = 0$ in Lemma 2.4, we obtain, for any real-valued $\varphi \in C_0^\infty(G)$, the following estimate:

(9) $$|\varphi|_{0,G}^2 \leq b_0^2 (k_0 t)^{2m} |\varphi|_{m,G}^2.$$

Furthermore, an appropriate modification of the proof of the global version of Gårding's inequality [2, Theorem 7.6] shows that there exist constants $c_1 \in (0, \infty)$ and $c_3 \in [0, \infty)$ such that, for any real-valued $\varphi \in C_0^\infty(G)$, we have

(10) $$F_G[L; \varphi] \geq c_1 E_0 \|\varphi\|_{m,G}^2 - c_3 |\varphi|_{0,G}^2.$$

The constant c_1 depends on m and n; the constant c_3 depends on $\sup\left\{ |A_{\alpha\beta}(x)| : x \in \Omega, |\alpha| \leq m, |\beta| \leq m \right\}$, m, n, E_0 and the modulus of continuity for the principal coefficients $A_{\alpha,\beta}(x)$ $(|\alpha| = m, |\beta| = m, x \in \Omega)$.

From (9) and (10) we deduce that, for any real-valued $\varphi \in C_0^\infty(G)$, we have:

(11) $$F_G[L; \varphi] \geq \left[b_0^{-2} c_1 E_0(k_0 t)^{-2m} - c_3 \right] |\varphi|_{0,G}^2.$$

It is now clear that $\lambda_0(L;G)$ is well-defined.

The estimate (11) also implies that the operator $\tilde{L}$ generated by L is lower semibounded, hence the results of [12, Section 11] and the fact that $C_0^\infty(G)$ is dense in $H_0^m(G)$ imply that there exists a real-valued function $u_0 \in H_0^m(G)$ such that

(12) $$F_G[L; u_0] = \lambda_0(L;G)$$

and, for all $\psi \in C_0^\infty(G)$, we have

(13) $$\int_G u_0 \cdot (L - \lambda_0 I) \psi = 0.$$

It follows, from (13) and a known regularity theorem [6, Theorem 5. II, p.36], that u_0 is also in $C^{2m}(G)$ and satisfies (8).

LEMMA 2.6. (A Comparison Principle for Eigenvalues)   Let $\{\Omega_t : t > 0\}$ be a family of bounded, open subsets of Ω such that:

(H1)   For each $t \in (0, \infty)$, the open set Ω_t has thickness t and at most a

              finite number of holes, and its Lebesgue measure depends continuously

              on t;

(H2)   If $0 < t_1 < t_2$, then $\Omega_{t_1} \neq \Omega_{t_2}$ and $\Omega_{t_1} \subset \Omega_{t_2}$.

For any $t \in (0, \infty)$, let

(14) $$\mu_0(t) := \lambda_0(L;\Omega_t),$$

where λ_0 is as in Lemma 2.5.   Then the function $\mu_0 : (0, \infty) \to R^1$ is continuous and nondecreasing, and we have

(15) $$\lim_{t \to 0+} \mu_0(t) = +\infty.$$

PROOF.   Using (14), (7), the definition of $F_G[L; \varphi]$ (see (6)), the mean-value theorem for integrals, the continuity hypotheses in Lemma 2.5 on the functions $A_{\alpha,\beta}$ $(|\alpha| \leq m, |\beta| \leq m)$ and the continuity hypothesis in (H1), we see that $\mu_0(t)$ is continuous on $(0, \infty)$.

Furthermore, (14) and (7) imply that for any positive number t we have:

(16) $\mu_0(t) = \inf \left\{ F_{\Omega_t}[L;\varphi] : |\varphi|_{0,\Omega_t} = 1, \; \varphi \in C_0^\infty(\Omega_t) \right\}.$

It follows, from (16) and the hypothesis (H2), that if $0 < t_1 < t_2$, then $\mu_0(t_1) \geq \mu_0(t_2)$.

Hence, $\mu_0(t)$ is nondecreasing on $(0, \infty)$.

Finally, (16) and (11) imply that for any positive number t we have:

(17) $\mu_0(t) \geq b_0^{-2} c_1 E_0 (k_0 t)^{-2m} - c_3.$

From (17) we immediately obtain (15), since b_0, c_1, E_0, k_0 and c_3 do not depend on t,

and since b_0, c_1, E_0 and k_0 are positive.

REMARK 2.7. Our next result, Theorem 2.8, extends [10, Theorem 2.4] by replacing

the differential operator M (defined in (4)) by the more general differential operator

L. Thus, Theorem 2.8 is also a generalization of known one-dimensional results due to

Leighton and Nehari [11, Theorem 3.6] and the author [8, Theorem 4.3]. These known

results and our new result give sufficient conditions under which the existence of a

nontrivial oscillatory solution of an elliptic or ordinary differential equation of order four

or higher implies that the equation is also nodally oscillatory. We note that (22), one

of the hypotheses of Theorem 2.8, is satisfied if a certain symmetric matrix S is

negative semidefinite throughout Ω. The entries of the matrix S consist of the

coefficients $A_{\alpha\beta}(x)$ $\left(|\alpha| \leq m, \; |\beta| \leq m, \; x \in \Omega \right)$ arranged in an appropriate manner. To

construct S, we follow [4] (with minor modifications): Let Z_+^n denote the set of all

n-tuples $\alpha := (\alpha(1), ..., \alpha(n))$ of nonnegative integers, and let $|\alpha| := \sum_{j=1}^{n} \alpha(j)$. For each

p in $\{0, 1, ..., m\}$, let N_p denote the number of distinct elements in the set

$I(p,n) := \left\{ \alpha \in Z_+^n : |\alpha| = p \right\}$, and let $\sigma_p : \{1, 2, ..., N_p\} \to I(p, n)$ be a bijection. Note

that $N_p = \dfrac{(p + n - 1)!}{p!(n-1)!}$, and that

(18) $\displaystyle\sum_{|\alpha|,|\beta|=0}^{m} A_{\alpha\beta} D^\alpha \varphi \, D^\beta \varphi \; = \sum_{p,q=0}^{m} \sum_{|\alpha|=p} \sum_{|\beta|=q} A_{\alpha\beta} D^\alpha \varphi \, D^\beta \varphi$

$\displaystyle = \sum_{p,q=0}^{m} \sum_{i=1}^{N_p} \sum_{j=1}^{N_q} A_{\sigma_p(i), \sigma_q(j)} D^{\sigma_p(i)} \varphi \, D^{\sigma_q(j)} \varphi.$

For each pair of integers p and q in $\{0,1,...,m\}$, let $S(p,q)$ be the $N_p \times N_q$ matrix whose entry in the i-th row and j-th column is $A_{\sigma_p(i),\sigma_q(j)}$. Then

(19)
$$S(p,q) = [S(q,p)]^T,$$

where the superscript "T" denotes "transpose", and

(20)
$$\sum_{p,q=0}^{m} \sum_{i=1}^{N_k} \sum_{j=1}^{N_k} A_{\sigma_p(i),\sigma_q(j)} D^{\sigma_p(i)} \varphi \, D^{\sigma_q(j)} \varphi = \sum_{p,q=0}^{m} [\Phi(p)]^T \, S(p,q) \, \Phi(q),$$

where, for each p in $\{0,1,...,m\}$,

(21)
$$\Phi(p) := \left(D^{\sigma_p(1)} \varphi,...,D^{\sigma_p(N_p)} \varphi \right)^T.$$

Finally, let S be the matrix whose (p, q)-th block is $S(p,q)$. From (18), (19), (20) and (21) we deduce that if the matrix S is negative semidefinite, then the hypothesis (22) will be satisfied. [It is known [4] that if the matrix S is negative semidefinite, then it will be so independently of the specific bijections σ_p ($0 \le p \le m$).]

THEOREM 2.8. Suppose that (3) has at least one nontrivial oscillatory solution in Ω, and that each G_k ($k \ge 1$) in Definition 1.2 has thickness $\tau_k \in (0, \infty)$ and has a family of nonempty open subsets G'_{k,t_k} (having thickness t_k, with $0 < t_k \le \tau_k$ and $G'_{k,\tau_k} := G_k$) which satisfy the hypotheses of Lemma 2.6 (with Ω_t replaced by G'_{k,t_k}). Suppose also that for any bounded, open set $G \subseteq \Omega$, and for any φ in $C_0^\infty(G)$, we have

(22)
$$F_G[L; \varphi] \le 0.$$

Then (3) is nodally oscillatory in Ω.

PROOF. The hypothesis (22) implies that for each positive integer k we have:

(23)
$$\inf\left\{ F_{G_k}[L;\varphi]: \varphi \in C_0^\infty(G_k), \|\varphi\|_{0,G_k} = 1 \right\} \le 0.$$

We recall also that for any bounded, open set $G \subseteq \Omega$ (see (7)),

$$\lambda_0(L;G) := \inf\left\{ |\varphi|_{0,G}^{-2} F_G[L;\varphi] : 0 \ne \varphi \in C_0^\infty(G) \right\}.$$

Moreover, it is easily verified that

(24)
$$\inf\left\{ F_G[L; \varphi] \|\varphi\|_{0,G}^{-2} : 0 \ne \varphi \in C_0^\infty(G) \right\}$$
$$= \inf\left\{ F_G[L; \varphi] : \varphi \in C_0^\infty(G), \|\varphi\|_{0,G} = 1 \right\}.$$

Furthermore, the proof of Lemma 2.5 shows that for each positive integer k we have

(25)
$$\inf\left\{F_{G_k}[L;\varphi]: \varphi \in C_0^\infty(G_k), \|\varphi\|_{0,G_k} = 1\right\}$$
$$\geq \inf\left\{\left[b_0^{-2}c_1 E_0(k_0\tau_k)^{-2m} - c_3\right]|\varphi|_{0,G}^2 : |\varphi|_{0,G} = 1\right\}$$
$$= b_0^{-2}c_1 E_0(k_0\tau_k)^{-2m} - c_3.$$

From (23), (7), (24) and (25) we see that for each positive integer k we have

(26)
$$0 \geq \lambda_0(L;G_k) \geq b_0^{-2}c_1 E_0(k_0\tau_k)^{-2m} - c_3.$$

Using (26), together with the arguments and conclusions of Lemma 2.6 and the definition of each of the families $\left\{G'_{k,t_k} : 0 < t_k \leq \tau_k\right\}$ $(k \geq 1)$, we see that, given any positive integer k, we can find s_k (in the interval $(0, \tau_k]$) and a nonempty open set $G'_{k,s_k} \subseteq G_k$ such that $\lambda_0(L;G'_{k,s_k}) = 0$. It follows from Lemma 2.5 that the equation $Lv_k = 0$ has a nontrivial solution v_k in $H_0^m\left(G'_{k,s_k}\right) \cap C^{2m}\left(G'_{k,s_k}\right)$.

But, by Definition 1.2, given any $r > 0$, there exists at least one G_k contained in the set $\left\{x \in \Omega : |x| > r\right\}$. From this fact and the last two sentences of the preceding paragraph it follows that, given any $r > 0$, there exists a nodal domain for L, namely, G'_{k,s_k}, contained in the region $\left\{x \in \Omega : |x| > r\right\}$. In other words, (3) is nodally oscillatory in Ω.

Acknowledgment. This work was supported by an Operating Grant from the Natural Sciences and Engineering Research Council of Canada.

REFERENCES

1.   R. A. Adams, Sobolev Spaces, Academic Press, New York, 1975.

2.   S. Agmon, Lectures on Elliptic Boundary Value Problems, Van Nostrand, Princeton, NJ, 1965.

3.   W. Allegretto, On the equivalence of two types of oscillation for elliptic operators, Pacific J. Math. 55 (1974), 319-328.

4.   W. Allegretto and C. A. Swanson, Comparison theorems for eigenvalues, Ann. Mat. Pura Appl. (4) 99 (1974), 81-107.

5.   T. M. Apostol, Mathematical Analysis, 2nd ed., Addison-Wesley, Reading, MA, 1974.

6.   G. Fichera, Linear Elliptic Differential Systems and Eigenvalue Problems, Lecture Notes in Mathematics No. 8, Springer-Verlag, Berlin-Heidelberg-New York, 1965.

7.   V. B. Headley, A monotonicity principle for eigenvalues, Pacific J. Math. 30 (1969), 663-668.

8.   V. B. Headley, Sharp nonoscillation theorems for even-order elliptic equations, J. Math. Anal. Appl. 120 (1986), 709-722.

9.   V. B. Headley, Weak and strong oscillation of even-order elliptic and ordinary differential equations, J. Math. Anal. Appl. 143 (1989), 379-393.

10.   V. B. Headley, Nodal oscillation and weak oscillation of elliptic equations of order 2m, Rocky Mountain J. Math. 20 (1990), 1003-1015.

11.   W. Leighton and Z. Nehari, On the oscillation of self-adjoint linear differential equations of the fourth order, Trans. Amer. Math. Soc. 89 (1958), 325-377.

12.   S. G. Mikhlin, The Problem of the Minimum of a Quadratic Functional, Holden-Day, San Francisco, 1965.

Positive Equilibria and Convergence in Subhomogeneous Monotone Dynamics

MORRIS W. HIRSCH University of California at Berkeley,
Berkeley, CA 94720

0   Introduction

The title of this conference, *Comparison Methods and Stability Theory*, is interesting: There are plenty of theories of Stability, but Comparison is too broad to be encompassed by a theory— there are only diverse methods! It is this very diversity— not limited to a fixed set of axioms, or a tightly defined collection of dynamical systems— that is responsible for both the wide applicability of comparison methods, and their diffuse nature.

Comparison methods are ways of systematically organizing and exploiting inequalities between trajectories in the same or different systems. Many of these stem from the classical Comparison Principle for second order quasilinear parabolic PDEs. For systems of ODEs the corresponding fundamental results are the comparison theorems of Müller and Kamke.

As the conference made clear, comparison methods are extremely useful. What is unified about the methods, and what makes them so valuable, is that they permit the use of geometry and topology to answer dynamical questions that are *crude* rather than *detailed*, and are *global* rather than *local*. Standard numerical techniques can calculate explicit trajectories with arbitrary accuracy; but comparison methods can exploit structural features of the equations to answer such questions such as: Is there a positive solution?

For a system of evolution equations that is not too complicated, dynamical systems theory provides a *solution flow* (more precisely, a local semiflow) in any of a number of function spaces. This can be viewed as *geometrifying* the problem, in that it permits the use of geometrical (incuding topological) reasoning. Even if the same results can be obtained more economically with purely analytical arguments, there is often a great virtue in being able to interpret the problem in geometrical terms. This often suggests images and diagrams that in turn lead to important intuitions that are difficult to perceive in a welter of equations.

A simple example: Many minimization theorems can be interpreted as saying that there is a unique pair of points minimizing the distance from a hyperplane to a disjoint closed strictly convex set— the latter statement being visually "obvious", and immediately suggestive of successive approximations for constructing the minimum.

Another example: The existence of a solution of an equation can often be reduced to finding a fixed point of a map. The Leray-Schauder theorem and the attending degree theory is often applicable, rendering not only the existence but often the stability of the fixed point readily ascertainable. The picture that goes along with the Leray-Schauder theorem is the same as for the Brouwer fixed point theorem— to a topologist the picture is suggestive and persuasive! In fact Leray-Schauder is derived from Brouwer; and Brouwer is a generalization of the visually obvious intermediate value theorem.

These examples show how geometrification can lead to a conceptual interpretation of analytic arguments which makes them more comprehensible, even visualizable.

The parabolic and Müller-Kamke comparison principles leads to further geometrification: It means that *the solution flow preserves the natural ordering of the function space*. This is a powerful tool for proving convergence of trajectories to equilibria. Its exploitation requires detailed knowledge of the geometry of the ordered vector space that has been chosen as the domain of the dynamics.

For PDEs this choice is often crucial for the success of comparison methods. It can be a difficult problem to reconcile the often conflicting assumptions needed to justify various desirable features— continuity of solutions in initial values, *a priori* estimates, regularity, strong ordering, normal ordering, etc.

In this article we discuss the dynamics of the class of *subhomogeneous* maps between ordered vector spaces: $f(\lambda x) \leq \lambda f(x)$ if $x \geq 0$ and $\lambda \geq 1$. When combined with monotonicity, this is a very powerful but not uncommon setting in which existence of fixed points and convergence of trajectories can be proved.

Basic definitions and examples are given in Sections 1 and 2, while Sections 3, 4 and 5 present the main results. Some important older results on monotone subhomogeneous maps are discussed in Section 6.

Convergence theorems for subhomogeneous strongly monotone flows in Euclidean spaces have been proved by Khanmy and Gabriel (1989). J. Keener (1993) obtained a fixed point theorem for certain subhomogeneous monotone maps in finite dimensional ordered vector spaces.

For various theories of strongly monotone flows, see the articles by Fusco and Oliva (1988, 1990); Hirsch (1982–1991); Mallet-Paret and Smith (1990); Matano (1986); Mierczynski (1991); Nadirashvili (1993); Smith (1987); Smith and Thieme (1990, 1991). For general results on semilinear parabolic equations see Henry (1981), Amann (1984, 1988, 1990).

The review article of Amman (1976) contains a great deal of basic material on monotone maps in functional analysis, while Amman (1976a) contains many applications to elliptic and parabolic equations. More recent results on monotone maps are found in Alikakos, Hess and Matano (1989); Dancer and Hess (1991); Hirsch (1985); Polačik (1989); Takáč (1990).

1   Ordered Topological Vector Spaces

Let X denote a topological vector space ordered by a closed convex cone X_+ such that if x and $-x$ are both in X_+ then $x = 0$. The interior of X_+ is denoted by X_{++}.

The basic order relation $y \geq x$ in X is defined to mean $y - x \in X_+$. Other relations are defined as follows:

- $y \succ x$ means $y \geq x, y \neq x$.

- $y \gg x$ means $y - x \in X_{++}$.

- $x \leq y$ means $y \geq x$; similarly for $\prec$ and $\ll$.

- For subsets $A, B \subset X$ we write $A \leq B$ if $a \leq b$ for all $a \in A$, $b \in B$; similarly for $\prec$ and $\ll$.

A subset S of X is *order convex* if $a \leq b \leq c$ and $a, c \in N$ implies $b \in N$. Such a set need not be convex.

If A, B are subsets of X and $A \leq B$, then we define the (generalized) *order interval*

$$[A, B] = \{x \in X : A \leq x \leq B\}.$$

This set is convex and order convex, and closed if A and B are compact.

A subset S of X is *order bounded* if there are nonempty compact subsets $A \leq B$ in X such that $S \subset [A, B]$.

If X_{++} is nonempty, then X_+ is a *solid* cone and X is *strongly ordered*.

The cone X_+ is *normal* provided the topology of X is defined by a norm $|| \cdot ||$ such that $0 \leq x \leq y$ implies $||x|| \leq ||y||$. Although normality is a common hypothesis, I do not use it for any of the main results.

EXAMPLE 1.1 Euclidean space $\mathbf{R}^n$ is given the *vector order* defined by the cone $\mathbf{R}^n_+$ of vectors with no negative components. There are many other ways to order $\mathbf{R}^n$, for example by the *light cone* $\{x \in \mathbf{R}^n : x_1 \geq \sqrt{x_2^2 + \cdots + x_n^2}\}$.

EXAMPLE 1.2 Let $\overline{\Omega} \subset \mathbf{R}^n$ be a nonempty smooth compact n-dimensional submanifold with interior Ω and boundary $\text{bd}\Omega$.

For $k = 0, 1$ let $C^k(\overline{\Omega})$ denote the Banach space of real-valued C^k maps on $\overline{\Omega}$, endowed with the natural (pointwise) order on functions, and one of the standard C^k norms. These ordered Banach spaces are strongly ordered, and normal for $k = 0$ but not for $k = 1$.

Let $C_0^k(\overline{\Omega}) \subset C^k(\overline{\Omega})$ denote the closed linear subspace of functions vanishing on $\text{bd}\Omega$. Then $C_0^1(\overline{\Omega})$ is strongly ordered, but $C_0^0(\overline{\Omega})$ is not. If $v \in C_0^1(\overline{\Omega})$, then $v \gg 0$ if and only if: $v(x) \geq 0$ for all $x \in \overline{\Omega}$ and $\partial v/\partial \eta < 0$, where η denotes the outward unit vector field normal to $\text{bd}\Omega$. Notice that it can happen that $v \gg 0$ even though $v(x) = 0$ for some or all $x \in rmBd\overline{\Omega}$! This is because we are using the C^1 topology.

EXAMPLE 1.3 Let M be a C^∞ compact Riemannian manifold. Let ξ be a nowhere vanishing vector field along the boundary $\text{bd}M$, transverse to $\text{bd}M$ and pointing out of M. For $k = 0, 1$ define $C_B^k(M)$ to be the ordered Banach space of C^k functions $u : M \to \mathbf{R}$ satisfying the following boundary condition if $\text{bd}M$ is nonempty:

$$Bu := bu|\text{bd}M + \delta\frac{\partial u}{\partial \xi} = 0;$$

where either $\delta = 0$ and $b = 1$ (Dirichlet boundary operator), or else $\delta = 1$ and $b : M \to \mathbf{R}$ is continuous and nonnegative (Neumann or regular oblique derivative boundary operator). Then $C_B^k(M)$ is strongly ordered, with $u \gg 0$ if and only if: $u > 0$ on $M \backslash \text{bd}M$ and $[\partial u/\partial \xi](p) < 0$ at every point $p \in \text{bd}M$ at which $u(p) = 0$.

EXAMPLE 1.4 Consider the ordered topological vector space $X = C_0^0(\overline{\Omega})$ of Example 1.2, which is normal but not strongly ordered. Fix a function $e \succ 0$, called an *order unit*, and define

$$C_e(\overline{\Omega}) = \{u \in X : \exists \alpha, \beta \in \mathbf{R} \text{ with } \alpha e \leq u \leq \beta e\}.$$

Define a new norm on $C_e(\overline{\Omega})$: For $u \in C_e(\overline{\Omega})$,

$$|u|_e = \inf\{\lambda \in \mathbf{R}_+ : u \in [-\lambda e, \lambda e]\}.$$

Giving $C_e(\overline{\Omega})$ its natural ordering as a function space, we obtain a normal, strongly ordered Banach space.

2   Monotone Subhomogenous Maps

A map f between subsets of ordered topological vector spaces is *monotone* if $x \leq y$ implies $f(x) \leq f(y)$. It is *strongly monotone* in case $x \prec y$ implies $f(x) \ll f(y)$; in this case X must of course be strongly ordered. If $x \leq y$ implies $f(x) \geq f(y)$ then f is *antimonotone*.

The map f is *subhomogeneous* $f(\lambda x) \leq \lambda f(x)$ whenever $x \in X_+$ and $\lambda \geq 1$. It is *strongly subhomogeneous* provided $f(\lambda x) \ll \lambda f(x)$ whenever $x \in X_{++}$ and $\lambda > 1$.

If $x \geq 0$ implies $f(x) \geq 0$, I say f *preserves positivity*.

If f maps order bounded sets into precompact sets then f is called *order compact*.

Any of these properties is ascribed to a flow $\Phi = \{\Phi_t\}$ provided it holds for each map Φ_t, $t > 0$.

EXAMPLE 2.1 (a) It is obvious but useful that linear maps are subhomogeneous. The pointwise limit of subhomogeneous maps is subhomogeneous. Antimonotone maps are subhomogeneous.

(b) A map $f : \mathbf{R} \to \mathbf{R}$ is subhomogeneous if and only if $f(s)/s$ is nonincreasing for $s > 0$. If f is differentiable it is subhomogeneous if and only if $f'(s) \leq f(s)/s$ for $s > 0$.

(c) Suppose f, g are both subhomogeneous, and the composition $f \circ g$ is defined. Then $f \circ g$ is subhomogeneous provided g preserves positivity.

(d) The sum of subhomogeneous maps is subhomogeneous, and strongly subhomogeneous if one of the summands is.

EXAMPLE 2.2 Suppose E, F are vector spaces of real-valued functions on some set Ω, each ordered as usual by the cone of functions ≥ 0. Let $f : E_+ \to F_+$ be a map having the form $[f(u)](x) = u(x)g(u(x))$, $(u \in E_+, x \in \Omega)$, where $g : \mathbf{R} \to \mathbf{R}$ is nonincreasing. Then f is subhomogeneous.

EXAMPLE 2.3 Recall the function space $C_B^k(M)$ from Example 1.3 If $g : \mathbf{R}_+ \to \mathbf{R}_+$ is C^k, nondecreasing, then the map

$$f : C_B^k(M) \to C_B^k(M), \ f(u) : x \mapsto u(x)g(u(x))$$

is monotone and subhomogeneous.

EXAMPLE 2.4 Let $\Omega \subset \mathbf{R}^n$ be as in Example 1.2, with bdΩ a smooth manifold of differentiability class $C^{2+\mu}, 0 < \mu < 1$.

Let B denote a Dirichlet or Neumann boundary operator as in Example 1.3, taking $M = \overline{\Omega}$. Let A be a second order differential operator of the form

$$A = \sum_{i,j} a_{ij} D_i D_j + \sum a_i D_i + a$$

where $D_i = \partial/\partial x_i$, the coefficients a_{ij}, a_i, a are C^μ (i.e., μ-Hölder) functions on $\overline{\Omega}$, and the matrices $[a_{ij}(x)]$, $x \in \overline{\Omega}$ are positive definite. Assume $a > 0$ in the Neumann case $(b = 0)$.

It is well known that there is a linear *solution operator* $S := S_{A,B}$ on $C^0(\overline{\Omega})$ such that for every $v \in C^\mu(\overline{\Omega})$, the unique solution of the elliptic boundary-value problem

$$Au = v, \tag{1}$$
$$Bu = 0 \tag{2}$$

is given by $u = Sv$. Moreover S is strongly monotone as a map from $C^0(\overline{\Omega})$ to $C_e(\overline{\Omega})$ (see Example 1.4), where e is the unique solution of the boundary value problem

$$Au = 1,$$
$$Bu = 0.$$

This strong monotonicity means that for every $u \succ 0$ there are positive constants α, β such that $\alpha u \leq Su \leq \beta u$.

EXAMPLE 2.5 Let the differential operators A, B be as above, with corresponding solution operator $S := S_{A,B}$. For a locally Lipschitz map $h : C^1(\overline{\Omega}) \times \mathbf{R} \to \mathbf{R}$, define the corresponding *Nemytskii operator* $H := H_h$ by:

$$[H(u)](x) = h(x, u(x)), \ (u \in C^0(\overline{\Omega}), \ x \in \overline{\Omega}).$$

Consider the semilinear elliptic boundary value problem:

$$Au = H(u), \tag{3}$$
$$Bu = 0. \tag{4}$$

Let T denote the composition $S \circ H : C^0(\overline{\Omega}) \to C^0(\overline{\Omega})$. It is well-known that (3), (4) are equivalent to the *fixed point problem*

$$Tu = u, \ u \in C^0(\overline{\Omega}).$$

The map T is known to be strongly monotone (by the strong maximum principle). Since S takes values in $C_e(\overline{\Omega})$, we can consider T as a strongly monotone operator on $C_e(\overline{\Omega})$. If the map $\mathbf{R} \to \mathbf{R}$, $y \mapsto h(x,y)$ is subhomogeneous for each fixed $x \in \overline{\Omega}$, then it is easy to see that T is subhomogeneous.

EXAMPLE 2.6 Let M be a compact Riemannian manifold and B a boundary operator as in Example 1.2. Denote by A a second order elliptic operator on M, expressed in local coordinates as A in Example 2.4. Let $h : M \times \mathbf{R} \to \mathbf{R}$ be locally Lipschitz. Consider the parabolic initial-boundary value problem:

$$\frac{\partial u}{\partial t} = Au + h(x,u), \ (x \in \overline{\Omega}, \ t > 0); \tag{5}$$

$$u(x,t) = 0, \ (x \in \mathrm{bd}\Omega, \ t \geq 0) \tag{6}$$

$$u(x,0) = v(x), \ (x \in \overline{\Omega}). \tag{7}$$

It is known that there are solution flows $S^k = \{S_t^k\}_{t \geq 0}$ in C_0^k, $k = 0,1$, so that the solution to (5), (6), (7)is given by by $u(x,t) = [S_t^k v](x)$.

These flows are monotone (by the maximum principle), S^1 is strongly monotone (by the strong maximum principle), and S^0 is order compact (see e. g. Hirsch (1988).) When h is subhomogeneous in the second variable then these flows are subhomogeneous.

EXAMPLE 2.7 Let $G : \mathbf{R}^n \to \mathbf{R}^n$ be a C^1 vector field which is *quasimonotone* (also called *cooperative*): the off-diagonal entries in the Jacobian matrices $Dg(x)$ are nonnegative. Then the solution flow of the differential equation $dx/dt = G(x)$ is monotone, and strongly monotone if these matrices are irreducible except at a countable subset of $\mathbf{R}^n$. If also G is [strongly] subhomogeneous, then the flow is [strongly] subhomogeneous; see Theorem 3.1 below.

EXAMPLE 2.8 Dancer and Hess (1991) consider semilinear parabolic time-periodic equations in a smooth bounded domain Ω, with time-periodic Dirichlet or Neumann boundary conditions:

$$\frac{\partial u}{\partial t} = A(t)u + h(x,t,u),$$

$$B(t)u = 0$$

where the operator

$$A(t) := \sum_{j,k} a_{jk} D_j D_k + \sum_j a_j D_j u + a$$

and the boundary operator $B(t)$ have coefficients depending continuously on (x,t) and periodically on t, and the map $h : \overline{\Omega} \times \mathbf{R} \times \mathbf{R} \to \mathbf{R}$ is likewise periodic in t.

They show, under reasonable assumptions too long to quote here, that there is a solution flow in the Cartesian product of a circle with an order interval $V = [u, v] \subset Y^\beta, 0 < \beta < 1$. Here $Y^\beta \subset C^{1+\mu}(\overline{\Omega})$, $0 < \mu \leq 1$, is a fractional power space induced by the operator pair $(A(0), B(0))$ in $Y = L^p(\Omega), p > n$. They prove that

(i) the corresponding Poincaré map $P : V \to V$ is strongly monotone; and

(ii) $P(V)$ has compact closure in Y^β.

It is easy to see that P is subhomogeneous provided the nonlinearity $h(x, t, y)$ is subhomogeneous in $y \in \mathbf{R}$ for each pair $x \in \Omega, t \in \mathbf{R}$.

Compare Theorem 3.2 for a related result in an abstract setting.

3   Subhomogeneous Monotone Flows

Khanmy and Gabriel (1989) introduced monotone flows generated by subhomogeneous vector fields in $\mathbf{R}^n$ and obtained convergence results. In this section I show under mild assumptions that the flow of a subhomogeneous, locally quasimonotone vector field in a Banach space is monotone and subhomogeneous.

Let E be a Banach space and $E_0 \hookrightarrow E$ another Banach space identified with a linear subspace of E by a continuous monotone injection having dense range. Consider a differential equation

$$\frac{du}{dt} = Au + g(u), \ t > 0 \tag{8}$$

where $A : E_0 \to E$ is a linear map which generates an analytic semigroup $\{e^{tA}\}$ (e. g. a bounded operator), and $g : E_0 \to E$ is a locally Lipschitz map, generally nonlinear. The corresponding integral equation is

$$u(t) = e^{tA}u(0) + \int_0^t e^{(t-s)A}g(u(s))ds; \tag{9}$$

Assume there is a solution flow Φ to these equations (Henry 1981). This means $\Phi_t(u) \in E_0$ for $t > 0, u \in E$ and $d\Phi_t(u)/dt = g(u(t))$ for $u \in E, t > 0$; and $\Phi_t(u)$ is jointly continuous in (t, u).

THEOREM 3.1 *Besides the foregoing assumptions, suppose:*

(a) E_0 *is ordered and E is strongly ordered;*

(b) e^{tA} *is monotone for all $t \geq 0$;*

(c) $g : E_0 \to E$ is subhomogeneous and $g(0) \geq 0$;

(d) g is locally quasimonotone: For every $u \in E_0$ there exists a real number $c > 0$ such that $g + cI$ is monotone in some neighborhood of u (where I is the inclusion of E_0 in E).

Then $\Phi_t|E_+$ for $t > 0$ is monotone and subhomogeneous. It is strongly subhomogeneous if g is strongly subhomogeneous, and strongly monotone if e^{tA} is strongly monotone for all $t > 0$.

Before beginning the proof I recall the conditions under which the solution flow is monotone or strongly monotone, ignoring subhomogeneity.

A well-known condition equivalent to monotonicity of e^{tA}, $t \geq 0$ is that the resolvents $R(c, A) = (cI - A)^{-1}$ are monotone for all sufficiently large scalars $c > 0$. A sufficient condition for strong monotonicity is that E be one of the function spaces $C_B^m(M)$ defined in Example 2.3 above, and A be a second-order, uniformly strongly elliptic differential operator (e. g. the Laplacean); this follows from the standard maximum principles for parabolic equations. It is not sufficient that the resolvents be strongly monotone (Hirsch 1988). From the integral equation (9) one can see that the solution flow $\{\Phi_t\}$ to Equation (8) is monotone under assumptions $(a), (b)$ and (d), and strongly monotone provided e^{tA} is strongly monotone for all $t > 0$.

Proof of Theorem 3.1 For any $v \in E_0$, we use assumption (d) to choose $c > 0$ so large that $g + cI$ is monotone in a neighborhood of v. We then replace g by $g + cI$ and A by $A - cI$ and observe that all the hypotheses are still valid. Henceforth we assume g is monotone in a neighborhood N of v.

From (9) we see that for sufficiently small $t > 0$ we can express $\Phi_t|N$ (possibly after shrinking N) as the uniform limit as $k \to \infty$ of nonlinear operators $\Phi_{k,t}$ defined by the formula

$$\Phi_{k+1,t}v = e^{tA}v + \int_0^t e^{(t-s)A} g(\Phi_{k,s}v)ds \tag{10}$$

with $\Phi_{0,t}v = v$.

By induction on k we shall prove that for all $k \in \mathbf{N}$, $\Phi_{k,s}$ is subhomogeneous and monotone, and maps points of E_+ into E_+.

The inductive hypothesis that e^{rA} is monotone for all $r > 0$, together with the inductive assumption applied to the right hand side of Equation 10, expresses $\Phi_{k+1,t}$ as a monotone map. Therefore $\Phi_{k+1,t}$ is monotone. Since $g(0) \geq 0$ it follows that $\Phi_{k+1,t}v \geq 0$ when $v \geq 0$.

To complete the induction we exploit the inductive assumption that $\Phi_{k,s}$ is subhomomogeneous for all s.

Assume $v \geq 0$ and fix the neighborhood N as above. Given k, for sufficiently small $t > 0$ and $\lambda > 1$ sufficiently close to 1, we have $\Phi_{k,t}\lambda v \in N$. Therefore by monotonicity of $g|N$ and the inductive hypotheses we have for $0 \leq s \leq t$: $g(\Phi_{k,s}\lambda v) \leq g(\lambda\Phi_{k,s}v)$, using the fact that

$\Phi_{k,s}$ preserves positivity. Since g also preserves positivity and is subhomogeneous, it follows that $g(\Phi_{k,s}(\lambda v)) \leq \lambda g(\Phi_{k,s}v)$. Applying (10) to λv we therefore get:

$$\begin{aligned} \Phi_{k+1,t}(\lambda v) &\leq \lambda e^{tA}v + \int_0^t e^{(t-s)A}\lambda g(\Phi_{k,s}v)ds \\ &= \lambda\Phi_{k+1,t}v, \end{aligned}$$

completing the induction.

Since a limit of subhomogeneous maps is subhomogeneous it follows that Φ_t is subhomogeneous in N for sufficiently small t. By looking at a compact interval in the ray through v we see that Φ_t is subhomogeneous.

Now Φ_s for any $s > 0$ is the composition of a finite number of maps, each of the form Φ_r with $0 < r < t$. Since a composition of monotone, subhomogeneous, positivity preserving maps is subhomogeneousand positivity preserving, this shows Φ_s is subhomogeneous and positivity preserving for all $s \geq 0$. The same kind of argument shows that Φ_s is strongly subhomogeneous whenever g is.

When A generates a strongly monotone semigroup then strong monotonicity of Φ_t now follows from the integral equation (9). **QED**

The foregoing proof extends to nonautonomous systems of the type

$$\frac{du}{dt} = A(t)u + g(t,u),\ t > 0 \tag{11}$$

where each $A(s) : E_0 \to E$ is linear and continuous and generates (for each fixed $s \geq 0$) a monotone analytic semigroup $\{e^{tA(s)}\}_{t \geq 0}$ in E; and $g : \mathbf{R} \times E_0 \to E$ is locally Lipschitz in u and locally quasimonotone in u, where "locally" refers to neighborhoods of (t,u). Assume that $e^{tA(s)x}$ is continuous in (t,s,x). There is a solution process Ψ composed of maps $\Psi_{s,t} : E \to E$, indexed by pairs $s \in \mathbf{R}, t \geq 0$, such that for $v \in E$ the solution to (11) with initial value $u(s) = v$ is $t \mapsto \Psi_{s,s+t}u$. As before, assume E_0 is ordered and E is strongly ordered.

THEOREM 3.2 *Besides the foregoing, assume that for each $t \geq 0$ the map $u \mapsto g(t,u)$ is subhomogeneous, and that $g(t,0) \geq 0$. Then each map $\Psi_{s,t}|E_+$ is monotone and subhomogeneous. It is strongly subhomogeneous if g_t is strongly subhomogeneous for all $t > 0$, and strongly monotone if $e^{tA(s)}$ is strongly monotone for all $t > 0$, $s > 0$.*

The proof, similar to that of Theorem 3.1, is left to the reader.

REMARK 3.3 The preceding theorem applies in particular to the case where $A(t)$ and $g(t,u)$ have period $\omega > 0$ in t. It then follows that *the period (or Poincaré) map $\Psi_{0,\omega} : E \to E$ has the monotonicity and subhomogeneity specified in Theorem 3.1.*

4 Existence of Fixed Points Without Solidity or Normality

In Sections 4 and 5 the following standing assumptions are made:

Hypotheses 4.0

(a) X denotes an ordered topological vector space.

(b) $N \subset X_+$ is a nonempty, closed, order convex set.

(c) $f : N \to X_+$ is a continuous, monotone, subhomogeneous map.

(d) $K = f(K) \subset N$ is a nonempty compact invariant set.

It then follows that each iterate f^k is continuous, monotone and subhomogeneous on its domain, and $f^k(K) = K$.

The point of the following theorem is to prove the existence of a fixed point for a monotone homogeneous map f under fairly minimal assumptions, and to make a crude estimate of its location. There are no assumptions of solidity or normality of X. I assume here that f is order compact, however. While most of the theorems in Section 5 do not assume this, they do have "strong" hypotheses on f. All the results presuppose the existence of a compact invariant set.

A similar result with slightly different assumptions was proved by R. Nussbaum in his interesting study, *Hilbert's Projective Metric and Iterated Nonlinear Maps*, Memoirs of the American Mathematical Society 75 (1988), No. 391, Theorem 4.1.

THEOREM 4.1 *Besides Hypothesis 4.0, assume:*

(a) *X is a Banach space.*

(b) *$f(N)$ has compact closure in X.*

(c) *There exist real numbers $\beta \geq 1 \geq \alpha \geq 0$ such that*

$$K \subset [\alpha K, \beta K] \subset N.$$

Then f has a fixed point. Moreover:

(d) *if $u, v \in X$ are such that $u \leq f^n(\alpha K)$ and $v \geq f^n(\beta K)$ for all n, then there is fixed point in $[u, v]$.*

Proof   The strategy is the standard one of finding a positively invariant compact convex set $C \subset N$, and then applying the Schauder-Tychonoff fixed point theorem. Since f may be nonlinear, and N may be nonconvex, we have to exploit subhomogeneity to find such a C.

Fix an arbitrary $y \in K$. I claim that for all $n \in \mathbf{N}$, $f^n(\alpha y)$ is defined and contained in N, and

$$\alpha f^n(y) \leq f^n(\alpha y) \leq K. \tag{12}$$

We proceed by induction. Suppose this holds for a particular $n = k - 1 \geq 0$, the case $n = 0$ following from hypothesis (c).

Then $f^k(\alpha y)$ is defined. Since $\alpha y \leq K$ and f^k is monotone, $f^k(\alpha y) \leq f^k(K) = K$. By subhomogeneity of f^k we have $f^k(\alpha y) \geq \alpha f^k(y)$. Since the latter is in αK, it is in N. Thus $f^k(\alpha y) \in [\alpha f^k(y), K]$, and this order interval is contained in N by order convexity of N. This completes the induction.

Since the closure of $f(N)$ is compact and N is closed, it follows that the orbit of αy lies in a compact subset of N. This implies that the omega-limit set $\omega(\alpha y)$ is a nonempty compact invariant subset of N, and $\omega(\alpha y) \leq K$.

A similar proof shows that $\omega(\beta y)$ is a nonempty compact invariant subset of N, and $\omega(\beta y) \geq K$.

The set $B = [\omega(\alpha y), \omega(\beta y)]$ is therefore a nonempty closed convex subset of N, and the closure D of $f(B)$ is a compact positively invariant subset of B. Because X is a Banach space, the closed convex hull C of D is compact; and it is easy to see that C is positively invariant. The Schauder-Tychonoff theorem now implies f has a fixed point in B.

Notice that (12) implies that $u \leq f^n(\alpha y)$ for all n. Similarly $v \geq f^n(\beta y)$. Therefore the orbits of αy and βy lie in the closed set $[u, v]$; so therefore do their omega limit sets. This shows that $B \subset [u, v]$, whence $[u, v]$ contains a fixed point.   **QED**

5   Subhomogeneous Maps in Strongly Ordered Spaces

In this section X is a strongly ordered topological vector space.

Note that in the following theorem no compactness assumptions are made concerning f, except the existence of a compact invariant set. Similar results under different hypotheses were obtained by P. Takáč (1990).

THEOREM 5.1 *In addition to Hypotheses 4.0, assume that one of the following two conditions holds:*

(a) *f is monotone and strongly subhomogeneous;*

(b) *f is strongly monotone and subhomogeneous.*

Let $K = f(K)$ be a nonempty compact invariant set, $K \subset N \cap X_{++}$. Then there are fixed points $p, q \in K$ such that $p \leq K \leq q$. Moreover the set of fixed points has the form $Jp = \{\rho p : \rho \in J\}$ where $J \subset \mathbf{R}_+$ is a closed interval (possibly infinite, possibly degenerate) containing 1.

COROLLARY 5.2 *Every minimal set— in particular every periodic orbit— in X_{++} consists of a fixed point.*

The proof of Theorem 5.1 is based on the following simple fact:

LEMMA 5.3 *If $p \in K$ and $p \leq K$ then $p = f(p)$.*

Proof $f(p) \leq p$ because f is monotone and $f(K) = K$, so $f(p) \leq f(K) = K$, and $p \in K$. And $f(p) \geq p$ because $f(p) \in K$ and $K \geq p$. **QED**

This shows more generally that if a monotone map f has an invariant set $K = f(K)$ which contains its greatest lower bound (or least upper bound) p, then p is a fixed point.

Proof of Theorem 5.1. Define sets:

$$M = \{x \in X_{++} : x \leq K\},$$

$$S = \{x \in M : y \gg x \Rightarrow y \notin M\}.$$

Notice that M is positively invariant $(f(M) \subset M)$. And no two points of S are related by $\gg$; we say S is $\gg$-*balanced*.

For each $x \in K$ let R_x denote the ray through x (issuing from the origin).

Because X is strongly ordered and K is a compact subset of X_{++}, it is easy to see that $R_x \cap M$ is a closed interval whose upper endpoint is the intersection $S \cap R_x$. Call this point of intersection Tx.

It can be shown that the resulting map $T : K \to S$ is continuous. The proof depends on the fact that the natural map from E_0 to the space of oriented lines through the origin is continuous; it also utilizes compactness of K and the $\gg$-balanced property of S.

It follows that there is a positive continuous function $\alpha : K \to \mathbf{R}$ such that $Tx = \alpha(x)x$, with $0 < \alpha(x) \leq 1$. Let μ be the maximum value of α. Then $0 < \mu \leq 1$.

Conclusion (i) of Theorem 5.1 follows from Lemma 5.3 once we prove $\mu = 1$. We proceed by contradiction: Assume $0 < \mu < 1$.

Fix $a \in K$ with $\alpha(a) = \mu$. Then $Ta = \mu a \in S$. By subhomogeneity we have $f(\mu a) \geq \mu f(a)$.

By maximality of μ we have $\mu f(a) \geq \alpha(f(a))f(a) = Tf(a)$. Thus we have $Tf(a) \leq f(\mu a)$. Moreover, by monotonicity of f and invariance of K we get $f(\mu a) \leq K$. Thus

$$Tf(a) \leq f(\mu a) \leq K. \tag{13}$$

The key observation is this: *If f is strongly subhomogeneous then the first order relation in (13) is $\ll$. If f is strongly monotone then the second order relation is $\ll$.* But in either case it follows that there exists $y \gg Tf(a)$ such that $y \in M$, contradicting $Tf(a) \in S$. Therefore it must be that $\mu = 1$.

Now suppose ρp is fixed for some $\rho > 1$. Then νp is fixed for all $\nu \in [1, \rho]$. For $f(\nu p) \leq \nu f(p) = p$; but also $f(\nu p) = f((\nu/\rho)\rho p) \geq (\nu/\rho)f(\rho p) \geq (\nu/\rho)\rho p = \nu p$, proving $f(\nu p) \geq \nu p$. So νp is fixed. This and a similar argument for σp, $0 < \sigma < 1$, shows:

There is a closed interval $J \subset \mathbf{R}_+$ such that λp is fixed if and only if $\lambda \in J$.

Notice that the proof of this used only subhomogeneity.

Now let z be any fixed point. Applying what has been already proved to the compact set $\{p, z\}$, we see that $z \in Jp$. This completes the proof of Theorem 5.1.

CONJECTURE 5.4 The conclusion of Theorem 5.1 is true even if $0 \in K$, provided that $K \backslash 0 \subset X_{++}$.

THEOREM 5.5 *Assume that f is monotone and strongly subhomogeneous and suppose there exists a compact nonempty invariant set $K \subset X_{++}$. Then K consist of a single fixed point $p \gg 0$, and every trajectory with compact closure in X_{++} converges to p.*

Proof By Theorem 5.1 there are fixed points $p, \lambda p \gg 0$ such that $p \leq K \leq \lambda p$, where $\lambda \geq 1$. But strong subhomogeneity implies that if $\lambda \neq 1$, then $f(\lambda p) \neq p$. Thus $K = \{p\}$. The convergence statement follows by taking K to be any compact omega limit set in X_{++}. **QED**

THEOREM 5.6 *Assume f is strongly monotone and subhomogeneous. If x has compact orbit closure in X_{++} then $f^n(x)$ converges to a fixed point.*

Proof By Theorem 5.1 the set of fixed points is has the form Jp for some fixed point p and some closed interval $J \subset [0, \infty)$ containing 1. Consider any $x \in N$ having compact orbit closure. There is a maximal number $\alpha \geq 0$ and a minimal number $\beta \geq \alpha$ such that the omega limit set L of x lies the order interval $[\alpha p, \beta p]$.

I claim $\alpha = \beta$. For if L is not a singleton, L cannot contain αp or βp: An omega limit set for a strongly monotone flow cannot contain two points ordered by $\prec$ (Hirsch 1988). Thus if $\alpha \neq \beta$ we assume $\alpha p \prec L \prec \beta p$. By invariance of L and strong monotonicity we get

$$f(\alpha p) \ll L \ll f(\beta p)$$

Now either $\beta \succ J$ or $\alpha \prec J$ or $[\alpha, \beta] \subset J$. Suppose $\beta \succ J$. Above I showed that this implies $f(\beta p) \leq \beta p$. But $\beta \succ J$ also implies, by strong monotonicity:

$$f(\beta p) \gg f(Jp) = Jp.$$

Since $p \in J$ this gives $f(\beta p) \gg \beta p$, a contradiction. If $\alpha \prec J$ the argument is similar.

If $[\alpha, \beta] \subset J$ then $\alpha p \prec L \prec \beta p$. But strong monotonicity now implies

$$\alpha p = f(\alpha p) \ll f(L) = L \ll f(\beta p) = \beta p.$$

But this implies there exist $\alpha' > \alpha$, $\beta' < \beta$ such that $L \subset [\alpha' p, \beta' p]$, contradicting the definitions of α and β. **QED**

The following result differs from Theorem 5.6 in that it assumes that N is a cone and there is a fixed point in X_{++}, and concludes that even orbits whose closures are not compact can have at most one limit point, which must be a fixed point:

THEOREM 5.7 *Assume f is strongly monotone and subhomogeneous and suppose there is a fixed point $p \gg 0$. Suppose also that N is a cone. Then every nonempty omega limit set consists of a single fixed point.*

Proof First I show that every omega limit set is order bounded. Let $L \subset N$ be the nonempty omega limit set of some point $u \in N$. Choose $\mu > 1$ so large that $\mu p \geq u$. Since N is a cone, $\mu p \in N$. Then monotonicity of f^n shows:

$$f^n(\mu p) \geq f^n(u) \text{ if } n \geq 0.$$

Now $f^n(\mu p) \leq \mu f^n(p) = \mu p$, by subhomogeneity. It follows that $\mu p \geq f^n(u)$. Therefore $\mu p \geq L$, so $L \subset [0, \mu p]$.

Let $\nu \geq 0$ be the smallest number such that $L \leq \nu p$; then $\nu > 0$.

If $\nu p \in L$ then L contains its supremum νp; in this case strong monotonicity and an argument similar to that in the proof of Theorem 5.6 shows that $L = \nu p$.

There remains the possibility that $\nu p \succ L$. Then

$$\nu p = \nu f(p) \geq f(\nu p) \gg f(L) = L,$$

by subhomogeity and strong monotonicity. But if $\nu p \gg L$ then $\lambda p \gg L$ for some $\lambda < \mu$, contradicting minimality of μ. This completes the proof. **QED**

The foregoing proof that L is order bounded did not use *strong* monotonicity. Moreover the assumption of a fixed point $p \gg 0$ can be weakened to the existence of a nonempty compact invariant set $K \gg 0$:

THEOREM 5.8 *Assume N is a cone and f is monotone and subhomogeneous. If there exists a nonempty compact invariant set $K \gg 0$, then every orbit is order bounded.*

Proof Fix $u \in N$. Choose $\mu > 1$ and $z \in N$ such that

$$z \geq \mu K \geq u.$$

For every natural number $n \geq 1$ and every $y \in K$ we have $\mu f^n(y) \in \mu K$, whence $z \geq \mu f^n(y)$. Therefore:

$$z \geq \mu f^n(y) \geq f^n(\mu y) \geq f^n(u).$$

This shows that the orbit of u lies in $[0, z]$. **QED**

THEOREM 5.9 *Assume N is a cone and f is strongly monotone, subhomogeneous and order compact. If there exists a nonempty compact invariant set $K \gg 0$, then every trajectory converges to a fixed point.*

Proof Let Y denote the orbit closure of some point $u \in N$. By Theorem 5.8, Y is order bounded. Since f is order compact, Y is compact. Theorem 5.6 now implies that the trajectory of u converges. **QED**

There are no analogues to the preceding Theorems if f is merely subhomogeneous and monotone, as can be seen by the following:

EXAMPLE 5.10 Give $\mathbf{R}^3$ its light cone ordering of Example 1.1. Define a flow Φ of global period 1 by rotating points around the x_1-axis at angular velocity 2π. The set of fixed points is the x_1-axis.

Fix an irrational number $\tau > 0$. Then the map $\Phi_\tau : \mathbf{R}^3 \to \mathbf{R}^3$ is monotone and subhomogeneous. Every orbit of Φ_τ is compact, but there are no periodic points off the x_1 axis. Except for fixed points, no orbits converge to periodic points.

6 Nonexpansiveness and Finiteness of Limit Sets

In this paper I have emphasized the use of "strong" assumptions: Except in Section 4, the theorems treat maps that are strongly monotone and homogeneous, or are monotone and strongly homogeneous. But there is a very deep theory of maps that are assumed to be merely monotone and homogeneous. For completeness I review some of these results. I am grateful to R. Nussbaum for informing me about this theory.

Assume X is a strongly ordered topological vector space. In this case the *order topology* on X is defined, having as a neighborhood base the family of all *open order intervals*

$$[[a, b]] = \{x \in X : a \ll x \ll b\}.$$

It is not hard to see that *if a map $f : A \to B$ is monotone and continuous, where A, B are subsets of ordered topological vector spaces, then f is continuous when A and B are given the topology induced from the order topologies.*

If $e \gg 0$ then the norm $|\cdot|_e$ defined as in Example 1.4 is a norm for the order topology. But for the order topology on X_{++} there is a more interesting metric, the *parts metric* d_p of Thompson (1963):

$$d_p(x,y) = \inf\{\rho \in \mathbf{R}_+ \; : \; e^{-\rho}x \ll y \ll e^{\rho}x\}.$$

For the case where $X = \mathbf{R}^n$ and $X_+ = \mathbf{R}_+^n$, the metric space (X_{++}, d_p) is isometric to $\mathbf{R}^n$ with the metric $d_{\max}$ induced by the norm $|y|_{\max} = \max |y_i|$. The isometry

$$\mathbf{R}_{++}^n \to \mathbf{R}^n : x \mapsto y, \; y_i = \log x_i$$

is a monotone map.

More generally, suppose X_+ is a *polyhedral cone* $P \subset \mathbf{R}^n$, defined as the subset where $m := m(P) \geq n$ linear functions on $\mathbf{R}^n$ are simultaneously nonnegative. Then X_{++} with the parts metric is isometric to a subset of $(\mathbf{R}^m, d_{\max})$.

For the following discussion assume $N \subset X_{++}$ is order convex, and $f : N \to X_{++}$ is monotone and subhomogenous.

It is easy to prove:

LEMMA 6.1 $d_p(f(x), f(y)) \leq d_p(x,y)$ *for all* $x, y \in N$.

This important property of monotone subhomogenous maps in solid cones is called *nonexpansiveness*. Nonexpansiveness of a map is a very powerful condition. It implies, for example, that there are no unstable fixed points— every fixed point is Liapunov stable.

Therefore a subhomogeneous monotone map in a polyhdral cone $P \subset \mathbf{R}^n$ has the same dynamics as a nonexpansive map in a subset of $(\mathbf{R}^m, d_{\max})$, $m = m(P)$.

Less obvious is the remarkable fact that *the action of f map on any nonempty compact omega limit set L is a d_p-isometry* (Dafermos and Slemrod 1973). This implies that *the map $f|L : L \to L$ embeds in a compact abelian group Γ of d_p-isometries acting transitively on L, and the cyclic subgroup generated by f is dense in Γ.*

Thus L is homeomorphic to a quotient group of Γ, in such a way that f acts by translation.

Example 5.10 shows that despite this very restrictive dynamics on compact limit sets, there are no general convergence theorems for monotone subhomogenous maps in arbitrary strongly ordered Banach spaces.

But it turns out that the geometry of the positive cone is crucial here: For the vector order in $\mathbf{R}^n$, or in fact for any order in $\mathbf{R}^n$ defined by a polyhedral cone $P \subset \mathbf{R}^n$, *every compact omega limit set of a monotone homogeneous map f is finite, with cardinality bounded above by an explicit function of the number of faces $m(P) = m$.* See D. Weller (1987), and R. D. Nussbaum (1990) who obtains the bound

$$2^m (m!)(\frac{1}{\ln 2})^m,$$

and conjectures that this can be improved to 2^m.

Nussbaum (see also R. C. Sine 1979) shows that if a flow in $\mathbf{R}^n$ is nonexpansive for a polyhedral norm, but not necessarily monotone, then every nonempty compact omega limit set is a point.

References

[1] N. D. Alikakos, P. Hess, and H. Matano (1989), *Discrete order preserving semigroups and stability for periodic parabolic differential equations*, J. Diff. Equations **82**, 322–341.

[2] H. Amann (1976), *Nonlinear operators in ordered Banach spaces*, Nonlinear Operators and the Calculus of Variations (New York) (J. P. Gossez, E. J. Lami-Dozo, J. Mawhin, and L. Waelbroeck, eds.), Summer School Held in Bruxelles 8-19 September 1975, Springer-Verlag.

[3] H. Amann (1976a), *Fixed point equations and nonlinear eigenvalue problems in ordered Banach spaces*, SIAM Review **18**, 620–704.

[4] H. Amann (1984), *Existence and regularity for semilinear parabolic evolution equations*, Ann. Scuola Norm. Sup. Pisa Cl. Sci. (4) **11**, 593–676.

[5] H. Amann (1988), *Dynamic theory of quasilinear parabolic equations. I. Abstract evolution equations*, Nonliner Anal. **12**, 895–919.

[6] H. Amann (1990), *Dynamic theory of quasilinear parabolic equations. II. Reaction-diffusion systems*, Differential Integral Equations **3**, 13–75.

[7] C. M. Dafermos and M. Slemrod (1973), *Asymptotic behaviour of nonlinear contraction semigroups*, J. Functional Analysis **13**, 97–106.

[8] E. N. Dancer and P. Hess (1991), *Stability of fixed points for order-preserving discrete-time dynamical systems*, J. reine angewandte Math. **419**, 125–139.

[9] G. Fusco and W. M. Oliva (1988), *Jacobi matrices and transversality*, Proc. Royal Soc. Edinburgh **109A**, 231–243.

[10] G. Fusco and W .M. Oliva (1990), *Transversality between invariant manifolds of periodic orbits for a class of monotone dynamical systems*, J. Dynamics and Diff. Equations **2**, 1–17.

[11] D. Henry (1981), *Geometric theory of semilinear parabolic equations*, Lecture Notes in Mathematics, vol. 840, Springer-Verlag.

[12] M. W. Hirsch (1982), *Systems of differential equations that are competitive or cooperative I: Limit sets*, Siam J. Math. Anal. **13**, 167–179.

[13] M. W. Hirsch (1984), *The dynamical systems approach to differential equations*, Bull. Amer. Math. Soc. **11**, no. 1, 1–64.

[14] M. W. Hirsch (1985), *Systems of differential equations that are competitive or cooperative II: Convergence almost everywhere*, Siam J. Math. Anal. **16**, 432–439.

[15] M. W. Hirsch (1985a), *Attractors for discrete-time monotone dynamical systems in strongly ordered spaces*, Geometry and Topology. Lecture Notes in Mathematics 1167 (New York) (J. Alexander and J. Harer, eds.), Proceedings of Special Year at University of Maryland, Springer-Verlag, pp. 141–153.

[16] M. W. Hirsch (1988), *Systems of differential equations that are competitive or cooperative III: Competing species*, Nonlinearity **1**, no. 1, 51–71.

[17] M. W. Hirsch (1988a), *Stability and convergence in strongly monotone dynamical systems*, J. reine angewandte Math. **383**, 1–53.

[18] M. W. Hirsch (1989a), *Systems of differential equations that are competitive or cooperative V: Convergence in three-dimensional systems.*, Journal of Differential Equations **80**, 94–106.

[19] M. W. Hirsch (1990), *Systems of differential equations that are competitive or cooperative IV: Structural stability in three-dimensional systems.*, SIAM Journal of Mathematical Analysis **21**, 1225–1234.

[20] M. W. Hirsch (1991), *Systems of differential equations that are competitive or cooperative VI: A C^r closing lemma for 3-dimensional systems*, Ergodic Theory and Dynamical Systems **11**, 443–454.

[21] J. P. Keener (1993), *The Perron-Frobenius theorem and the ranking of football teams*, SIAM Review **35**, no. 1, 80–93.

[22] C. Khanmy and J. Gabriel (1989), *On convergence in cooperative differential systems having at most two equilibria*, Tech. report, University of Lausanne.

[23] J. Mallet-Paret and H. L. Smith (1990), *The Poincaré-Bendixson theorem for monotone cyclic feedback systems*, J. Dynamics and Diff. Equations **2**, 367–421.

[24] H. Matano (1986), *Strongly order preserving local semi-dynamical systems— theory and applications*, Semigroups and Applications, Vol. I (H. Brézis, M. G. Crandall, and F. Kappel, eds.), Longman Scientific and Technical, Research Notes in Mathematics No. 141, pp. 178–85.

[25] J. Mierczyński (1991), *On monotone trajectories*, Proc. Amer. Math. Soc. **113**, 537–544.

[26] N. S. Nadirashvili (1993), *Uniformly monotone dynamical systems*, Proc. Royal. Soc. Edinburgh **123A**, 59–74.

[27] R. D. Nussbaum (1990), *Omega limit sets of nonexpansive maps: finiteness and cardinality estimates*, Differential and Integral Equations **3**, 523–540.

[28] P. Polàčik (1989), *Convergence in smooth strongly monotone flows defined by semilinear parabolic equations*, J. Diff. Equations **79**, 89–110.

[29] P. Polàčik (1989a), *Domains of attraction of equilibria and monotonicity properties of convergent trajectories in parabolic systems admitting strong comparison principles*, J. Reine Angewandte Math. **4400**, 32–56.

[30] R. C. Sine (1979), *On nonlinear contraction semigroups in sup norm spaces*, Nonlinear Analysis **3**, 885–890.

[31] H. L. Smith and H. R. Thieme (1990), *Quasiconvergence and stability for strongly order preserving semiflows*, SIAM J. Math. Anal. **21**, 673–692.

[32] H. L. Smith and H. R. Thieme (1991), *Convergence for strongly order preserving semiflows*, SIAM J. Math. Anal. **22**, 1081–1101.

[33] H. L. Smith (1987), *Monotone semiflows generated by functional differential equations*, J. Diff. Equations **66**, 420–442.

[34] H. L. Smith (1988), *Systems of ordinary differential equations which generate an order-preserving flow*, SIAM Review **30**, 87–111.

[35] P. Takáč (1990), *Asymptotic behavior of discrete-time semigroups of sublinear, strongly increasig mappings with applications in biolgy*, Nonlinear Analysis, Theory, Methods & Applications **14**, 35–42.

[36] P. Takáč (1990a), *Convergence to equilibrium on invariant d-hypersurfaces for strongly increasing discrete-time semigroups*, J, Math. Anal. Appl. **148**, 223–244.

[37] P. Takáč (1992), *Large-time behavior of a time-periodic cooperative system of reaction-diffusion equations depending on parameters*, SIAM J. Math. Anal. **23**, 387–411.

[38] A. C. Thompson (1963), *On certain contraction mappings in a partially ordered vector space*, Proc. Amer. Math. Soc. **14**, 438–443.

[39] D. Weller (1987), *Hilbert's metric, part metric, and selfmappings of a cone*, Ph.D. thesis, University of Bremen.

Blowup of Solution for the Heat Equation with a Nonlinear Boundary Condition

Bei Hu, Department of Mathematics, University of Notre Dame, Notre Dame, IN 46556,

Hong-Ming Yin, Department of Mathematics, University of Notre Dame, Notre Dame, IN 46556

1. INTRODUCTION

In this presentation we shall report some recent progress about the profile near the blowup time for the solution of the heat equation:

$$u_t = \Delta u \quad \text{for } x \in \Omega,\ t > 0, \tag{1.1}$$

subject to the following Neumann boundary and initial conditions:

$$\frac{\partial u}{\partial n} = f(u) \geq 0, \qquad x \in \partial\Omega,\ t > 0, \tag{1.2}$$

$$u(x,0) = u_0(x) \geq 0, \qquad x \in \Omega, \tag{1.3}$$

where Ω is a bounded domain in R^n and $\frac{\partial}{\partial n}$ the outward normal derivative on $\partial\Omega$.

The first author is partially supported by US National Science Foundation Grant DMS 90-24986 and DMS 92-24935; the second author is partially supported by NSERC of Canada.

It is known for a long time (cf. [13]) that the problem does not have a global solution in
time, if

$$\int^{\infty} \frac{1}{f(s)f'(s)} ds < \infty,$$

and $u_0(x)$ is suitably large. However, there are many important and interesting questions
which have been open for some years. For examples, how does the solution approach the
blowup time? Where is the hot spot located (blowup set)? For an equation with a nonlinear
source

$$u_t - \Delta u = f(u),$$

these problems have been studied by a lot of authors (cf.[1],[7]-[8], [10] for examples). For
(1.1)-(1.3), in one space dimension as well as a radial symmetric domain in R^n, the questions
were answered recently in [6], under certain monotonicity assumptions on the initial value.
An improvement, besides other results, was given in [2] where the monotonicity condition
was removed. For several space dimensions, the problem is much more challenging and there
is no result except some partial answer obtained in a recent paper [15].

The first result in this paper is concerned with blowup set. The examples in [12] indicate
that the blowup may occur in the interior of the domain if the heat supply through the
boundary is fast enough (with exponential rate). We prove that the blowup only occurs on
the boundary. The second main result deals with the blowup rate for $f(u) = u^p$ for $p > 1$. It
is proved that under certain conditions on the initial data and p,

$$c(T - t)^{-1/[2(p-1)]} \le \max_{\Omega} u(x,t) \le C(T - t)^{-1/[2(p-1)]},$$

where T is the blowup time.

The key step for the proof of the first result is to construct suitable auxiliary functions.
The second result is derived by analyzing the local behavior near a blowup point. Because
of the limitation of space, we shall only give the outline of their proofs. More detailed proofs
will be published elsewhere.

2. BLOWUP IN FINITE TIME

Throughout this paper, we shall use C and c to denote various generic constants, if there is
no confusion. A solution of (1.1)-(1.3) is always understood in the classical sense.

THEOREM 1: For any nonzero, nonnegative initial data $u_0(x)$, the solution of the system
(1.1)-(1.3) blows up in a finite time, if

$$f(u) \ge 0, \ f'(u) \ge 0, \ f''(u) \ge 0, \ \int_{z_0}^{\infty} \frac{1}{f(s)} ds < \infty,$$

for any $z_0 > 0$.

Proof: Local existence is clear. By the maximum principle, $\inf_{x \in \Omega} u(x, \varepsilon) > 0$ (for small $\varepsilon > 0$). Replacing $t = 0$ by $t = \varepsilon$ if necessary, we may assume without loss of generality that $\inf_{x \in \Omega} u_0(x) = c > 0$. Take $v(x)$ such that

$$\inf_{x \in \Omega} \Delta v > 0,$$

$$\frac{\partial v}{\partial n} = f(v) \quad \text{for } x \in \partial \Omega,$$

$$\frac{c}{2} \le v(x) \le c \quad \text{for } x \in \Omega.$$

The existence of such a $v(x)$ can be obtained, for example, by the variational method. Then the comparison principle implies that $u(x, t) \ge \varphi(x, t)$, where $\varphi(x, t)$ is the solution of (1.1)-(1.3) with the initial condition $\varphi(x, 0) = v(x)$. Clearly, $\varphi_t(x, t) \ge 0$, by maximum principle. As a consequence, $\varphi(x, t) \ge \varphi(x, 0) \ge c/2$. Let $\psi(x, t) = \varphi_t(x, t) - \delta f(\varphi(x, t))$, then a direct calculation shows that

$$\frac{\partial \psi}{\partial t} - \Delta \psi \ge 0 \quad \text{for } x \in \Omega, \ t > 0,$$

$$\frac{\partial \psi}{\partial n} = f'(u)\psi \quad \text{for } x \in \partial \Omega, \ t > 0.$$

$\varphi_t(x, 0) = \Delta v(x) > 0$ implies $\psi(x, 0) \ge 0$ if δ is small enough. It follows that $\psi(x, t) \ge 0$, which implies that $\varphi_t \ge \delta f(u)$. Thus $\varphi(x, t)$ blows up in finite time, and $u(x, t)$ must blow up at a finite time.

3. BLOWUP RATE AND BLOWUP SET

Suppose that T is the blowup time. We first derive the blowup rate for $f(u) = u^p$ with $p > 1$.

Theorem 2: Let Ω be a bounded domain in R^n such that $\partial \Omega \in C^{2+\alpha}$ for some $0 < \alpha < 1$.
(i)

$$\max_{x \in \overline{\Omega}} u(x, t) \ge \frac{C}{(T - t)^{1/[2(p-1)]}}. \tag{3.1}$$

(ii) Suppose that $1 < p < \infty$ for $n = 2$ and $1 < p < \dfrac{n-1}{n-2}$ for $n \ge 3$ and that the initial value $u_0 \in C^2(\overline{\Omega})$ satisfies

$$u_0 \ge 0, \quad \Delta u_0 \ge 0 \quad \text{for } x \in \Omega,$$

$$\frac{\partial u_0}{\partial n} = u_0^p \quad \text{for } x \in \partial \Omega.$$

Then

$$\max_{x\in\overline{\Omega}} u(x,t) \le \frac{C}{(T-t)^{1/[2(p-1)]}}. \tag{3.2}$$

Proof: By applying a similar argument as that in [2] and [14], one can derive the lower bound from the integral representation of solution. To derive the upper bound, we consider the equations for the functions u and u_t and use maximum principle to obtain

$$u(x,t) \ge 0, \quad u_t(x,t) \ge 0.$$

Thus the function $M(t) = \max_{x\in\overline{\Omega}} u(x,t)$ is monotone nondecreasing and $M(t) \to \infty$ as $t \to T - 0$. The maximum principle implies that $M(t) = \max_{x\in\partial\Omega} u(x,t)$.

We now take $T/2 < t^* < T$, and let $M^* = M(t^*)$. Take any point $x^* \in \partial\Omega$ such that $M(t^*) = u(x^*,t^*)$ (there may be more than one choice of such x^* for each t^*) and introduce the rescaled function

$$\varphi_\lambda(y,s) = \frac{1}{M^*} u(\lambda Ry + x^*, \lambda^2 s + t^*) \quad \text{for } y \in \overline{\Omega_\lambda}, \quad -\frac{T}{2\lambda^2} \le s \le 0,$$

where $\Omega_\lambda = \{y; \lambda Ry + x^* \in \Omega\}$ and R is a rotation operator such that $(-1,0,0,\cdots,0)$ is the exterior normal vector of $\partial\Omega_\lambda$ at 0. We choose λ such that

$$\lambda(M^*)^{(p-1)} = 1.$$

Then φ_λ solves

$$\frac{\partial\varphi_\lambda}{\partial s} = \Delta_y\varphi_\lambda \quad \text{for } y \in \Omega_\lambda, \quad -\frac{T}{2\lambda^2} \le s \le 0,$$

$$\frac{\partial\varphi_\lambda}{\partial n} = \varphi_\lambda^p \quad \text{for } y \in \partial\Omega_\lambda, \quad -\frac{T}{2\lambda^2} \le s \le 0,$$

$$\varphi_\lambda(0,0) = 1,$$

$$0 \le \varphi_\lambda(y,s) \le 1, \quad (\varphi_\lambda)_s(y,s) \ge 0.$$

We claim the following key estimate: there exist $c > 0$, $\delta > 0$ (independent of the choices of x^* and the rotation R) such that

$$\frac{\partial\varphi_\lambda}{\partial s}(0,0) \ge c \quad \text{for } T - \delta < t^* < T. \tag{3.3}$$

The proof is quite technical. The idea is as follows. Assume it is not true, we choose a sequence of λ which approches zero and then use the compactness to obtain a positive solution of the steady-state equation. This would contradict the uniqueness results in Section 4. We shall omit the detail here.

Rewrite (3.3) as

$$\frac{\partial u}{\partial t}(x^*,t^*) \ge c(M^*)^{2p-1} \quad \text{for } T - \delta < t^* < T.$$

For each $h > 0$,

$$\frac{M(t^* + h) - M(t^*)}{h} = \frac{M(t^* + h) - u(x^*, t^*)}{h} \geq \frac{u(x^*, t^* + h) - u(x^*, t^*)}{h}.$$

Letting $h \to 0+$, (noticing that $M(t)$ is Lipschitz continuous), we obtain

$$M'(t^*) \geq u_t(x^*, t^*) \geq cM^{2p-1}(t^*).$$

Integrating the above equation, we conclude the estimate (3.2).

Next, we shall prove that the blowup will occur only at the boundary of the domain.

Theorem 3: Suppose that the function $u(x, t)$ is continuous on the domain $\overline{\Omega} \times [0, T)$ and satisfies

$$u_t = \Delta u \quad \text{for } (x, t) \in \Omega \times [0, T),$$
$$u \leq \frac{C}{(T - t)^q} \quad \text{for } (x, t) \in \partial\Omega \times [0, T), \quad \text{for some } q > 0.$$

Then for any $\Omega' \subset\subset \Omega$,

$$\sup\{u(x, t); \ (x, t) \in \Omega' \times [0, T)\} < \infty.$$

Proof: By approximating the domain from inside if necessary, we may assume without loss of generality that $\partial\Omega$ is smooth, say C^2.

Let $d(x) = \text{dist}(x, \partial\Omega)$ and

$$v(x) = d^2(x) \quad \text{for } x \in N_\varepsilon(\partial\Omega),$$

where $N_\varepsilon(\partial\Omega) = \{(x \in \Omega, d(x) < \varepsilon\}$. Since $\partial\Omega$ is C^2, the function $v(x)$ is in $C^2(\overline{N_\varepsilon(\partial\Omega)})$ if ε is small enough. Clearly,

$$\Delta v - \frac{(q + 1)|\nabla v|^2}{v} = 2 - 4(q + 1) \quad \text{on } \partial\Omega.$$

Since $v \in C^2(\overline{N_\varepsilon(\partial\Omega)})$,

$$\Delta v - \frac{(q + 1)|\nabla v|^2}{v} \geq -4(q + 1) \quad \text{in } \overline{N_{\varepsilon_0}(\partial\Omega)}$$

if ε_0 is small enough. We next extend $v(x)$ to a function on $\overline{\Omega}$ such that $v \in C^2(\overline{\Omega})$ and $v \geq c_0 > 0$ on $\overline{\Omega \setminus N_{\varepsilon_0}(\partial\Omega)}$. Then

$$\Delta v - \frac{(q + 1)|\nabla v|^2}{v} \geq -C^* \quad \text{on } \overline{\Omega}$$

for some $C^* > 0$. Set

$$w(x,t) = \frac{C_1}{[v(x) + C^*(T-t)]^q}.$$

Then

$$w_t - \Delta w = \frac{C_1 q}{[v(x) + C^*(T-t)]^{(q+1)}} \left(C^* + \Delta v - \frac{(q+1)|\nabla v|^2}{v + C^*(T-t)} \right) > 0.$$

Take C_1 to be large enough so that $w(x,0) \geq u(x,0)$ and $C_1 \geq (C^*)^q$. Then the maximum principle implies that $w(x,t) \geq u(x,t)$, and

$$\sup\{u(x,t); (x,t) \in \Omega' \times [0,T]\} \leq C_1 \sup \left\{ \frac{1}{v^q(x)}; x \in \Omega' \right\} < \infty.$$

Corollary: Assume that $u_0(x) \geq 0$ and $\Delta u_0(x) \geq 0$. Then no blowup will occur in the interior of the domain Ω.

Proof: By the assumption, we see that $u_t \geq 0$. Moreover, the maximum principle implies that $u_t(x,\varepsilon) > 0$ for a small ε. As in the proof of Theorem 1, we can easily establish $u_t \geq \delta f(u)$ for $t \geq \varepsilon$ for small δ and ε. It then follows that for $f(u) = u^p$ or $f(u) = e^u$:

$$\max_{x \in \bar\Omega} u(x,t) \leq \frac{C}{(T-t)^{1/(p-1)}},$$

or

$$\max_{x \in \bar\Omega} u(x,t) \leq C ln \frac{1}{T-t}.$$

Consequently, the result follows from Theorem 3.

4. SOME RESULTS FOR ELLIPTIC EUQATIONS

In this section we prove some nonexistence results. In addition to the application in proving Theorem 2, it is of independent interest. We first consider the case $n = 2$.

Theorem 4: Suppose that $\varphi(y) = \varphi(y_1, y_2)$ satisfies

$$\Delta \varphi = 0 \quad \text{for } 0 < y_1 < \infty, -\infty < y_2 < \infty,$$
$$-\frac{\partial \varphi}{\partial y_1} \geq 0 \quad \text{for } y_1 = 0,$$
$$0 \leq \varphi(y) \leq 1.$$

Then

$$\varphi(y) \equiv \text{constant}.$$

Proof: Take $y_0 \in [0, \infty) \times (-\infty, \infty)$. For any $0 < \varepsilon < 1$, $0 < \delta < 1$, we construct the auxiliary function

$$\psi(y) = \varepsilon \log \left(\frac{|y - y_0|^2}{\delta^2} \right) + C_\delta \, ,$$

where $C_\delta = \max_{|y - y_0| = \delta} [\varphi(y_0) - \varphi(y)]$. that $\psi = C_\delta$ on $\{|y - y_0| = \delta\}$, $\psi(y) > 1 \geq \varphi(y_0) - \varphi(y)$ on $|y - y_0| = e^{1/\varepsilon}$. A direct calculation also shows that $-\psi_{y_1}(0, y_2) \geq 0$. Thus by maximum principle,

$$\varphi(y_0) - \varphi(y) \leq \psi(y) \quad \text{in the region } \{y_1 > 0, \, \delta < |y - y_0| < e^{1/\varepsilon}\}.$$

Letting $\varepsilon \to 0+$ and then $\delta \to 0+$, we conclude $\varphi(y_0) - \varphi(y) \leq 0$ for any y and y_0.

Theorem 5: Suppose that $n \geq 3$, $p < \dfrac{n-1}{n-2}$ and

$$\Delta \varphi = 0 \quad \text{for } 0 < y_1 < \infty, -\infty < y_k < \infty \ (2 \leq k \leq n),$$
$$-\frac{\partial \varphi}{\partial y_1} = \varphi^p \quad \text{for } y_1 = 0,$$
$$\varphi(y) \geq 0.$$

Then

$$\varphi(y) \equiv 0.$$

Proof: Assume for the contrary that $\varphi \not\equiv 0$, then we claim that

$$\varphi(y) \geq \frac{c}{|y|^{n-2}} \quad \text{for } y_1 > 0, 1 \leq |y| < \infty, \tag{4.1}$$

for some $c > 0$. In fact, if we take $c = \min_{|y|=1} \varphi(y)$, then by maximum principle, $c > 0$. The function $w(|y|) = \dfrac{c}{|y|^{n-2}}$ satisfies

$$\Delta w = 0 \quad \text{for } 0 < y_1 < \infty, 1 \leq |y| < \infty,$$
$$-\frac{\partial w}{\partial y_1} = 0 \leq -\frac{\partial \varphi}{\partial y_1} \quad \text{for } y_1 = 0,$$
$$w(y) \leq \varphi(y) \quad \text{on } |y| = 1.$$

Therefore by maximum principle,

$$w(|y|) - w(R) \leq \varphi(y) \quad \text{for } 0 < y_1 < \infty, 1 \leq |y| < R.$$

Letting $R \to \infty$, we obtain (4.1).

In order to study the behavior of $\varphi(y)$ near ∞, we introduce the Kelvin's inversion:

$$\psi(z) = |y|^{n-2}\varphi(y), \quad z = \frac{y}{|y|^2}.$$

The function $\psi(z)$ may have a singularity at 0, it satisfies the equations:

$$\Delta\psi = 0 \quad \text{for } 0 < z_1 < \infty, -\infty < z_k < \infty \ (2 \leq k \leq n), \tag{4.2}$$

$$-\frac{\partial\psi}{\partial z_1} = |z|^{-\alpha}\psi^p \quad \text{for } z_1 = 0, \ |z| > 0, \quad \text{where } \alpha = n - p(n-2). \tag{4.3}$$

Recalling (4.1), we have

$$\psi(z) \geq c > 0 \quad \text{for } 0 \leq z_1 \leq 1, \ 0 < |z| \leq 1. \tag{4.4}$$

For any $\varepsilon > 0$, we take a smooth cut-off function $\zeta(z)$ such that

$$\zeta(z) = 0 \quad \text{for } |z| \leq \varepsilon \text{ and } |z| \geq 4\varepsilon,$$

$$\zeta(z) = 1 \quad \text{for } 2\varepsilon \leq |z| \leq 3\varepsilon,$$

$$0 \leq \zeta(z) \leq 1, \quad |\nabla\zeta(z)| \leq \frac{C}{\varepsilon}.$$

Multiplying the equation (4.2) with $\zeta^2\psi^{-1}$ and integrating over $\Omega = \{z_1 > 0, 0 < |z| < 1\}$, we obtain

$$\int_\Omega \zeta^2\frac{|\nabla\psi|^2}{|\psi|^2}dz + \int_{\{z_1=0\}} \zeta^2|z|^{-\alpha}\psi^{p-1}dS$$

$$= 2\int_\Omega \frac{\zeta}{\psi}\nabla\zeta\nabla\psi dz$$

$$\leq \frac{1}{2}\int_\Omega \zeta^2\frac{|\nabla\psi|^2}{|\psi|^2}dz + 2\int_\Omega |\nabla\zeta|^2dz.$$

Hence

$$\int_{\{z_1=0\}} \zeta^2|z|^{-\alpha}\psi^{p-1}dS \leq 2\int_\Omega |\nabla\zeta|^2dz,$$

which implies that

$$\varepsilon^{n-1}\varepsilon^{-\alpha} \leq C\varepsilon^n\frac{1}{\varepsilon^2}.$$

Noticing that $\alpha > 1$, we obtain a contradiction if ε is small enough.

5. OPEN QUESTIONS

The restriction on p in Theorem 5 may not be optimal. We conjecture that the uniqueness holds for $1 < p < \frac{n}{n-2}$ if $n > 2$. This would imply that the estimate (3.2) holds if $p \in (1, \frac{n}{n-2})$.

The second open question is to eliminate the monotonocity condition of $u(x,t)$ with respect to t in Theorem 2 and Theorem 3. For Theorem 3, one only need to show that the solution grows no faster than $\frac{1}{(T-t)^q}$ for some $q > 0$.

It is not difficult to see that most results in this paper can be extended to the following eqution

$$u_t = \triangle u - u^q$$

associated with the same initial and boundary conditions (1.2)-(1.3).

REFERENCES

1. J. Bebernes and D. Eberly, Mathematical Problems from Combustion Theory, Springer-Verlag, New York, 1989.

2. J. M. Chadam and H. M. Yin, A diffusion equation with localized chemical reactions, IMA preprint series No. 970, 1992, to appear in Proceedings of the Edinburgh Mathematical Society.

3. M. Chipot, M. Fila and P. Quittner, Stationary solutions, blowup and convergence to stationary solutions for semilinear parabolic equations with nonlinear boundary conditions, Acta Math. Univ. Comenianae, LX(1991), 35-103.

4. B. Gidas and J. Spruck, Global and local behavior of positive solutions of nonlinear elliptic equations, Comm. Pure and Applied Math., 34(1981), 525-598.

5. M. Fila, Boundedness of global solutions for the heat equation with nonlinear boundary conditions, Comment. Math. Univ. Carolinae, 30(1989), 479-484.

6. M. Fila and P. Quitter, The blowup rate for the heat equation with a nonlinear boundary condition, Mathematical Methods in the Applied Sciences, 14(1991), 197-205.

7. A. Friedman and B. McLeod, Blowup of positive solutions of semilinear heat equations, Indiana Univ. Math. J. 34(1985), 425-477.

8. Y. Giga and R. V. Kohn, Characterizing blowup using similarity variables, Indiana Univ. Math. J., 36(1987), 425-447.

9. H. A. Levine and L. E. Payne, Nonexistence Theorems for the heat equation with nonlinear boundary conditions and for the porous medium equation backward in time, J. of Diffs. Eqs., 16(1974), 319-334.

10. W. Liu, The blowup rate of solutions of semilinear heat equations, J. Diff. Eqns., 77(1989), 104-122.

11. C. V. Pao, Nonlinear Elliptic and Parabolic Systems.

12. A. A. Samarskii, On new methods of studying the asymptotic properties of parabolic equations, Proceedings of the Steklov Institute of Mathematics, 158(1983),165-176.

13. W. Walter, On existence and nonexistence in the large of solutions of parabolic differential equations with a nonlinear boundary condition, SIAM J. Math. Anal., 6(1974), 85-90.

14. F. B. Weissler, Existence and non-existence of global solutions for a semilinear heat equation', Israel J. of Maths., 38(1981), 29-40.

15. H. M. Yin, Blowup versus global solvability for a class of nonlinear parabolic equations, preprint, University of Toronto, 1992.

On the Existence of Extremal Solutions for Impulsive Differential Equations with Variable Time

SAROOP KAUL   Department of Mathematics and Statistics, University of Regina, Regina, CANADA   S4S 0A2

1 INTRODUCTION

In this paper we continue the study of the existence of extremal solutions for scalar impulsive differential equations (IDE) with variable time begun in [1,6]. In [1,6] we considered the problem starting with the extremal solutions of the associated differential equation without impulses. Here, however, we establish the existence of such solutions directly using Peano's method.

To see how an extremal solution leads to a comparison principle and stability criteria see [1].

The following discussion is restricted to maximal solutions. The treatment of minimal solutions is similar. The proofs of the following results are long and will appear elsewhere [2].

2 UPPER SOLUTIONS

Consider the IVP for the IDE

$$\begin{cases} u' = g(t,u), & t \neq \tau(u) \\ \Delta u = \psi(u), & t = \tau(u) \\ u(\beta_0^+) = u_0, \end{cases} \tag{2.1}$$

where $g \in C[E,R]$, $E = R_+ \times \Omega$, Ω an open subset of R, $(\beta_0, u_0) \in E$. Assume that $\tau \in C^1[\Omega, R_+]$, $\Psi \in C[\Omega, \Omega]$ and $h(u) = u + \Psi(u) \in \Omega$ for $u \in \Omega$.

We call

$$\begin{cases} u' = g(t, u) \\ u(\beta_0) = u_0 \end{cases} \qquad (2.2)$$

the IVP associated with (2.1).

Let

$$S = \{(t, u) : \tau(u) = t \text{ for } u \in \Omega\}$$
$$E_0 = \{(t, u) : 0 \leq t < \tau(u) \text{ for } u \in \Omega\}$$
$$E_1 = \{(t, u) : t \geq \tau(u)\}$$

and $B_0 = \partial E_0 - S$ and $B_1 = \partial E_1 - S$, where ∂ is the boundary operator in Ω.

DEFINITION (2.3)   A function $\omega \in PC_1^1[\,[\beta_0, \alpha), \Omega]$, $\alpha > \beta_0$, is said to be an upper solution of (2.1) on $[\beta_0, \alpha)$ if

$$\begin{cases} \omega'(t) \geq g(t, u), \ \ t \neq \tau(u) \\ \omega(t^+) \geq \omega(t) + \psi(\omega(t)), \ t = \tau(u) \\ \omega(\beta_0^+) \geq u_0. \end{cases} \qquad (2.3.1)$$

THEOREM (2.4)   Suppose $\tau(u_0) > \beta_0$ and

$$\frac{\partial \tau(u)}{\partial u} \cdot g(t, u) < 1 \qquad (2.4.1)$$

on E. Then (2.1) has an upper solution u(t) on a maximal interval $[\beta_0, \beta_1)$ such that, either,
I. u(t) exhibits the pulse phenomenon on $[\beta_0, \beta_1)$, or
II. u(t) hits the surface S a finite number of times [possibly never], and either
II(i): $\lim_{t \to \beta_1} u(t) \in B_0$, or
II(ii): $\lim_{t \to \beta_1} u(t) \in B_1$

Proof: We use Peano's technique to establish this existence result.

DEFINITION (2.5)   The IVP (2.2) has infinite scape time in E_0 with respect to S, if for any solution $u(t, \beta_0, u_0)$ of (2.2) in E_0, $(\beta_0, u_0) \in E_0$, whose maximal interval of existence $[\beta_0, \beta_1)$ in E_0 is finite, $\lim_{t \to \beta_1} u(t, \beta_0, u_0)$ is a point in S.

THEOREM (2.6)   Suppose $\tau(u)$ is bounded on Ω and (2.2) has infinite scape time in E_0. Then there exists an $\epsilon_0 > 0$ such that any upper solution u(t) of (2.2) in E_0 over a maximal interval $[\beta_0, \beta_1)$ satisfying

$$u(\beta_0) = u_0 + \epsilon_0, \qquad (2.6.1)$$

$$\tau(u(t)) > t, \qquad (2.6.2)$$

and

$$g(t, u(t)) < D_+ u(t) \leq g(t, u(t)) + \epsilon_0 \qquad (2.6.3)$$

on $[\beta_0, \beta_1)$, satisfies

$$\lim_{t \to \beta_1} u(t) \quad \text{is a point of } S. \tag{2.6.4}$$

3 EXTREMAL SOLUTIONS

In this section we consider an IVP for the IDE,

$$\begin{cases} u' = g(t, u), & t \neq \tau_k(u) \\ \triangle u = \psi_k(u), & t = \tau_k(u) \\ u(\beta_0^+ = u_0, & \beta_0 \geq 0 \end{cases} \tag{3.1}$$

where E, g and Ω are as in Section 2. Furthermore, for each $k = 1, 2, \cdots, \psi_k \in C^1[\Omega, R_-]$ such that $h_k(u) = u + \psi_k \in \Omega$ for any $u \in \Omega$; also $\tau_k(u) < \tau_{k+1}(u)$ and $\tau_k(u) \to \infty$ as $k \to \infty$, and for each k, $\tau_k(u)$ is bounded by T_k. Define $\tau_0(u) = 0$.

We wish to show that a maximal solution for (3.1) exists. To do that we first give conditions under which there exists an upper solution of (3.1) which intersects each surface $S_k : t = \tau_k(u)$ exactly once. Next we show, given any such solution, how to get another such which lies "below" it. This allows us to build a decreasing sequence of upper solutions, each intersecting a surface exactly once, which can be shown to converge to a solution of (3.1) which is its maximal solution. Define

$$E_m = \{(t, u) : \tau_m(u) \leq t < \tau_{m+1}(u), u \in \Omega\}.$$

THEOREM (3.1) Let $\tau_1(u_0) > \beta_0$. Assume that,

(i) $\dfrac{\partial \tau_k}{\partial u} g(t, u) < 1$, $(t, u) \in E$, $k = 1, 2, \cdots$;

(ii) $\tau_k(u) < \tau_{k+1}(h_k(u))$, $u \in \Omega$;

(iii) $\tau_k(u)$ is an increasing function of u for each $k = 1, 2, \cdots$;

(iv) On each set E_m the associated IVP (2.2) for $u(t_0) = v_0$ with $(t_0, v_0) \in E_m$ has infinite scape time.

Then (3.1) has an upper solution

$$u(t) = u_k(t), \; \beta_k \leq t < \beta_{k+1}$$

for $k = 0, 1, 2, \cdots$, such that,

(a) $\tau_k(u_k(t)) < t < \tau_{k+1}(u_k(t))$, $\beta_k \leq t < \beta_{k+1}$, and

(b) $\tau_{k+1}(u_k(\beta_{k+1})) = \beta_{k+1}$.

THEOREM (3.2) Assume all the hypotheses of Theorem (3.1). Assume further that

(v) $\dfrac{\partial \tau_k(h_k^{-1}(u))}{\partial u} g(t, u) < 1$

and $v_0 \in \Omega$, $v_0 < u_0$ and $\tau(v_0) > \beta_0$. Then, given an upper solution $u(t)$ of (3.1) as in Theorem (3.1) satisfying (3.1.1), there exists an upper solution $v(t)$ of (3.1) such that $v(\beta_0) = v_0$ and $v(t) < u(t)$ for all t.

Before stating the next theorem we need to define a new function of $u \in \Omega$.

$$H_k^n(u) = h_n \circ \cdots \circ h_k(u) \text{ for } n > k$$

$$H_n^n(u) = h_n(u).$$

Since each h_j is an increasing function of u so is H_k^n.

THEOREM (3.3)   Assume all the hypothese of Theorem (3.1). Assume, furthermore, that for $n \geq k$,

(vi) $\dfrac{\partial}{\partial u} H_k^n(u) g(t, u) \geq g(t, H_k^n(u)).$

Suppose $v_0 \in \Omega$, $v_0 < u_0$ and $\tau_1(v_0) > \beta_0$. Then given an upper solution $u(t)$ of (3.1) as in Theorem (3.1) there exists an upper solution $v(t)$ of (3.1) such that $v(\beta_0) = v_0$ and $v(t) < u(t)$ for all t.

THEOREM (3.4)   Assume all the hypotheses of Theorem (3.1). Assume further either

(a) $\dfrac{\partial \tau_k(h_k^{-1}(u))}{\partial u} \, g(t, u) < 1$

or

(b) $\dfrac{\partial H_k^n(u)}{\partial u} \, g(t, u) \geq g(t, H_k^n(u))$

for all $n \geq k$. Then (3.1) with the initial condition $u(\beta_0^+) = v_0$ has a maximal solution.

REFERENCES

1. Kaul, S, Lakshmikantham, V., Leela, S. Extremal solutions, comparison principle and stability criteria for impulsive differential equations with variable times, (to appear).

2. Kaul, S.K. On the existence of extremal solutions for impulsive differential equations with variable time, accepted for publication by *JNA-TMA*.

3. Lakshmikantham, V., Bainov, D.D., and Simeonov, P.S. (1989). *Theory of Impulsive Differential Equations*, World Scientific, Singapore.

4. Lakshmikantham, V. and Leela, S. (1969). *Differential and Integral Inequalities*, Vol. I., Academic Press, New York.

5. Lakshmikantham, V.; Leela, S., and Martynyuk, A.A. (1989). *Stability Analysis of Nonlinear Systems*, Marcel Dekker, New York.

6. Lakshmikantham, V., Leela, S., and Kaul, S., Comparison principle for impulsive differential equations with variable times and stability theory, (to appear).

7. Rozhko, V.F. (1975). Lyapunov stability in discontinuous dynamical systems. *Diff. Uravm.* **11**, 1105-1012.

Global Asymptotic Stability of Competitive Neural Networks

SEMEN KOKSAL, Department of Mathematics, Bradley University, Peoria, IL 61625

1. INTRODUCTION

Differential systems as models of artificial neural networks have been surveyed by Grossberg in [4]. Two kinds of dynamic processes can occur in neural networks: The change of the activation of neurons and the change of the strength of the synaptic connections (weights) between neurons with respect to time. The stability of first kind has been considered previously by Hopfield [6], Hirsch [5], Matsuka [10], and Cohen and Grossberg [2]. The stability of the latter has been investigated by Dong [3]. In each of these investigations, a single Lyapunov function was used to describe the stability requirements.

A stability property of a differential system may be considered as a family of properties depending on certain parameters. Consequently, when we employ a single Lyapunov function to establish a given stability property, the Lyapunov function chosen is forced to play the same role for every choice of these parameters. However, if we utilize a family of Lyapunov functions instead of one, it is natural to expect that each member of the family may satisfy weaker assumptions. This observation provides the motivation for the use of vector Lyapunov functions. The concept of vector Lyapunov functions was introduced by Bellman [1], and has been extended by Lakshmikantham et al. [9].

In [7], Koksal and Sivasundaram have investigated practical stability properties of Hopfield type neural networks by employing the method of vector Lyapunov functions. Also, they have used the extension of LaSalle's invariance principle to the vector Lyapunov functions to study the global asymptotic stability of the same network. This work has been extended to study the stability properties of continuous Hopfield networks with variable weights by Koksal and Fausett [8].

In the present paper, we use the extension of LaSalle's invariance principle to the method of vector Lyapunov functions to study global asymptotic stability on n-dimensional competitive dynamical systems that can be written in the form

$$x_i' = a_i(x_i)[b_i(x_i) - \sum_{k=1}^{n} c_{ik}d_k(x_k)] \equiv h_i(X) \tag{1}$$

for $i = 1, 2, \ldots, n$.

2. THE EXTENSION OF LASALLE'S INVARIANCE PRINCIPLE AND ITS APPLICATION

We consider the autonomous differential system

$$X' = f(X), \quad X(0) = X_0 \tag{2}$$

where $f \in C[\Omega^*, R^n], \Omega^*$ being an open set in R^n and $f(0) = 0$. Let Ω be any arbitrary set in R^n such that $\bar{\Omega} \subset \Omega^*$, where $\bar{\Omega}$ is the closure of Ω.

DEFINITION 1:   A point $y \in R^n$ is said to be a positive limit point of $X(t)$ if there exists a sequence $\{t_n\}$ such that $t_n \to \infty$ and $X(t) \to y$ as $n \to \infty$.

DEFINITION 2:   A solution $X(t)$ of equation (2) is said to be compact if $X(t)$ is contained in a compact set relative to Ω^*, that is, $X(t)$ is bounded for $t \geq t_0$ and has no positive limit points on the boundary of Ω^*.

The extension of LaSalle's invariance principle in terms of the vector Lyapunov functions is the following.

THEOREM 1: *Assume that*

(i) $V \in C'[\Omega^*, R^n]$, $V'_j(x) = \frac{\partial V_j}{\partial X} f(X) \leq 0$ *for each* $j \in S(X)$, $X \in \Omega$ *where*

$$S(X) = [j : V_j(X) = V_0(X)] \text{ and } V_0(X) = \max_j V_j(X);$$

(ii) $E = [X : V'_i(X) = 0 \text{ for some } i \in S(X), X \in \Omega \cap \Omega^*];$

(iii)   *L is the largest invariant set in E.*

Then every compact solution for $t \geq 0$ *of* (2) *that remains in* Ω *tends to L as* $t \to \infty$.

The following theorem gives sufficient conditions for the vector Lyapunov function V to satisfy the assumptions of Theorem 1.

THEOREM 2: *Assume that*

(i) $V \in C'[\Omega^*, R^n];$

(ii)   *for each* $X \in \Omega$, *there exists a vector* $u(X)$ *with* $u(X) \neq 0$ *such that for each* $i \in S(X)$,

$$V_i(X) \geq V_j(X), \quad j = 1, 2, \ldots, n.$$

implies that

(a) $\frac{\partial V_i(X)}{\partial x_j} u_j(X) \geq 0 \ \ i \neq j,$

(b) $\left(\frac{\partial V(X)}{\partial X} u(X) \right)_j \leq 0,$

(c) $\frac{f_i(X)}{u_i(X)} \geq 0$ *and* $\frac{f_i(X)}{u_i(X)} \geq \frac{f_j(X)}{u_j(X)},$ $\ j = 1, 2, \ldots, n.$

Then V is the desired vector Lyapunov function in Theorem 1.

For the proofs of Theorems 1 and 2, see [9]. We shall now discuss our result as an application of the above theorems.

THEOREM 3: *(Global asymptotic stability) Assume that in Equation* (1),

(i)   *The right hand side function* $h_i(X) \equiv 0$ *when* $X = 0$ *for all* $i = 1, 2, \ldots, n;$

(ii) $a_i(x_i)$, $b_i(x_i)$ *and* $d_i(x_i)$ *are differentiable functions for* $i = 1, 2, \ldots, n;$

(iii)

$$m_{ii}(X) - \tilde{m}_{ii}(X) + \sum_{\substack{j=1 \\ j \neq i}}^{n} |m_{ij}(X) - \tilde{m}_{ij}(X)| < 0 \text{ if } x_i^2 > x_j x_i$$

where

$$m_{ij}(X) = \int_0^1 \frac{\partial F_i(sX)}{\partial x_j} ds, \quad F_i(X) = a_i(x_i) b_i(x_i)$$

and

$$\tilde{m}_{ij}(X) = \int_0^1 \frac{\partial G_i(sX)}{\partial x_j} ds, \quad G_i(X) = a_i(x_i) \sum_{j=1}^n c_{ij} d_j(x_j)$$

then the system (1) *is globally asymptotically stable.*

PROOF: Equation (1) can be written in matrix-vector form as

$$X' = A(X)[b(X) - Cd(X)] \equiv H(X)$$

where

$$A(X) = diag(a_1(x_1), \ldots, a_n(x_n))$$

$$X \quad = [x_1, x_2, \ldots, x_n]^T$$

$$b(X) \quad = [b_1(x_1), \ldots, b_n(x_n)]^T$$

$$d(X) \quad = [d_1(x_1), \ldots, d_n(x_n)]^T$$

$$C \quad = \begin{bmatrix} c_{11} & \cdots & c_{1n} \\ \cdots & \cdots & \cdots \\ c_{n1} & \cdots & c_{nn} \end{bmatrix}$$

From (i) and (ii), by using the mean value theorem for integrals, we obtain

$$H(X) - H(Y) = \left[\int_0^1 \frac{\partial H}{\partial x}(sX + (1-s)Y)ds \right](X - Y)$$

and by setting $Y = 0$, for each i we have

$$x_i' = \sum_{j=1}^n [m_{ij}(X) - \tilde{m}_{ij}(X)]x_j \quad i = 1, 2, \ldots, n$$

where m_{ij} and $\tilde{m}_{ij}$ are given in (iii).

Now we let $V_i(X) = x_i^2$ and choose

$$u_i(X) = [m_{ii}(X) - \tilde{m}_{ii}(X) + \sum_{j=1}^n |m_{ij}(X) - \tilde{m}_{ij}(X)|]x_i \text{ if } i \in S(X).$$

Then,

$$V_i'(X) = 2x_i x_i'$$

$$= 2x_i[(m_{i1}(X) - \tilde{m}_{i1}(X))x_1 + \ldots + (m_{in}(X) - \tilde{m}_{in}(X))x_n]$$

$$\leq 2x_i[m_{ii}(X) - \tilde{m}_{ii}(X) + \sum_{\substack{j=1 \\ i \neq j}}^n |m_{ij}(X) - \tilde{m}_{ij}(X)|]x_i$$

$$< 0 \text{ for } X \neq 0.$$

When $X = 0$, $V_i'(X) = 0$; therefore it is clear that $E = L = \{0\}$. Hence it follows from Theorems 1 and 2, using $V_0(X) = \max\limits_{j \in S(X)} V_j(X)$ and $\Omega = \Omega^* = R^n$, the trivial solution of the system (1) is globally asymptotically stable.

3. CONCLUSION

Cohen and Grossberg, in their work [2], used LaSalle's invariance principle in terms of a single Lyapunov function and showed that the tendency of the trajectories of (1) to approach equilibrium points is dependent on the symmetry of the matrix $\| c_{ij} \|$ of interaction coefficients. Also, for their result it was necessary to assume that the functions $a_i(\xi)$ and $d_i(\xi)$ are positive for $\xi \in (-\infty, \infty)$. By using the extension of LaSalle's invariance principle to the method of vector Lyapunov functions, we have shown under certain assumptions that the symmetry condition on the interaction coefficients and the positivity of the functions $a_i(\xi)$ and $d_i(\xi)$ are not the necessary conditions for the global asymptotic stability of the trivial solution of the system (1).

REFERENCES

[1]   Bellman, R., Vector Lyapunov functions. *SIAM J. Control and Optimization* 1 (1962),32-34.

[2]   Cohen, M.A., and S. Grossberg, Absolute stability of global pattern formation and parallel memory storage by competitive neural networks, *IEEE Trans. Systems, Man, and Cybernetics* **SMC-13** (1983), 815-826.

[3]   Dong, D.W., *Dynamic Properties of Neural Networks*,   Ph.D. Dissertation, California Institute of Technology (1991).

[4]   Grossberg, S., Nonlinear neural networks: Principles, mechanisms, and architectures, *Neural Networks* 1 (1988), 17-61.

[5]   Hirsch, M.W., Convergent activation dynamics in continuous time networks, *Neural Networks* 2 (1989), 331-349.

[6]   Hopfield, J.J., Neurons with graded response have collective computational properties like those of two-state neurons', *Proc. National Academy of Sciences, USA* 81 (1984), 3088-3092.

[7]   Koksal, S., and S. Sivasundaram, Stability properties of the Hopfield type neural networks, *Dynamics and Stability of Systems* Vol.8, No.3 (1993).

[8]   Koksal, S., and D.W. Fausett, Stability analysis of neural networks using vector Lyapunov functions, *Proc. of First World Congress of Nonlinear Analysts* (to appear).

[9]   Lakshmikantham,V.,V. Matrosov, and S. Sivasundaram, *Vector Lyapunov Functions and Stability Analysis of Nonlinear Systems,* Kluwer Academic Publishers, The Netherlands (1989).

[10] Matsuka, K., Stability conditions for nonlinear continuous neural networks with asymmetric connections weights', *Neural Networks* 5 (1992),495-500.

A Graph Theoretical Approach to Monotonicity with Respect to Initial Conditions

H. KUNZE* and D. SIEGEL† Department of Applied Mathematics, University of Waterloo, Waterloo, Ontario, Canada, N2L 3G1

1 INTRODUCTION

Mathematical conditions for monotone and order preserving flows with respect to an orthant are well known (see [3]). However, conditions for a single component of a system of ordinary differential equations to be monotone with respect to changes in a single initial value of some component in the system have not been developed. We present a graph theoretical approach to this question. An equivalent graphical condition for a flow to be order preserving with respect to an orthant will be given. Sufficient graphical conditions will be given for more general monotonicity results, including a result guaranteeing strictly positive or negative derivatives of components with respect to initial conditions.

We consider the following general system of ordinary differential equations:

$$\dot{\tilde{x}} = \tilde{f}(\tilde{x}), \ \tilde{x} = (x_1, \ldots, x_n) \in \Omega, \ \Omega \subset R^n, \tag{1}$$

where Ω is open, connected and convex and $\tilde{f} \in C^1(\Omega)$. Let $D\tilde{f}$ denote the Jacobian matrix

*Research of this author supported by a Natural Sciences and Engineering Research Council of Canada Post-graduate Scholarship.

†Research of this author supported by a Natural Sciences and Engineering Research Council of Canada Individual Research Grant.

of $\tilde{f}$, namely

$$(D\tilde{f})_{ij} = f_{i,j} = \frac{\partial f_i}{\partial x_j}.$$

For $S \subset \Omega$, suppose that

$$H(\tilde{f}, S) \qquad f_{i,j}(\tilde{x}) \geq 0 \;\; \text{or} \;\; f_{i,j}(\tilde{x}) \leq 0, \forall\, \tilde{x} \in S \subset \Omega, \forall\, i \neq j.$$

Under this hypothesis, we can associate with the matrix $D\tilde{f}$ a signed digraph $G(\tilde{f}, S)$. The vertices are labeled $v_1, \ldots, v_n$, where vertex v_i is associated with solution component x_i, and the edges are constructed in the following way:

G.i) If $f_{j,i} \geq 0$, $\forall\, \tilde{x} \in S$, $i \neq j$, and $f_{j,i} > 0$ at some point of S, a positive edge, labeled e_{ij}, directed from vertex v_i to vertex v_j is drawn in the graph.

G.ii) If $f_{j,i} \leq 0$, $\forall\, \tilde{x} \in S$, $i \neq j$, and $f_{j,i} < 0$ at some point of S, a negative edge, labeled e_{ij}, directed from vertex v_i to vertex v_j is drawn in the graph.

G.iii) If $f_{j,i} = 0$, $\forall\, \tilde{x} \in S$, $i \neq j$, no edge from vertex v_i to vertex v_j is drawn in the graph.

1.1   Notation and Terminology

The following concepts will be required in the upcoming graphical discussion.

There is a *path* between two vertices, v_i and v_j, $i \neq j$, if a series of edges, ignoring direction, connects the vertices. The edges and vertices in a path need not be distinct. There is a *directed path* from v_i to v_j, $i \neq j$, if a series of edges directed from v_i to v_j connects the vertices.

A *cycle* is a closed path and a *directed cycle* is a closed directed path. A *simple cycle* is a cycle consisting of distinct edges and vertices.

The *length* of a path or cycle is the total number of edges comprising it.

The *sign* of a path or cycle is the product of the signs of the edges comprising it.

We say that the ordered vertex pair (v_i, v_j), $i \neq j$, is

(i) *positively (negatively) consistently di-connected* if all directed paths from v_i to v_j are positive (negative) and at least one such path exists,

(ii) *not di-connected* if there is no directed path from v_i to v_j, and

(iii) *inconsistently di-connected* if there is a directed path of each sign from v_i to v_j.

These are the only three possibilities for any distinct vertex pair. We say the vertex pair (v_i, v_i) is *inconsistently di-connected* if v_i is part of a negative directed cycle. Otherwise, we say that (v_i, v_i) is *positively consistently di-connected*.

The ordered vertex pair (v_i, v_j) is said to be *consistent* if (v_i, v_j) is either consistently di-connected or not di-connected.

For example, consider the graph in Figure 1, where positive edges are represented by solid lines and negative edges are represented by dashed lines. In this graph, there are paths between any two vertices. There are directed paths between many pairs of vertices. However, there is no directed path from v_2 to any other vertex. The 4-cycle described by $v_1 e_{13} v_3 e_{35} v_5 e_{54} v_4 e_{41} v_1$ is positive while the 3-cycle described by $v_1 e_{13} v_3 e_{32} v_2 e_{12} v_1$ is negative.

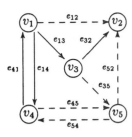

Figure 1: A sample graph

There are no negative directed cycles. There are many directed paths from v_1 to v_2. In particular, the directed path described by $v_1 e_{12} v_2$ is negative and the directed path described by $v_1 e_{13} v_3 e_{35} v_5 e_{52} v_2$ is positive; so, (v_1, v_2) is inconsistently di-connected. (v_1, v_5), for example, is negatively consistently di-connected and (v_2, v_4), for example, is not di-connected.

2   MONOTONE, ORDER PRESERVING, AND CONSISTENT FLOWS

The flow φ for (1) is the mapping $\varphi_t : \Omega \mapsto \Omega$ where $\varphi_t(\tilde{x}) = (\varphi_t^1(\tilde{x}), \ldots, \varphi_t^n(\tilde{x}))$ is the solution to (1) with initial value $\tilde{x}$.

We define the positive interval of existence $I(\tilde{x}) = \{t \geq 0 : \varphi_t(\tilde{x}) \in \Omega\}$ and the positive semi-trajectory $\Gamma(\tilde{x}) = \{\varphi_t(\tilde{x}) : t \in I(\tilde{x})\}$.

We say that φ is a *monotone flow* if $\tilde{w}_0 \leq \tilde{v}_0 \Rightarrow \varphi_t(\tilde{w}_0) \leq \varphi_t(\tilde{v}_0)$, $\forall t \in I(\tilde{w}_0) \cap I(\tilde{v}_0)$.

We say that φ is an *order preserving flow* with respect to an orthant if $\exists$ an $n \times n$ matrix

$$P = diag[P_i] = diag[(-1)^{m_1}, \ldots, (-1)^{m_n}], \ m_i \in \{0, 1\}, \tag{2}$$

such that $P\tilde{w}_0 \leq P\tilde{v}_0 \Rightarrow P\varphi_t(\tilde{w}_0) \leq P\varphi_t(\tilde{v}_0)$, $\forall t \in I(\tilde{w}_0) \cap I(\tilde{v}_0)$.

In [3], H. Smith showed that the system (1) induces a monotone flow if and only if $f_{i,j}(\tilde{x}) \geq 0, \forall \tilde{x} \in \Omega, \forall i \neq j$. One direction of the proof follows from Kamke's Theorem.

He also showed that (1) induces an order preserving flow (with respect to an orthant) if and only if $\exists$ an $n \times n$ matrix P as in (2) such that $P_i f_{i,j} P_j \geq 0, \forall i \neq j$.

We see that an order preserving flow with P the identity matrix is a monotone flow and that we have a monotone flow if and only if all edges in $G(\tilde{f}, \Omega)$ are positive.

Our first Theorem gives a graph theoretical equivalent to the above condition for an order preserving flow. We will need two simple observations before presenting this result.

LEMMA 1 *Every negative cycle contains a negative simple cycle.*

Proof: The proof is by induction on the length of the negative cycle. Suppose the negative cycle has length 2. Then it must consist of one edge of each sign and the result follows.

Suppose the result is true for negative cycles of length $\leq m$. We consider a negative cycle of length $m + 1$. If the negative cycle has no repeated vertices, the result follows. If the negative cycle has a repeated vertex, then the negative cycle is a union of two cycles, one of each sign. The negative cycle in this union must have length $\leq m - 1$, and therefore must contain a negative simple cycle by the induction hypothesis. $\square$

LEMMA 2 *If $\tilde{f}$ induces an order preserving flow and the graph $G(\tilde{f}, \Omega)$ has a positive (negative) directed edge labeled e_{ij} from vertex v_i and v_j, $i \neq j$, then $P_i P_j = 1$ (-1) where $P = diag[P_i]$ is the matrix associated with the order preserving flow.*

Proof: Suppose e_{ij} has positive sign. Then $f_{j,i} \geq 0$ in Ω and $f_{j,i} > 0$ at some point of Ω. Since $\tilde{f}$ gives an order preserving flow, there is a matrix P as in (2) such that $P_j f_{j,i} P_i \geq 0$. Evaluating at the point where $f_{j,i} > 0$, we must have $P_j P_i > 0$ and hence $P_i P_j = 1$. The other case is argued in the same way.   □

THEOREM 1 *Under hypothesis $H(\tilde{f}, \Omega)$, the system (1) induces an order preserving flow with respect to an orthant if and only if there are no negative simple cycles in the graph $G(\tilde{f}, \Omega)$.*

Proof: If $\tilde{f}$ induces an order preserving flow, there is a matrix P as in (2). Suppose that the graph $G(\tilde{f}, \Omega)$ has a negative simple cycle with vertices $v_{l_1}, \ldots, v_{l_k}$, where v_{l_i} is connected to $v_{l_{i+1}}$, $1 \leq i \leq k$, with the convention $v_{l_{k+1}} = v_{l_1}$. By Lemma 2, $P_{l_i} P_{l_{i+1}}$ is the sign of the (undirected) edge between v_{l_i} and $v_{l_{i+1}}$. $(P_{l_1} P_{l_2})(P_{l_2} P_{l_3}) \cdots (P_{l_k} P_{l_1}) = -1$ since the cycle is negative. This is a contradiction because $(P_{l_1} P_{l_2})(P_{l_2} P_{l_3}) \cdots (P_{l_k} P_{l_1}) = (P_{l_1} P_{l_2} \cdots P_{l_k})^2 = 1$.

Next, we prove that if $G(\tilde{f}, \Omega)$ has no negative simple cycles then we must have an order preserving flow. The graph $G(\tilde{f}, \Omega)$ consists of connected components $G_1(f, \Omega), \ldots, G_p(f, \Omega)$. Since the variables corresponding to the vertices in two different subgraphs do not interact, we need only consider a connected subgraph of $G(\tilde{f}, \Omega)$, say $G_1(f, \Omega)$ with n_1 vertices.

Choose any vertex v_1 in $G_1(f, \Omega)$. Since there are no negative simple cycles, every vertex in the subgraph is connected to v_1 by paths of only one sign. If not, then v_1 is connected to v_k, say, by paths of both sign. Hence, v_1 is part of a negative cycle (combining the two paths) and Lemma 1 applies, giving a contradiction. We define the disjoint sets

$$\mathcal{Q} = \{v_k : v_1 \text{ and } v_k \text{ are only connected by positive paths}\}, \text{ and}$$
$$\mathcal{R} = \{v_k : v_1 \text{ and } v_k \text{ are only connected by negative paths}\}.$$

In order to avoid a simple contradiction, vertices in $\mathcal{Q}$ can only be connected to each other by positive paths, vertices in $\mathcal{R}$ can only be connected to each other by positive paths, and vertices in $\mathcal{Q}$ can only be connected to vertices in $\mathcal{R}$ by negative paths. We relabel the vertices in $G_1(f, \Omega)$ so that $v_2, \ldots, v_q \in \mathcal{Q}$ and $v_{q+1}, \ldots, v_{n_1} \in \mathcal{R}$. Hence,

$$\begin{aligned}
f_{i,k} &\geq 0, \quad 1 \leq i, k \leq q, \ i \neq k, \\
f_{i,k} &\geq 0, \quad q+1 \leq i, k \leq n_1, \ i \neq k, \\
f_{i,k} &\leq 0, \quad 1 \leq i \leq q, \ q+1 \leq k \leq n_1, \\
f_{i,k} &\leq 0, \quad 1 \leq k \leq q, \ q+1 \leq i \leq n_1.
\end{aligned}$$

For this subsystem, Df has the sign pattern

$$\left(\frac{\partial f_i}{\partial x_j} \right) = \left(\begin{array}{ccc|ccc} \cdot & & + & & & \\ & \cdot & & & & - \\ + & & \cdot & & & \\ \hline & & & \cdot & & + \\ & & - & & \cdot & \\ & & & + & & \cdot \end{array} \right)$$

$$\underbrace{}_{q} \quad \underbrace{}_{n_1 - q}$$

where '+' means the corresponding partial derivative is non-negative and '−' means it is non-positive. Hence, choosing the matrix $P = diag[1, \ldots, 1, -1, \ldots, -1]$, where P has q entries of 1 and $n_1 - q$ entries of -1, means that $P(Df)P$ has non-negative off-diagonal entries. Thus, this subsystem gives an order preserving flow.   □

Although a version of Theorem 1 was stated informally in [3], we were unable to locate a careful statement or proof in the literature.

As we will be focusing from now on on the signs of partial derivatives of components with respect to initial conditions we can drop the convexity assumption on Ω. Furthermore, it will suffice to only make assumptions on $G(\tilde{f}, \Gamma(\tilde{x}))$ rather than $G(\tilde{f}, \Omega)$. The following Theorem shows how to determine the signs of partial derivatives with respect to initial conditions when $G(\tilde{f}, \Gamma(\tilde{x}))$ has no negative simple cycles.

THEOREM 2 *Suppose that hypothesis $H(\tilde{f}, \Gamma(\tilde{x}))$ holds and that $G(\tilde{f}, \Gamma(\tilde{x}))$ has no negative simple cycles. If all paths connecting v_i and v_j in $G(\tilde{f}, \Gamma(\tilde{x}))$ are positive then*

$$\frac{\partial \varphi_t^j}{\partial x_i}(\tilde{x}) \geq 0 \ and \ \frac{\partial \varphi_t^i}{\partial x_j}(\tilde{x}) \geq 0,$$

$\forall \ t \in I(\tilde{x})$. *If all paths connecting v_i and v_j in $G(\tilde{f}, \Gamma(\tilde{x}))$ are negative, then the derivatives are both non-positive.*

Proof: If all paths connecting v_i and v_j in $G(\tilde{f}, \Gamma(\tilde{x}))$ are of one sign then v_i and v_j must be part of a connected subgraph $G_1(f, \Gamma(\tilde{x}))$ which has no negative simple cycles. The argument used to prove Theorem 1 then applies and gives the result.   □

The following result shows that we can still draw conclusions on the signs of these partial derivatives when $G(\tilde{f}, \Gamma(\tilde{x}))$ has a negative simple cycle. This result makes use of the directions of the edges in $G(\tilde{f}, \Gamma(\tilde{x}))$.

THEOREM 3 *Suppose hypothesis $H(\tilde{f}, \Gamma(\tilde{x}))$ holds. If (v_i, v_i) is consistent in $G(\tilde{f}, \Gamma(\tilde{x}))$ {i.e. v_i is not part of a negative directed cycle in $G(\tilde{f}, \Gamma(\tilde{x}))$}, then*

$$\frac{\partial \varphi_t^i}{\partial x_i}(\tilde{x}) > 0, \ \forall \ t \in I(\tilde{x}).$$

Furthermore, if (v_i, v_j), $i \neq j$, is consistent in $G(\tilde{f}, \Gamma(\tilde{x}))$ then, $\forall \ t \in I(\tilde{x})$,

$$\frac{\partial \varphi_t^j}{\partial x_i}(\tilde{x}) \begin{cases} = 0 \ if \ (v_i, v_j) \ is \ not \ di\text{-}connected \ in \ G(\tilde{f}, \Gamma(\tilde{x})) \\ \geq 0 \ if \ (v_i, v_j) \ is \ positively \ consistently \ di\text{-}connected \ in \ G(\tilde{f}, \Gamma(\tilde{x})) \\ \leq 0 \ if \ (v_i, v_j) \ is \ negatively \ consistently \ di\text{-}connected \ in \ G(\tilde{f}, \Gamma(\tilde{x})) \end{cases}.$$

Proof: Suppose (v_1, v_j) is consistent in $G(\tilde{f}, \Gamma(\tilde{x}))$. For any vertex v_k, (v_1, v_k) is either consistently di-connected, inconsistently di-connected or not di-connected. Define the disjoint sets

$$\begin{aligned} \mathcal{Q}_1 &= \{v_k : (v_1, v_k) \ is \ positively \ consistently \ di\text{-}connected \ in \ G(\tilde{f}, \Gamma(\tilde{x}))\}, \\ \mathcal{Q}_2 &= \{v_k : (v_1, v_k) \ is \ negatively \ consistently \ di\text{-}connected \ in \ G(\tilde{f}, \Gamma(\tilde{x}))\}, \\ \mathcal{R} &= \{v_k : (v_1, v_k) \ is \ not \ di\text{-}connected \ in \ G(\tilde{f}, \Gamma(\tilde{x}))\}, \ and \\ \mathcal{S} &= \{v_k : (v_1, v_k) \ is \ inconsistently \ di\text{-}connected \ in \ G(\tilde{f}, \Gamma(\tilde{x}))\}. \end{aligned}$$

We relabel the vertices so that $v_1, \ldots, v_{q_1} \in \mathcal{Q}_1$, $v_{q_1+1}, \ldots, v_{q_2} \in \mathcal{Q}_2$, $v_{q_2+1}, \ldots, v_r \in \mathcal{R}$, and $v_{r+1}, \ldots, v_n \in \mathcal{S}$. With the corresponding relabelling of x_i, $1 \le i \le n$, the system (1) takes on the form

$$
\begin{aligned}
\dot{x}_i &= f_i(x_1, \ldots, x_{q_2}, x_{q_2+1}, \ldots, x_r), & 1 \le i \le q_2, \\
\dot{x}_i &= f_i(x_{q_2+1}, \ldots, x_r), & q_2 + 1 \le i \le r, \\
\dot{x}_i &= f_i(x_1, \ldots, x_{q_2}, x_{q_2+1}, \ldots, x_r, x_{r+1}, \ldots, x_n), & r+1 \le i \le n.
\end{aligned}
$$

Since (v_1, v_j) is consistent, either

(i) $q_2 + 1 \le j \le r$. Then

$$
\frac{\partial \dot{\varphi}_t^j}{\partial x_1}(\tilde{x}) = 0, \ \forall\, t \in I(\tilde{x}).
$$

Or

(ii) $1 \le j \le q_2$. Then,

$$
\begin{aligned}
\frac{\partial \dot{\varphi}_t^j}{\partial x_1}(\tilde{x}) &= \sum_{k=1}^{q_2} f_{j,k} \frac{\partial \varphi_t^k}{\partial x_1}(\tilde{x}) + \sum_{k=q_2+1}^{r} f_{j,k} \frac{\partial \varphi_t^k}{\partial x_1}(\tilde{x}) \\
&= \sum_{k=1}^{q_2} f_{j,k} \frac{\partial \varphi_t^k}{\partial x_1}(\tilde{x}).
\end{aligned}
\tag{3}
$$

We can write (3) as $\dot{Y}(t) = MY(t)$ where Y is a q_2 vector and M is a $q_2 \times q_2$ matrix with $Y_j = \frac{\partial \varphi_t^j}{\partial x_1}$ and $M_{jk} = f_{j,k}$. Choosing

$$
P = diag[P_i] = diag[1, \ldots, 1, -1, \ldots, -1],
\tag{4}
$$

where P has q_1 entries of 1 and $q_2 - q_1$ entries of -1, means that PMP has non-negative off-diagonal entries. Note that $P_1 = 1$.

We let $Z = PY$. Since $Y(0) = E$, then $Z(0) = E$, where $E = (1, 0, \ldots, 0)^T$. We will prove that $Z \ge 0$, $\forall\, t \in I(\tilde{x})$, which by the construction of Z gives the remainder of the second result (with $i = 1$). Now,

$$
\dot{Y} = P\dot{Z} = MY = MPZ.
$$

So, we have

$$
\dot{Z} = (PMP)Z.
\tag{5}
$$

Since $(PMP)Z$ is quasimonotone nondecreasing, $Z(t) \ge 0$, $\forall\, t \ge 0$, by Kamke's Theorem.

We still need to prove the first result. Take $j = 1$ in the above setup. We will show that $Z_1(t) > 0$, $\forall\, t \in I(\tilde{x})$. From (5), we have

$$
\dot{Z} + \lambda Z = (PMP + \lambda)Z = \tilde{M}Z,
$$

where λ is chosen large enough so that $\tilde{M} = PMP + \lambda I \ge 0$. Solving for Z in terms of the right hand side, we get

$$
Z(t) = e^{-\lambda t} E + \int_0^t e^{\lambda(s-t)} \tilde{M}(Z(s)) Z(s) ds.
\tag{6}
$$

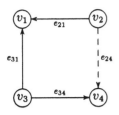

Figure 2: Consistent graph with corresponding flow that is not order preserving.

We can immediately conclude from (6) that $Z_1(t) > 0, \forall\, t \in I(\tilde{x})$.   □

We say that (1) gives a *consistent flow* if, for each i and j and $\forall\, \tilde{x} \in \Omega$, either

$$\frac{\partial \varphi_t^j}{\partial x_i}(\tilde{x}) \geq 0, \ \forall\, t \in I(\tilde{x}), \ \text{ or } \ \frac{\partial \varphi_t^j}{\partial x_i}(\tilde{x}) \leq 0, \ \forall\, t \in I(\tilde{x}).$$

If (v_i, v_j) is consistent in $G(\tilde{f}, S)$ for each i and j, we say that $G(\tilde{f}, S)$ is consistent. Thus, by Theorem 3, we get the following Corollary.

COROLLARY *If (1) gives a consistent graph $G(\tilde{f}, \Omega)$, then (1) gives a consistent flow.*

Consider the example graph $G(\tilde{f}, \Omega)$ in Figure 2. The graph consists of a single negative cycle and therefore does not correspond to an order preserving flow. However, there are no inconsistently di-connected vertices. This is a consistent graph and, by Theorem 3 we can immediately state the sign pattern for the matrix of partial derivatives of solution components with respect to initial conditions, namely,

$$\left(\frac{\partial \varphi_t^i}{\partial x_j}(\tilde{x}) \right) = \begin{pmatrix} + & 0 & 0 & 0 \\ + & + & 0 & - \\ + & 0 & + & + \\ 0 & 0 & 0 & + \end{pmatrix},$$

where a '+' ('−','0') means that the corresponding partial derivative is non-negative (non-positive, identically zero) and a '+' entry on the diagonal means the corresponding partial derivative is positive.

We can also use Theorem 3 to give partial results on the signs of partial derivatives with respect to initial values. Returning to the example graph in Figure 1, we see that (v_i, v_2) is inconsistently di-connected for $i \neq 2$. We can predict the following sign pattern for the matrix of partial derivatives of solution components with respect to initial conditions:

$$\left(\frac{\partial \varphi_t^i}{\partial x_j}(\tilde{x}) \right) = \begin{pmatrix} + & ? & + & + & - \\ 0 & + & 0 & 0 & 0 \\ + & ? & + & + & - \\ + & ? & + & + & - \\ - & ? & - & - & + \end{pmatrix},$$

where a '+' ('−','0') means that the corresponding partial derivative is non-negative (non-positive, identically zero), a '+' entry on the diagonal means the corresponding partial derivative is positive, and a '?' means we cannot predict the sign of the corresponding partial derivative.

The condition that $G(\tilde{f}, \Omega)$ be consistent is a sufficient, but not a necessary, condition for a consistent flow as the following example illustrates.

EXAMPLE: Consider

$$\dot{x}_1 = f_1 = \begin{cases} x_2^2, & x_2 \ge 0 \\ 0, & x_2 < 0 \end{cases} , \quad \dot{x}_2 = f_2 = \begin{cases} 0, & x_2 \ge 0 \\ -x_1 x_2^2, & x_2 < 0 \end{cases} .$$

One can check that $f_{1,2} \ge 0$ and $f_{2,1} \le 0$, $\forall \tilde{x} \in \Omega = R^2$, so the graph, $G(\tilde{f}, \Omega)$, for this system is as follows:

$$\widehat{v_1} \xleftarrow[e_{21}]{\overset{e_{12}}{\dashrightarrow}} \widehat{v_2}$$

The vertices are inconsistently di-connected and Theorem 3 does not apply. Yet, solving the system gives:

$$\varphi_t^1(x_1, x_2) = \begin{cases} (x_2)^2 t + x_1, & x_2 \ge 0 \\ x_1, & x_2 < 0 \end{cases} , \quad \varphi_t^2(x_1, x_2) = \begin{cases} x_2, & x_2 \ge 0 \\ \frac{x_2}{1 + x_1 x_2 t}, & x_2 < 0 \end{cases} .$$

This gives the following sign pattern for partial derivatives with respect to initial conditions:

$$\left(\frac{\partial \varphi_t^i}{\partial x_j}(\tilde{x}) \right) = \begin{pmatrix} + & + \\ - & + \end{pmatrix}$$

where a '+' ('−') means that the corresponding partial derivative is non-negative (non-positive) and a '+' entry on the diagonal means the corresponding partial derivative is positive. By definition, this is a consistent flow. This result could be obtained by combining the conclusions of Theorem 3 for each of the half-planes $x_2 \ge 0$ and $x_2 < 0$.

3   A STRICT SIGN RESULT

The graph $G(\tilde{f}, \tilde{x}_1)$, where $\tilde{x}_1$ is some point on $\Gamma(\tilde{x})$, can tell us when we have strict positivity or negativity of partial derivatives with respect to initial conditions.

THEOREM 4 *Suppose hypothesis $H(\tilde{f}, \Gamma(\tilde{x}))$ holds. For some $t_1 \in I(\tilde{x})$, let $\tilde{\varphi}_{t_1}(\tilde{x}) = \tilde{x}_1$. Then, $\forall t \ge t_1$ ($\forall t > 0$ if $t_1 = 0$), $t \in I(\tilde{x})$, $i \ne j$,*

$$\frac{\partial \varphi_t^j}{\partial x_i}(\tilde{x}) \begin{cases} > 0 \ \text{if } (v_i, v_j) \text{ is positively consistently di-connected in } G(\tilde{f}, \Gamma(\tilde{x})) \text{ and } G(\tilde{f}, \tilde{x}_1) \\ < 0 \ \text{if } (v_i, v_j) \text{ is negatively consistently di-connected in } G(\tilde{f}, \Gamma(\tilde{x})) \text{ and } G(\tilde{f}, \tilde{x}_1) \end{cases} .$$

Proof: The proof begins in a similar way to that of Theorem 3. Suppose (v_1, v_j), $j \ne 1$, is consistently di-connected in $G(\tilde{f}, \Gamma(\tilde{x}))$ and in $G(\tilde{f}, \tilde{x}_1)$, with $t_1 > 0$. Define the disjoint sets

$$\begin{aligned}
\mathcal{Q}_1 &= \{v_k : (v_1, v_k) \text{ is positively consistently di-connected in } G(\tilde{f}, \Gamma(\tilde{x})) \text{ and } G(\tilde{f}, \tilde{x}_1)\}, \\
\mathcal{Q}_2 &= \{v_k : (v_1, v_k) \text{ is negatively consistently di-connected in } G(\tilde{f}, \Gamma(\tilde{x})) \text{ and } G(\tilde{f}, \tilde{x}_1)\}, \\
\mathcal{R}_1 &= \{v_k : (v_1, v_k) \text{ is positively consistently di-connected in } G(\tilde{f}, \Gamma(\tilde{x})) \text{ and not} \\
&\qquad \text{di-connected in } G(\tilde{f}, \tilde{x}_1)\}, \\
\mathcal{R}_2 &= \{v_k : (v_1, v_k) \text{ is negatively consistently di-connected in } G(\tilde{f}, \Gamma(\tilde{x})) \text{ and not} \\
&\qquad \text{di-connected in } G(\tilde{f}, \tilde{x}_1)\}, \text{ and} \\
\mathcal{S} &= \{v_k : (v_1, v_k) \text{ is inconsistently di-connected in } G(\tilde{f}, \Gamma(\tilde{x})) \text{ or } (v_1, v_k) \text{ is not} \\
&\qquad \text{di-connected in } G(\tilde{f}, \Gamma(\tilde{x}))\}.
\end{aligned}$$

We relabel the vertices so that $v_1, \ldots, v_{q_1} \in \mathcal{Q}_1$, $v_{q_1+1}, \ldots, v_{r_1} \in \mathcal{R}_1$, $v_{r_1+1}, \ldots, v_{q_2} \in \mathcal{Q}_2$, $v_{q_2+1}, \ldots, v_{r_2} \in \mathcal{R}_2$, and $v_{r_2+1}, \ldots, v_n \in \mathcal{S}$. Performing the corresponding relabelling of x_i, $2 \leq i \leq n$, puts the system (1) in the form

$$
\begin{aligned}
\dot{x}_i &= f_i(x_1, \ldots, x_{r_2}), & 1 \leq i \leq r_2, \\
\dot{x}_i &= f_i(x_1, \ldots, x_n), & r_2 + 1 \leq i \leq n.
\end{aligned}
$$

We will show that for $t \geq t_1$,

$$
\frac{\partial \varphi_t^j}{\partial x_1}(\tilde{x})
\begin{cases}
> 0, & 1 < j \leq q_1 \\
< 0, & r_1 + 1 \leq j \leq q_2
\end{cases}.
$$

For $1 \leq i \leq r_2$, we proceed in as in the proof of Theorem 3. We have,

$$
\frac{\partial \dot{\varphi}_t^i}{\partial x_1}(\tilde{x}) = \sum_{k=1}^{r_2} f_{i,k} \frac{\partial \varphi_t^k}{\partial x_1}(\tilde{x}). \tag{7}
$$

We can write (7) as $\dot{Y}(t) = MY(t)$ where Y is an r_2 vector and M is an $r_2 \times r_2$ matrix with $Y_i = \frac{\partial \varphi_t^i}{\partial x_1}$ and $M_{ik} = f_{i,k}$. Choosing

$$
P = diag[P_i] = diag[1, \ldots, 1, -1, \ldots, -1], \tag{8}
$$

where P has r_1 entries of 1 and $r_2 - r_1$ entries of -1, means that PMP has non-negative off-diagonal entries.

Let $Z = PY$. By Theorem 3, we know that $Z_k \geq 0$, $\forall\, t \in I(\tilde{x})$, $k \neq 1$, and that $Z_1 > 0$, $\forall\, t \in I(\tilde{x})$. We will prove that $Z_j(t) > 0$, $\forall\, t \geq t_1$, $1 < j \leq q_1$ and $r_1 + 1 \leq j \leq q_2$. Then the conclusion of the theorem would follow. As in the proof of Theorem 3, we have

$$
\dot{Z} = (PMP)Z \Rightarrow \dot{Z} + \lambda Z = (PMP + \lambda)Z = \tilde{M}Z,
$$

where λ is chosen large enough so that $\tilde{M} = PMP + \lambda I \geq 0$. Solving for Z in terms of the right hand side, we get

$$
Z(t) = e^{-\lambda t} E + \int_0^t e^{\lambda(s-t)} \tilde{M}(Z(s)) Z(s) ds. \tag{9}
$$

The proof will now proceed by induction on the length of the shortest directed path in $G(\tilde{f}, \tilde{x}_1)$ from v_1 to v_j. Suppose that a shortest directed path from v_1 to v_j in $G(\tilde{f}, \tilde{x}_1)$ has length 1. Then

$$
\begin{aligned}
Z_j(t) &= \int_0^t e^{\lambda(s-t)} \sum_{l=1}^{r_2} \tilde{M}_{jl} Z_l \, ds \\
&\geq \int_0^t e^{\lambda(s-t)} \tilde{M}_{j1} Z_1 \, ds.
\end{aligned} \tag{10}
$$

Since $\tilde{M}_{j1}(t_1) = P_j f_{j,1}(\tilde{x}_1) P_1 > 0$, we conclude that $Z_j(t) > 0$ for $t \geq t_1$.

Now suppose the result is true if the shortest directed path has length m. We consider the case when the shortest directed path from v_1 to v_j in $G(\tilde{f}, \tilde{x}_1)$ has length $m+1$. Suppose the intermediate vertices are $v_{k_1}, \ldots, v_{k_m}$, with v_{k_1} adjacent to v_1, v_{k_l} adjacent to $v_{k_{l+1}}$,

$1 \leq l \leq m - 1$, and v_{k_m} adjacent to v_j. Note that each k_l satisfies either $1 < k_l \leq q_1$ or $r_1 + 1 \leq k_l \leq q_2$. Then

$$
Z_j(t) = \int_0^t e^{\lambda(s-t)} \sum_{l=1}^{r_2} \tilde{M}_{jl} Z_l ds
$$

$$
\geq \int_0^t e^{\lambda(s-t)} \tilde{M}_{jk_m} Z_{k_m} ds. \tag{11}
$$

Again, $\tilde{M}_{jk_m}(t_1) = P_j f_{j,k_m}(\tilde{x}_1) P_{k_m} > 0$. In order for a shortest directed path from v_1 to v_j to have length $m + 1$, a shortest directed path from v_1 to v_{k_m} must have length m. Hence, $Z_{k_m}(t) > 0$, $t \geq t_1$, by the induction hypothesis. Thus, $Z_j(t) > 0$, $t \geq t_1$. The proof by induction is now complete.

The case $t_1 = 0$ is argued in exactly the same way. □

Theorem 4 requires some knowledge of the solution trajectory in order to be useful. The following Corollary gives a result which does not require any information about the solution trajectory.

COROLLARY *If (v_i, v_j) is positively (negatively) consistently di-connected in $G(\tilde{f}, \Omega)$ and in $G(\tilde{f}, \tilde{x})$, $\forall \tilde{x} \in \Omega$, then, $\forall t \in I(\tilde{x}) \backslash \{0\}$,*

$$
\frac{\partial \varphi_t^j}{\partial x_i}(\tilde{x}) > 0 \quad \left(\frac{\partial \varphi_t^j}{\partial x_i}(\tilde{x}) < 0 \right).
$$

A particular case of this Corollary is stated by M. Hirsch in [1] and in [2]: *If $\tilde{f}$ is a cooperative vector field and $D\tilde{f}(\tilde{x})$ is also irreducible for all $\tilde{x}$, then $\{\phi_t\}$ has positive derivatives.*

A cooperative vector field is one that satisfies $f_{i,j}(\tilde{x}) \geq 0$, $\forall \tilde{x} \in \Omega$, $\forall i,j, i \neq j$. $D\tilde{f}(\tilde{x})$ is irreducible means that $G(\tilde{f}, \tilde{x})$ is strongly connected, i.e. there is a directed path between any two distinct vertices. It should be pointed out that the proofs given in [1] and [2] are incorrect (this has been confirmed by M. Hirsch). A formula similar to (9) was introduced in [1] but there is no induction argument as presented above.

References

[1] M.W. Hirsch, Differential equations and convergence almost everywhere in strongly monotone semiflows, Contemporary Mathematics, **17** (1983), 267–285.

[2] M.W. Hirsch, Systems of differential equations that are competitive or cooperative II: Convergence almost everywhere, SIAM J. Math. Anal., **16** (1985), 423–439.

[3] H. Smith, Systems of ordinary differential equations which generate an order preserving flow, SIAM Rev., **30** (1988), 87–113.

On the Stabilization of Uncertain Differential Systems

A.B. Kurzhanski, Moscow State University, Moscow, Russia

1   Introduction

The Problem of ensuring the stability of motion [1] under incomplete information on the system model is one of the topics dealt with by the present theory of feedback control [2, 3, 4].

This paper deals with the problem of stabilizing a time–variant linear system

$$\dot{x} = A(t)x + u + v \tag{1.1}$$

with an unknown but bounded input

$$v = v(t)$$

through the selection of feedback controls

$$u = U(t, x)$$

with magnitude bounds. The respective levels of the restrictions on the unknown input and on the control may allow either to asymptotically stabilize the system (uniformly in v) if the restriction on u prevails in some sense over v (with no matching conditions presumed, in general) or only to "practically" stabilize it when one can ensure that the trajectory $x(t)$ of the system is kept, after some instant of time τ, within a prescribed ellipsoidal set $\mathcal{E}$ centered in the origin, whatever is the disturbance $v = v(t)$. In the latter case the control u should not necessarily prevail over v, but the solvability of the problem will depend on the behaviour of the solution tube to a certain evolution equation of the "funnel" type. Therefore the emphasis would also be on a finite–time horizon control synthesis problem.

The specific points of this paper are such that

(a) the system is *time–variant* with continuous coefficients $A(t)$.

217

(b) the controls u and inputs v are both *bounded* by continuous set–valued functions with values in convex compact sets, namely

$$u \in \mathcal{P}(t), \quad v \in \mathcal{Q}(t), \tag{1.2}$$

where $\mathcal{P}(t) \in \mathrm{comp}\, R^n$, $\mathcal{Q}(t) \in \mathrm{comp}\, R^n$, $\{0\} \in \mathcal{P}(t)$, $\{0\} \in \mathcal{Q}(t)$ and $\mathrm{comp}\, R^n$ stands for the variety of convex compact subsets of R^n,

(c) the conventional "matching conditions"

$$\mathcal{P}(t) = k\mathcal{Q}(t), \quad 0 < k < 1,$$

are *not presumed*.

We further suppose that the sets $\mathcal{P}(t)$, $\mathcal{Q}(t)$ are ellipsoidal–valued, namely

$$\mathcal{P}(t) = \mathcal{E}(0, P(t)), \quad \mathcal{Q}(t) = \mathcal{E}(0, Q(t)), \tag{1.3}$$

where symbol $\mathcal{E}(a, P)$ stands for the ellipsoid

$$\mathcal{E}(a, P) = \{x : (x - a)'P^{-1}(x - a) \le 1\}$$

and P^{-1} is either the inverse of the symmetrical matrix $P > 0$ or the generalized (Moore–Penrose) inverse if $P \ge 0$, with $a \in R^n$.

2   The Stabilization Problem

Given equation (1.1) with constraints (1.2), (1.3) on u, v, we first wish, by selecting a feedback control strategy $u = U(t, x)$ ($U(t, x)$ may be set–valued), to "stabilize" the system in the following sense.

Given an ellipsoidal set

$$\mathcal{E}(0, T) = \{x : (x, T^{-1}x) \le 1\}, \quad T > 0,$$

find a ("practically") stabilizing control strategy multivalued map $u = U(t, x) \subseteq \mathcal{E}(0, P(t))$, such that for any initial position $\{\tau, x(\tau)\}$ there exists an instant of time $\sigma > \tau$ such that the solution tube $X(t, \tau, x(\tau)) = X[t]$ to the differential inclusion

$$\dot{x} \in A(t)x + U(t, x) + \mathcal{Q}(t) \tag{2.1}$$

would satisfy the property

$$X[t] \subseteq \mathcal{E}(0, T), \quad t \ge \sigma.$$

Let us first fix the instant σ. Then this develops into the solution of the following two problems:

Problem 1-(a) *(The terminal control problem under uncertainty)* is to ensure the inclusion

$$X[\sigma] \in \mathcal{E}(0, T).$$

Problem 1-(b) *(The problem of "viability under counteraction")* is to ensure the inclusion

$$X[t] \subseteq \mathcal{E}(0, T), \quad t \ge \sigma.$$

It is natural that the given reasoning is true, provided $x(\tau) \in W[\tau]$ where $W[\tau]$ is the set of states from which the solution to Problem 1-(a) does exist at all. Let us therefore start with the description of the solvability domains and the control strategies involved.

3 Solvability Tubes. The Synthesizing Strategies.

Starting with the terminal control Problem 1-(a), instants t, σ given, define $W[\tau]$ as the set of all states $x = x(\tau)$ from which the solution to Problem 1-(a) does exist.

As indicated in [5], the set–valued function $W[t]$, $\tau \leq t \leq \sigma$ (the "solvability tube") satifies an evolution equation of the "funnel type" which is

$$\lim_{\epsilon \to +0} \epsilon^{-1} h_+(W(t-\epsilon) - \epsilon \mathcal{Q}(t), \, W(t) - \epsilon \mathcal{P}(t)) = 0, \tag{3.1}$$

$$\tau \leq t \leq \sigma$$

with boundary condition

$$W[\sigma] = \mathcal{E}(0, T). \tag{3.2}$$

Here

$$h_+(W', W'') = \min\{\alpha | W' \subseteq W'' + \alpha S(0), \ \alpha > 0\},$$

$$S(0) = \{s : (s, s) \leq 1\}$$

is the *Hausdorff semidistance* for sets W', W'', so that if $h_-(W', W'') = h_+(W'', W')$, then

$$h(W', W'') = \max\{h_+(W', W''), \ h_-(W', W'')\}$$

is the conventional *Hausdorff distance* for W', $W'' \in \text{comp } R^n$.

The solution $W[t]$ to (3.1) depends on σ, $\mathcal{E}(0, T)$, so that $W[t] = W(t, \sigma, \mathcal{E}(0, T))$.

The control strategy $U(t, x)$ for solving Problem 1-(a) may now be defined as the following set–valued map

$$U(t, x) = \{u : \max_v \{dV(t, x)/dt \mid (2.1)\} \leq 0, \ v \in \mathcal{E}(0, Q(t))\} \tag{3.3}$$

where

$$V(t, x) = h_+^2(x, W[t]).$$

As indicated in [5], the solution to (3.1), (3.2) is given with int $W[\tau] \neq \emptyset$ by Pontriagin's "alternated integral".

The following proposition is true, whose proof is due to [5].

Theorem 2.1. *Once the solvability tube $W[t] \neq \emptyset$ for $\tau \leq t \leq \sigma$, the strategy $U(t, x)$ gives a solution to Problem 1-(a) for any initial state $\{\tau, x\}$, provided $x \in W[\tau]$.*

Lemma 2.1 *The set–valued map $U^*(t, x)$ is continuous in t and upper semicontinuous in x.*

This is due to the continuity of $W[t]$ and ensures the existence of solutions to the differential inclusion (2.1), $u = U$.

Assuming $x \in \mathcal{E}(0, T)$, σ fixed, let us now discuss Problem 1-(b). Take a time-instant $\sigma_k = k\sigma$ with parameter $k > 1$. Define $W_k[\sigma]$ to be the set of all states $x^* \in R^n$ for which the folowing problem is resolvable:
there exists a strategy $U(t, x)$ such that the solution tube $X(t, \tau, x^*) = X[t]$ to the differential inclusion

$$\dot{x} \in A(t)x + U(t, x) + \mathcal{Q}(t)$$

$$x(\tau) = x^*$$

satisfies the condition

$$X(t, \tau, x^*) \subseteq \mathcal{E}(0, T), \quad \sigma \leq t \leq k\sigma.$$

As indicated in [5], the set $W_k[t] = W(t, k\sigma, \mathcal{E}(0, T))$ satisfies the following equation of the "funnel" type[1]:

$$\lim_{\epsilon \to 0} \epsilon^{-1} h_+ \left(W(t - \epsilon) - \epsilon \mathcal{Q}(t), \ (W(t) - \epsilon \mathcal{P}(t)) \cap \mathcal{E}(0, T) \right) = 0, \qquad (3.4)$$

$$W(k\sigma) = \mathcal{E}(0, T).$$

In order that the trajectory tube

$$X(t, \tau, x^*) \subseteq \mathcal{E}(0, T), \quad \sigma \le t \le k\sigma,$$

it is nessesary and sufficient [5] that $x^* \in W_k[\sigma]$. The respective strategy is given by

$$U_k(t, x) = \left\{ u : \ \max_v \{ dV_k(t, x)/dt \mid (1.1) \} \le 0, \ v \in \mathcal{E}(0, \mathcal{Q}(t)) \right\},$$

where $V_k(t, x) = h_+^2(x, W_k[t])$.

According to the definition of $W_k[t]$, we have

$$W_k[\sigma] \subseteq \mathcal{E}(0, T).$$

However, in order to connect the solution of Problem 1-(a) with the latter one, it is sufficient that one would have

$$W_k[t] = \mathcal{E}(0, T), \quad \sigma \le t \le k\sigma. \qquad (3.5)$$

Lemma 2.2. *Under condition (3.5), for any state $x^* \in W[\tau]$ there exists a strategy*

$$U^0(t, x) = \begin{cases} U(t, x), & \tau \le t \le \sigma \\ U_k(t, x), & \sigma \le t \le k\sigma \end{cases}$$

such that the solution tube

$$X^0[t] = X(t, \sigma, x^0)$$

to the differential inclusion (2.1) does satisfy the condition

$$X^0[t] \subseteq \mathcal{E}(0, T), \quad \sigma \le t \le k\sigma$$

The proof of this assertion follows from the definition of U, U_k.

Let us now investigate the fulfillment of (3.5). A sufficient condition for the latter would be that

$$\left\{ u : \ \max_v \frac{d}{dt} d^2(x, \mathcal{E}(0, T)) \mid (1.1) \le 0, \ v \in \mathcal{E}(0, \mathcal{Q}(t)) \right\} \ne \emptyset \qquad (3.6)$$

$$\forall t \in [\tau, k\tau], \quad \forall x \in \mathcal{E}(0, T).$$

If this is true, (3.5) always does hold so that $W_k[t] \ne \emptyset$, $t \ge \sigma$.

Further on, using the substitution $z = S(t, t_0)x$, where

$$\frac{dS(t, t_0)}{dt} = -S(t, t_0)A(t),$$

---

[1] Here $W[t]$ of (3.1) and $W_k[t]$ of (3.4) are actually the *maximal solutions* (with respect to inclusion) of the corresponding equations.

$$S(t_0, t_0) = I,$$

we may substitute system (1.1) for

$$\dot{z} = u^* + v^*,$$

where

$$u^* \in \mathcal{E}(0, P^*(t)), \ v^* \in \mathcal{E}(0, Q^*(t))$$

and

$$P^*(t) = S'(t, t_0) P(t) S(t, t_0), \quad Q^*(t) = S'(t, t_0) Q(t) S(t, t_0).$$

This indicates that without loss of generality we may assume $A(t) \equiv 0$ and keep the former notations (provided the constraints $\mathcal{E}(0, P(t))$, $\mathcal{E}(0, Q(t))$ are time–variant).

Assumption 2.1 *There exists an $r > 0$ such that the geometrical ("Minkowski") difference*

$$\mathcal{E}(0, P(t)) \dot{-} \mathcal{E}(0, Q(t)) \supseteq r \mathcal{E}(0, I), \quad r > 0.$$

(Recall that $\mathcal{E}_1 \dot{-} \mathcal{E}_2 = \{x : x + \mathcal{E}_2 \subseteq \mathcal{E}_1\}$).

Lemma 2.3 *Under Assumption 2.1 the set $U_k(t, x) \neq \emptyset$ and therefore $W_k[t] = \mathcal{E}(0, T)$, $t \in [\sigma, k\sigma]$, whatever are the matrix $T > 0$ and $k > 0$.*

Indeed, assume

$$d = d(x, \mathcal{E}(0, T)) = \max_{\|l\| \leq 1} \left\{ (l, x) - \frac{1}{2} (l, T^{-1} l)^{\frac{1}{2}} \right\} > 0$$

One may then observe that the respective maximizer is

$$l^0(x) = \alpha T x^0, \quad \alpha = \|T x^0\|^{-1} > 0, \tag{3.7}$$

where x^0 is the metric projection of x on $\mathcal{E}(0, T)$. Following the rules for differentiating a function of the "maximum" type, we come to

$$\frac{d \, d^2(x, \mathcal{E}(0, T))}{dt} \Big|_{(2.1)} = (l^0(x), \dot{x}) d = (l^0(x), u + v) d, \quad d > 0.$$

Therefore the set

$$U_*(t, x) = \left\{ u : \max_v \{ (x - x_\epsilon, u + v) \mid (1.1) \} \leq 0 \right\} =$$

$$= \left\{ u : \max_v \{ (l^0(x), u + v) | (1.1) \} \leq 0 \right\} =$$

$$= \left\{ u : (-l^0, u) - \rho(-l^0(x) | \mathcal{E}(0, Q(t))) \geq 0, \ u \in \mathcal{E}(0, P(t)) \right\} = \emptyset, \quad v \in \mathcal{E}(0, Q(t)),$$

due to Assumption 2.1. This is true since the convex hull of the function

$$f(l) = \rho(l | \mathcal{E}(0, P(t)) - \rho(l | \mathcal{E}(0, Q(t))$$

which is

$$f^{**}(l) = \text{co}\,(f)(l) = \rho(l | \mathcal{E}(0, P(t)) \dot{-} \mathcal{E}(0, Q(t))) \geq 0, \quad \forall l \in R^n.$$

Lemma 2.4 *The multivalued function $U_*(t, x)$ is continuous in t and upper semicontinuous in x.*

This again follows from the continuity, now of $W_k[t]$ in t, and ensures the existence of a solution to equation (2.1).

Taking $U_k(t, x) = U_*(t, x)$ we come to the assertion

Theorem 2.2 *Under Assumption 2.1 the strategy $U^0(t, x)$ ($U_k = U_*$) solves the Problems 1-(a), 1-(b) (with σ fixed), provided $x(\tau) \in W[\tau]$.*

Let us now discuss the overall solution to the Problem.

4 On the Solvability of the Stabilization Problem.

The solution schemes indicated in the previous Section presumed instant σ to be fixed. Thus with $\mathcal{E}(0,T)$ given, the solution to Problem 1-(a) does exist for a starting position $\{\tau, x\}$, provided $x \in W[\tau] = W(\tau, \sigma, \mathcal{E}(0,T))$. In other terms the solvability domain for Problem 1-(a) is nonvoid iff $W[\tau] \neq \emptyset$, where $W[t]$ is the solution to equation (3.1). The structure of equation (3.1) yields

Lemma 3.1

(i) *The sets $W[t]$ are symmetrical about the origin.*

(ii) *Once $W[\tau] \neq \emptyset$ we also have $W[t] \neq \emptyset$ for $\tau \leq t \leq \sigma$.*

(iii) *With $\mathcal{E}'(0,T) \subseteq \mathcal{E}(0,T)$ the respective set $W'[t] \subseteq W[t]$.*

For a given starting position $\{\tau, x\}$ the problem of practical stabilization 1-(a), 1-(b) is obviously resolvable if

(a) there exists an instant $\sigma > \tau$ such that $x \in W[\tau] = W(\tau, \sigma, \mathcal{E}(0,T))$.

(b) the set $W_k(\sigma, k\sigma, \mathcal{E}(0,T)) = \mathcal{E}(0,T)$ for any $k > 1$.

Condition (b) is ensured by (3.6), while a nonvoid tube $W(t)$, $\tau \leq t \leq \sigma$ exists for any nondegenerate ellipsoid $\mathcal{E}(0,T)$ if

$$\mathcal{E}(0, P(t)) \overset{\cdot}{-} \mathcal{E}(0, Q(t)) \neq \emptyset, \ \tau \leq t \leq \sigma.$$

The last condition could be strengthened to the level of Assumption 2.1.

Lemma 3.2 *Under Assumption 2.1 for any triplet $\{\tau, rS(0), \mathcal{E}(0,T)\}$, $r > 0$, there exists an instant $\sigma > \tau$, such that*

$$rS(0) \subseteq W(\tau, \sigma, \mathcal{E}(0,T)),$$

and therefore, a strategy $U_r(t,x)$ that solves Problem 1-(a) uniformly in $x(\tau) \in rS(0)$.

What directly follows from Lemma 2.2 is

Lemma 3.3 *Under Assumption 2.1 there exists a strategy $U_\infty(t,x)$ such that once $x(\sigma) \in \mathcal{E}(0,T)$ the solution tube $X[t] = X(t, \sigma, x(\sigma))$ to equation (2.1) satisfies the inclusion $X[t] \subseteq \mathcal{E}(0,T)$, $\sigma \leq t$.*

Combining strategies $U_r(t,x)$, $(\sigma \leq t \leq \sigma)$ and $U_\infty(t,x)$, $(\sigma \leq t)$ into one strategy $U^{\mathcal{E}}(t,x)$, we ensure a practical stabilization of system (2.1) within the tube $\mathcal{E}[t] = \mathcal{E}(0,T)$, $t \geq \sigma$, uniformly in $x = x(\tau) \in rS(0)$.

It is not difficult to demonstrate however that under Assumption 2.1 there exists a strategy $U_\infty^0(t,x)$, such that for any $r > 0$ and any state $x(\tau) \in rS(0)$, the trajectory tube $X[t] = X(t, \tau, x(\tau))$ satisfies the following conditions:

(a) for a certain pair $\sigma > 0$, $\mathcal{E}(0,T)$, we have $X[t] \subseteq \mathcal{E}(0,T)$

(b) $lim_{t \to \infty} h_+(X[t], \{0\}) = 0$,

which ensure asymtotic stability of the system (2.1) provided $U(t,x) = U_\infty^0(t,x)$.

It would make sense however to introduce some stabilizability conditions and synthesizing solutions that would be less precise, but simpler to check and to implement.

5 Ellipsoidal Approximations. The Comparison Principle.

As indicated above, once the overall solvability tube $\mathcal{W}[t]$ is given, namely

$$\mathcal{W}[t] = W[t], \quad \tau \le t \le \sigma,$$

$$\mathcal{W}[t] = \mathcal{E}(0, T), \quad \sigma \le t \le k\sigma,$$

the synthesizing (practically stabilizing) strategy $U^0(t, x)$ is defined through the equation

$$U^0(t, x) = \left\{ u : \max_v \{ dV(t, x)/dt | (1.1) \} \le 0, \ v(t) \subseteq \mathcal{E}(0, Q(t)) \right\},$$

where

$$V(t, x) = d^2(x, \mathcal{W}[t]).$$

The tube $\mathcal{W}[t]$ is the solution to the evolution equation (3.1) of the "funnel" type. This equation may be integrated with the aid of Pontriagin's multivalued "alternated integral", whose calculation is however a rather lengthy procedure.

We therefore suggest to apply the following *comparison principle*. In papers [6, 7] it was indicated that the tube $W[t]$, $t_0 \le t \le \sigma$, could be approximated, both internally and externally, by ellipsoidal–valued functions $\mathcal{E}_-(0, P_-(t)|p(t), S(t))$, $\mathcal{E}_+(0, P_+(t)|p(t), S(t))$. These functions depend upon continuous parametrizing functions $S(t)$, $p(t)$, (S – invertible symmetrical matrix, $p > 0$ – scalar), so that

$$\mathcal{E}_-(0, P_-(t)|p(t), S(t)) \subseteq W[t] \subseteq \mathcal{E}_+(0, P_+(t)|p(t), S(t))$$

and

$$\text{cl } \{ \cup \mathcal{E}_-(0, P_-(t)|p(t), S(t)) \} = W[t] = \{ \cap \mathcal{E}_+(0, P_+(t)|p(t), S(t)) \}$$

Here the centers $a_-(t) \equiv 0$, $a_+(t) \equiv 0$ and the matrix functions $P_-(t)$, $P_+(t)$ satisfy some explicitly defined ODE's, which are

$$\dot{P}_+(t) = -p^{-1}(t)P_+(t) - p(t)P_+(t) + \qquad (5.1)$$

$$+ S^{-1}(t)[S(t)P_+(t)S(t)]^{1/2}[S(t)Q(t)S(t)]^{1/2}S^{-1}(t) +$$

$$+ S^{-1}(t)[S(t)Q(t)S(t)]^{1/2}[S(t)P_+(t)S(t)]^{1/2}S^{-1}(t),$$

$$P_+(\sigma) = T,$$

$$\dot{P}_-(t) = p^{-1}P_-(t) + p(t)Q(t) - \qquad (5.2)$$

$$- S^{-1}(t)[S(t)P_-(t)S(t)]^{1/2}[S(t)P(t)S(t)]^{1/2}S^{-1}(t) -$$

$$- S^{-1}(t)[S(t)P(t)S(t)]^{1/2}[S(t)P_-(t)S(t)]^{1/2}S^{-1}(t),$$

$$P_-(\sigma) = T,$$

$$\tau \le t \le \sigma.$$

(the indicated properties are true under a qualification restriction of the type that int $W[t] \neq \emptyset$, $t_0 \leq t \leq \sigma$).

Let $\mathcal{E}_-[t] = \mathcal{E}(0, P_-[t])$ be a fixed internal approximation for $W[t]$ and

$$V_-(t, x) = d^2(x, \mathcal{E}_-[t]) = d^2.$$

Then obviously

$$V(t, x) \leq V_-(t, x)$$

and the condition $x^0 = x[t_0] \in \mathcal{E}_-[t_0]$ with

$$U^-(t, x) = \left\{ u : \max_v \frac{dV_-(t, x)}{dt} \Big| (1.1) \leq 0, \ v \in \mathcal{E}(0, Q(t)) \right\}$$

ensures the inclusions

$$x(t, t_0, x^0) \subseteq \mathcal{E}_-[t] \subseteq W[t], \ t_0 \leq t \leq \tau$$

(the respective proofs are given in [7]).

Let us discuss the strategy $U^-(t, x)$. Recall that

$$d_- = d(x, \mathcal{E}_-[t]) = \max \left\{ (l, x) - \frac{1}{2}(l, P_-[t]l)^{\frac{1}{2}} \big| (l, l) = 1 \right\}$$

where the maximizer

$$l_-^0(t, x) = \alpha P_-^{-1}[t] x^0,$$

$$\alpha = \|P^{-1}[t] x^0\|,$$

and x^0 is the metric projection of x on $\mathcal{E}_-[t]$.

The strategy $U^-(t, x)$ is then single–valued in the domain $d_- > 0$, namely

$$U^-(t, x) = u^-(t, x) =$$

$$= \left\{ u : \max(l_-^0(t, x)u) | u \in \mathcal{E}(0, P_-[t]) \right\} = \tag{5.3}$$

$$= P_-[t] l^0(t, x)(l^0(t, x), P_-[t] l_-^0(t, x))^{-1/2}$$

and is therefore an *analytical design*.

Under Assumption 2.1 with

$$\mathcal{W}[t] = \mathcal{E}(0, T), \quad \sigma \leq t \leq k\sigma$$

the set $\mathcal{W}[t]$ is already an ellipsoid and therefore, the strategy

$$U_k(t, x) = U_k^-(t, x) = \left\{ u : \max_v d\, d(x, \mathcal{E}(0, T))/dt \leq 0 | (1.1), \ v \in \mathcal{E}(0, Q(t)) \right\},$$

$(d(x, \mathcal{E}(0, T)) > 0)$ is also a design similar to (5.3) but with $l_-^0(t, x)$ substituted by $l^0(x)$ of (3.7).

With $d_- = d(x, \mathcal{E}_-[t]) = 0$, $d_{\mathcal{E}} = d(x, \mathcal{E}(0, T)) = 0$ the strategy U_- may be taken as $U_-(t, x) = \mathcal{E}(0, P(t))$.

The given scheme then automatically ensures the existence of solutions to the differential inclusion

$$\dot{x} \in U_-(t, x) + v.$$

However to keep the "design" requirement one may simply take $u_-(t,x) \equiv 0$ when $x \in$ int $W[t]$.

There will be some problem with the existence of a solution to the equation

$$\dot{x} = u_-(t,x) + v,$$

though, since $u_-(t,x)$ will require an additional definition on the surface of the tube $\mathcal{W}(t)$ that would potentially allow "sliding" modes or "chattering" controls. This may formally be done within the techniques of [8].

6 Conclusion.

This paper provides a general scheme and some simple sufficient conditions for the solvability of the problem of "practical stabilization" for time–variant systems with input uncertainty together with constructive schemes for the design of the respective control strategies. The ellipsoidal techniques mentioned here allow graphical representation of the solutions similar to those indicated in [6, 7].

References

[1] A. Liapounoff, *Problème Génerale de Stabilité de Mouvement.* Paris, 1891.

[2] N.N. Krasovski, *The Control of a Dynamic System.* Nauka: Moscow, 1986.

[3] G. Leitmann, *Deterministic Control of Uncertain Systems via a Constructive Use of Liapunov Stability Theory.* Proceed. 14-th IFIP Conf. Leipzig, 1989, Lecture Notes in Control and Information Sciences, 143, Springer–Verlag.

[4] D.D. Siljak, *Decentralized Control of Complex Systems.* Academic Press, New–York, 1991.

[5] A.B. Kurzhanski and O.I. Nikonov, *Funnel equations and multivalued integration problems for control synthesis.* in Perspectives in Control Theory, PSCT, vol. 2, Birkhäuser: Boston, 1990, pp. 143–153.

[6] A.B. Kurzhanski and I. Vályi, *Ellipsoidal techniques for dynamic systems: the problem of control synthesis.* Dynamics and Control, vol. 1, N 4, pp. 357–378, 1991.

[7] A.B. Kurzhanski and I. Vályi, *Ellipsoidal techniques for dynamics systems: control synthesis for uncertain systems.* Dynamics and Control, vol. 2, N 2, pp. 87–112, 1992.

[8] A.F. Filippov, *Differential Equations with Discontinuous Right-hand Side.* Nauka, Moscow, 1985.

Comparison Principle for Impulsive Differential Equations with Variable Times

V. LAKSHMIKANTHAM, Florida Institute of Technology, Department of Applied Mathematics, Melbourne, Florida

1. INTRODUCTION

In this paper, we discuss the existence of extremal solutions for impulsive differential equations (IDE) with variable times using a new approach, develop the necessary comparison result parallel to the one in ODE and apply it for the investigation of stability criteria. In the context of stability investigation, it is natural to consider the existence of a solution that meets each given barrier (hypersurface) exactly once i.e., the lack of pulse phenomenon. With this motivation, we also consider a result on existence of solutions which meet the given hypersurfaces only once and this result is a refinement of the known result in [5]. We do hope that the new idea of this paper will be of value in the study of qualitative behavior of solutions of IDE with variable times whose progress so far has been slow.

Before we proceed to discuss IDE with variable times, we shall present corresponding results for IDE with fixed times so that one can appreciate the difficulties involved.

2. SYSTEMS WITH IMPULSES AT FIXED TIMES

Let us first consider the impulsive differential system in which the impulse effects occur at fixed moments of time, namely,

$$
\left[
\begin{aligned}
x' &= f(t,x), & t &\neq t_k, \\
x(t^+) - x(t) &= I_k(x(t)), & t &= t_k, \\
x(t_0^+) &= x_0.
\end{aligned}
\right.
\qquad (2.1)
$$

It is well-known [4] that the theory of differential inequalities plays an important role in the qualitative and quantitative study of solutions of differential equations. It is natural to expect that the corresponding theory of impulsive differential inequalities would be equally useful in the investigation of impulsive differential equations. We shall begin by presenting basic impulsive differential inequalities. For proofs see [5].

Let PC denote the class of piecewise continuous functions from R_+ to R, with discontinuities of the first kind only at $t = t_k$, $k = 1, 2, \ldots$. We begin with the following simple result.

THEOREM 2.1:   *Assume that*
   (A_0) *the sequence* $\{t_k\}$ *satisfies* $0 \leq t_0 < t_1 < t_2 < \ldots$, *with* $\lim_{k \to \infty} t_k = \infty$;
   (A_1) $m \in PC[R_+, R]$ *and* $m(t)$ *is left-continuous at* t_k, $k = 1, 2, \ldots$;
   (A_2) *for* $k = 1, 2, \ldots, t \geq t_0$,

$$
D^+ m(t) \leq p(t) m(t) + q(t), \quad t \neq t_k,
$$

$$
m(t_k^+) \leq d_k m(t_k) + b_k,
$$

where $q, p \in C[R_+, R]$, $d_k \geq 0$ *and* b_k *are constants.*

Then,

$$m(t) \le m(t_0) \prod_{t_0 < t_k < t} d_k \, exp(\int_{t_0}^{t} p(s)ds) +$$

$$\sum_{t_0 < t_k < t} \left(\prod_{t_k < t_j < t} d_j exp(\int_{t_k}^{t} p(s)ds) \right) b_k$$

$$+ \int_{t_0}^{t} \prod_{s < t_k < t} d_k exp(\int_{s}^{t} p(\sigma)d\sigma)q(s)ds, \quad t \ge t_0.$$

To prove a general comparison result concerning impulsive differential inequalities, we need the notion of maximal solution of scalar impulsive differential equation

$$\begin{cases} u' = g(t,u), t \ne t_k, & u(t_0) = u_0, \\ u(t_k^+) = \psi_k(u(t_k)), & t_k > t_k > t_0 \ge 0 \end{cases} \tag{2.2}$$

where $g \in C[R_+ \times R, R]$, $\psi_k : R \to R$.

DEFINITION 2.1:   Let $r(t) = r(t, t_0, u_0)$ be a solution of (2.2) on $[t_0, t_0 + a)$. Then $r(t)$ is said to be the maximal solution of (2.2) if for any solution $u(t) = u(t, t_0, u_0)$ of (2.2) existing on $[t_0, t_0 + a)$, the inequality $u(t) \le r(t)$, $t \in [t_0, t_0 + a)$ holds.

One can prove the following comparison result. See [5].

THEOREM 2.2:   *Assume that (A_0) and (A_1) hold. Suppose that*
(i) $g \in C[R_+ \times R, R]$, $\psi_k : R \to R$, $\psi_k(u)$ *is nondecreasing in u and for each $k = 1, 2, \ldots$,*

$$\begin{cases} D_- m(t) \le g(t, m(t)), t \ne t_k, & m(t_0) \le u_0, \\ m(t_k^+) \le \psi_k(m(t_k)); \end{cases}$$

(ii) $r(t)$ *is the maximal solution of (2.2) existing on $[t_0, \infty)$.*
Then,

$$m(t) \le r(t), \quad t_0 \le t < \infty.$$

The impulsive system (2.1) will be considered under the following assumption:
(B_0) (i) $0 < t_1 < t_2 < \ldots < t_k < \ldots$, and $t_k \to \infty$ as $k \to \infty$;
  (ii) $f : R_+ \times R^n \to R^n$ is continuous in $(t_{k-1}, t_k] \times R^n$ and for each $x \in R^n$,
        $k = 1, 2, \ldots$, $\lim_{(t,y) \to (t_k^+, x)} f(t, y) = f(t_k^+, x)$ exists;
 (iii) $I_k : R^n \to R^n$.

Let $V : R_+ \times R^n \to R_+$. Then V is said to belong to class V_0 if
  (i) V is continuous in $(t_{k-1}, t_k] \times R^n$ and for each $x \in R^n$, $k = 1, 2, \ldots$,
        $\lim_{(t,y) \to (t_k^+, x)} V(t, y) = V(t_k^+, x)$ exists;

(ii) V is locally Lipschitzian in x.

For $(t,x) \in (t_{k-1}, t_k] \times R^n$, define

$$D^+V(t,x) = \lim_{h \to 0^+} sup \frac{1}{h}[V(t+h, x+hf(t,x)) - V(t,x)].$$

Having the comparison theorem 2.2 at our disposal, it is easy to discuss stability criteria of (2.1). We shall merely state a typical result that corresponds to the theory of differential equations without impulses.

Assume that $f(t,0) \equiv 0$ and $I_k(0) = 0$ for all k so that we have the trivial solution of (2.1). The following known result [1] offers sufficient conditions in a unified way for various stability criteria.

THEOREM 2.3:   *Assume that*
(i) $V: R_+ \times S(\rho) \to R_+$, $V \in V_0$,

$$D^+V(t,x) \le g(t, V(t,x)), \quad t \ne t_k,$$

where $g: R_+ \times R_+ \to R$, $g(t,0) \equiv 0$ and g satisfies $(B_0\ ii)$;
(ii)   *there exists a $\rho_0 > 0$ such that $x \in S(\rho_0)$ implies that $x + I_k(x) \in S(\rho)$ for all k and $V(t, x + I_k(x)) \le \psi_k(V(t,x))$, $t = t_k$, $x \in S(\rho_0)$, where $\psi_k: R_+ \to R_+$ is nondecreasing;*
(iii) $b(|x|) \le V(t,x) \le a(|x|)$ *on $R_+ \times S(\rho)$ where $a, b \in C[[0,\rho), R_+]$ such that $a(0) = b(0) = 0$ and $a(u), b(u)$ are increasing in u.*
Then the stability properties of the trivial solution of (2.2) imply the corresponding stability properties of the trivial solution of (2.1).

For several result for IDE with fixed times, see [5].

The following corollary of Theorem 2.3 is interesting in itself and shows the interplay of the impulses.

COROLLARY 2.1:   *The functions*
(i) $g(t,u) \equiv 0$, $\psi_k(u) = d_k u$, $d_k \ge 0$ *for all k, are admissible in Theorem 2.3 to yield $x \equiv 0$ of (2.1) is uniformly stable provided the infinite product $\prod_{i=1}^{\infty} d_i$ converges. In particular, $d_k = 1$ for all k, is admissible;*
(ii) $g(t,u) = \lambda'(t)u$, $\lambda \in C^1[R_+, R_+]$, $\psi_k(u) = d_k u$, $d_k \ge 0$ *for all k, are admissible in Theorem 2.3 to imply stability of $x \equiv 0$ of (2.1) provided $\lambda'(t) \ge 0$ and*

$$\lambda(t_{k+1}) + \ln d_k \le \lambda(t_k), \text{ for all } k; \tag{2.3}$$

(iii)   *the functions in (ii) are also admissible in Theorem 2.3 to assure asymptotic stability of $x \equiv 0$ of (2.1) if (2.3) is strengthened to*

$$\lambda(t_{k+1}) + \ln \alpha d_k \le \lambda(t_k), \text{ for all } k; \text{ where } \alpha > 1. \tag{2.4}$$

PROOF: The claim in (i) follows easily. To prove (ii) and (iii), we see that any solution $u(t, t_0, u_0)$ of

$$u' = \lambda'(t)u, \quad t \ne t_k,$$

$$u(t_k^+) = d_k u(t_k),$$

$$u(t_0^+) = u_0 \ge 0,$$

is given by $u(t, t_0, u_0) = u_0 \prod_{t_0 < t_k < t} d_k exp[\lambda(t) - \lambda(t_0)]$, $t \geq t_0$. Since $\lambda(t)$ is nondecreasing, it follows from (2.3) that

$$u(t, t_0, u_0) \leq u_0 exp[\lambda(t_1) - \lambda(t_0)], t \geq t_0,$$

provided $0 < t_0 < t_1$. Hence choosing $\delta = \frac{\epsilon}{2} exp[\lambda(t_0) - \lambda(t_1)]$ stability of the trivial solution $u = 0$ of (2.2) follows. If (2.4) holds, we get $u(t, t_0, u_0) \leq u_0 exp[\lambda(t_1) - \lambda(t_0)]\frac{1}{\alpha^k}$, $t_{k-1} < t \leq t_k$, from which $\lim_{t \to \infty} u(t, t_0, u_0) = 0$ results.

3. EQUATIONS WITH IMPULSES AT VARIABLE RATES

Let $\{S_k\}$ be a sequence of surfaces given by $S_k : t = \tau_k(x)$, $k = 1, 2, \ldots$, such that $\tau_k(x) < \tau_{k+1}(x)$. Then we have the following impulsive differential system

$$\begin{cases} x' = f(t, x), & t \neq \tau_k(x), \\ x(t^+) - x(t) = I_k(x(t)), & t = \tau_k(x), \end{cases} \quad k = 1, 2, \ldots. \tag{3.1}$$

Systems with variable moments of impulsive effect such as (3.1) offer more difficult problems compared to the systems with fixed moments of impulsive effect. For example, we note that the moments of impulsive effect for the system (3.1) depend on the solutions that is, $t_k = \tau_k(x(t_k))$, for each k. Thus, solutions starting at different points will have different points of discontinuity. Also, a solution may hit the same surface $t = \tau_k(x)$ several times and such a behavior is called "pulse phenomenon". In addition, different solutions may coincide after some time and behave as a single solution thereafter. This phenomenon is called "confluence". See [5] for illustrative example.

If one knows that the trivial solution for (3.1) exists and points of impulse of solutions are also known, then one can reduce the stability conditions to the study of IDE with fixed moments. For such results see [5].

Very recently, an attempt is made in [1, 2] to develop the comparison principle that is required in the setup of IDE with variable times and then apply it to study stability theory in a unified way corresponding to the theory of ODE (ordinary differential equations) without impulses [3, 4].

We shall prove the existence of extremal solutions for IDE with variable times using a new approach, develop the necessary comparison result parallel to the one in ODE and apply it for the investigation of stability criteria.

Consider the initial value problem for ODE

$$x' = f(t, x), \ x(t_0) = x_0 \tag{3.2}$$

and the IVP for the impulsive differential system with variable times

$$\begin{aligned} x' &= f(t, x), & t \neq \tau_k(x), \\ x(t^+) &= x(t) + I_k(x(t)), & t = \tau_k(x), \\ x(t_0^+) &= x_0, \end{aligned} \quad \Bigg\} k = 1, 2, \ldots, \tag{3.3}$$

where $f \in C[R_+ \times R^n, R^n]$, $I_k \in C[R^n, R^n]$, $\tau_k \in C^1[R^n, (0, \infty)]$, $\tau_k(x) < \tau_{k+1}(x)$ and $\tau_k(x)$ is bounded, for each $k = 1, 2, \ldots$. For convenience, we shall denote $x(t^+)$ by x^+ in the sequel.

Assuming the existence of solutions of (3.2), we can prove the existence of solutions of (3.3) which meet the given hypersurfaces (barriers) $t = \tau_k(x)$, $k = 1, 2, \ldots$, only once. We merely state such a result. See for a proof [4, 5].

THEOREM 3.1: *Assume that*
  (i)   *for any (t_0, x_0), IVP (3.1) has a solution on $[t_0, \infty)$;*

  (ii) $\dfrac{\partial \tau_k(x)}{\partial x} f(t, x) < 1$;

  (iii) $\left(\dfrac{\partial \tau_k}{\partial x} [x + s I_k(x)] \right) I_k(x) < 0$, $0 \leq s \leq 1$ *and* $\tau_{k+1}[x + I_k(x)] > \tau_k(x)$.

Then a solution of IVP (3.3) exists on $[t_0, \infty)$ and meets every surface $s_k : t = \tau_k(x)$ only once.

Next we shall discuss existence of extremal solutions.

Consider the scalar IVP for ODE

$$u' = g(t, u), \ u(t_0) = u_0, \tag{3.4}$$

and the corresponding IVP for IDE with variable times

$$\begin{cases} u' = g(t, u), & t \neq \gamma_k(u), \\ u(t^+) = u(t) + \psi_k(u(t)), & t = \gamma_k(u(t)) \\ u(t_0^+) = u_0, & k = 1, 2, \ldots, \end{cases} \tag{3.5}$$

where $g \in C[R_+ \times R, R]$, $\psi_k \in C[R, (-\infty, 0)]$, $\gamma_k \in C^1[R, (0, \infty)]$, $\gamma_k(u)$ is increasing in u, $\gamma_k(u) < \gamma_{k+1}(u)$ and $\gamma_k(u)$ is bounded, for each $k = 1, 2, \ldots$.

Note that, in view of the assumptions $\dfrac{\partial \gamma_k(u)}{\partial u} > 0$ and $\psi_k(u) < 0$, and hence we have

$$\left(\dfrac{\partial \gamma_k}{\partial u} [u + s \psi_k(u)] \right) \psi_k(u) < 0, \ 0 \leq s \leq 1, \tag{3.6}$$

which will be needed in this particular form in the following discussion.

THEOREM 3.2: *Assume that*
  (a) $r(t) = r(t, t_0, u_0)$ *and* $\rho(t) = \rho(t, t_0, u_0)$ *are the maximal and minimal solutions of (3.4) existing on $[t_0, \infty)$ for every (t_0, u_0);*

  (b) $\dfrac{\partial \gamma_k(u)}{\partial u} g(t, u) < 1$ *and* $\gamma_{k+1}[u + \psi_k(u)] > \gamma_k(u)$;

  (c) $h_k(u) \equiv u + \psi_k(u)$ *is increasing in u and* $\dfrac{\partial h_k(u)}{\partial u} g(t, u) \geq g(t, h_k(u))$.

Then the maximal and minimal solutions of (3.5) exist for $t \geq t_0$ and hit each curve $\sigma_k : t = \gamma_k(u)$ only once.

PROOF: Let $r(t) = r(t, t_0, u_0)$ be the maximal solution and $u(t) = u(t, t_0, u_0)$ be any solution of (3.4) on $[t_0, \infty)$ such that $t_0 < \gamma_1(u_0)$. Then, by the definition of the maximal solution, we have

$$u(t) \leq r(t), \ t_0 \leq t \leq s_1$$

where $s_1 > t_0$ is such that $s_1 = \gamma_1(u(s_1))$. Since $\gamma_1(u)$ is increasing and bounded, it is clear that $u(t)$ hits the barrier $\sigma_1 : t = \gamma_1(u)$ first. As a result, $r(t)$ meets the curve σ_1 at $t_1 \geq s_1$ so that $t_1 = \gamma_1(r(t_1))$. Since $\psi_1(u) < 0$, it follows that

$$u_1^+ = u(s_1^+) = u(s_1) + \psi_1(u(s_1)) < u(s_1) \leq r(s_1),$$

which implies by comparison Theorem [4]

$$u_1(t) \le r(t), s_1 \le t \le t_1$$

where $u_1(t) = u(t, s_1, u_1^+)$ is any solution of (3.4) on $[s_1, \infty)$ provided $u_1(t)$ meets σ_2 after t_1. If $u_1(t)$ meets σ_2 before t_1, then it is easy to see that (3.4) is clearly satisfied. We shall show that

$$u_1(t_1) \le r_1^+ = r(t_1^+) \equiv r(t_1) + \psi_1(r(t_1)). \tag{3.7}$$

For this purpose, let $u(t, \epsilon)$ be any solution of

$$u' = g(t, u) + \epsilon, \ u(t_0) = u_0 + \epsilon$$

for $\epsilon > 0$ sufficiently small and note that $\lim_{\epsilon \to 0} u(t, \epsilon) = r(t)$. Since $h_1(u)$ is increasing and $u(s_1) \le r(s_1) < u(s_1, \epsilon)$, we get

$$u_1^+ = u(s_1^+) = h_1(u(s_1)) < h_1(u(s_1, \epsilon)).$$

We claim that

$$u_1(t) < h_1(u(t, \epsilon)), \ s_1 \le t \le t_1.$$

If this is not true, there exists a t^* with $s_1 < t^* \le t_1$ such that

$$u_1(t^*) = h_1(u(t^*, \epsilon)) \text{ and } u_1(t) < h_1(u(t, \epsilon)), \ s_1 \le t \le t^*.$$

This implies

$$u_1'(t^*) \ge \frac{d}{dt}[h_1(u(t^*, \epsilon))].$$

Now, using assumption (c),

$$g(t^*, u(t^*)) \ge \left[\frac{\partial h_1}{\partial u}(u(t^*, \epsilon))\right](g(t^*, u(t^*, \epsilon)) + \epsilon)$$

$$\ge g(t^*, h_1(u(t^*, \epsilon))) + \epsilon \frac{\partial h_1(u(t^*, \epsilon))}{\partial u}$$

which leads to a contradiction since $\epsilon > 0$ and $\frac{\partial h_1}{\partial u} > 0$. Consequently, $u_1(t) < h_1(u(t, \epsilon))$, $s_1 \le t \le t_1$ and therefore, $u_1(t) \le h_1(r(t))$, $s_1 \le t \le t_1$. This proves $u_1(t_1) \le h_1(r(t_1)) = r(t_1^+) = r_1^+$, which is the desired inequality (3.7).

Let $r_1(t) = r(t, t_1, r_1^+)$ be the maximal solution of (3.4) on $[t_1, \infty)$. Because of (3.6), we obtain

$$\gamma_1(r_1^+) - \gamma_1(r(t_1)) = \int_0^1 \left(\frac{\partial \gamma_1}{\partial u}[r_1^+ s + (1 - s)r(t_1)]\right) ds \psi_1(r(t_1)) < 0$$

and therefore, $\gamma_1(r_1^+) < \gamma_1(r(t_1)) = t_1$. By (b), we also get $\gamma_2(u_1^+) > \gamma_1(u_1) = s_1$ and $\gamma_2(r_1^+) > \gamma_1[r(t_1)] = t_1$. Now, setting $p(t) = t - \gamma_1(r_1(t))$, we see that $p(t_1) = t_1 - \gamma_1(r_1^+)$ > 0 and by assumption (b), $p'(t) = 1 - \left[\frac{\partial \gamma_1}{\partial u}(r_1(t))\right]g(t, r_1(t)) > 0$, which implies

$p(t) > p(t_1) > 0$ for $t \ge t_1$. Thus, $t > \gamma_1(r_1(t))$ for $t \ge t_1$ showing that $r_1(t)$ does not meet the curve σ_1 again. Similarly, we can show that $u_1(t) = u(t, s_1, u_1^+)$ of (3.4) defined earlier does not hit σ_1 again.

Because of (3.7), it is clear that $u_1(t) \leq r_1(t), t \geq t_1$ where $u_1(t)$ and $r_1(t)$ are solutions of (3.4) on $[s_1, \infty)$ and $[t_1, \infty)$ respectively and

$$u(t) \leq r(t), \ t \geq t_0,$$

where

$$u(t) = \begin{cases} u(t), & t_0 \leq t \leq s_1, \\ u_1(t), & s_1 < t < \infty \end{cases}$$

and

$$r(t) = \begin{cases} r(t), & t_0 \leq t \leq t_1, \\ r_1(t), & t_1 < t < \infty. \end{cases}$$

We proceed, as before, with $u_1(t)$ and $r_1(t)$ to claim that

$$u_1(t) \leq r_1(t), \ t_1 \leq t \leq s_2$$

where s_2 is such that $s_2 = \gamma_2(u_1(s_2))$. Again, since $\gamma_2(u)$ is increasing and bounded, $u_1(t)$ meets the curve $\sigma_2 : t = \gamma_2(u)$ first and $r_1(t)$ meets σ_2 at $t_2 \geq s_2$ so that $t_2 = \gamma_2(r_1(t_2))$. We can now proceed successively and use similar arguments to conclude that

$$u_1(t) \leq r(t), t \geq t_0$$

where $u(t)$ is any solution of (3.5) on $[t_0, \infty)$ given by

$$u(t) = u(t, t_0, u_0) = \begin{cases} u(t, t_0, u_0), & t_0 \leq t \leq s_1, \\ u(t, s_1, u_1^+), & s_1 < t \leq s_2, \\ \dots \dots \\ u(t, s_k, u_k^+), & s_k < t \leq s_{k+1} \\ \dots \dots \end{cases}$$

Hence $r(t)$ is the desired maximal solution of (3.5) on $[t_0, \infty)$. The proof for the minimal solution of (3.5) is similar with appropriate modifications. This completes the proof. It is important to note that the maximal solution $r(t)$ of (3.5) is defined by

$$r(t) = r(t, t_0, u_0) = \begin{cases} r(t, t_0, u_0), & t_0 \leq t \leq t_1, \\ r(t, t_1, r_1^+), & t_1 < t \leq t_2, \\ \dots \dots & \dots \dots \\ r(t, t_k, r_k^+), & t_k < t \leq t_{k+1} \\ \dots \dots & \dots \dots, \end{cases}$$

where $r(t, t_k, r_k^+)$ is the maximal solution of (3.5) through (t_k, r_k^+).

We begin with the comparison result.

THEOREM 3.3: *Assume that* $m \in PC[R_+, R_+]$, $g \in C[R_+^2, R]$ *and*

$$\left. \begin{array}{ll} D^+ m \leq g(t, m), & t \neq \gamma_k(m), \\ m(t^+) \leq m(t) + \psi_k(m(t)), & t = \gamma_k(m) \\ m(t_0^+) \leq u_0 & \end{array} \right\} k = 1, 2, \dots,$$

where $\gamma_k \in C^1[R_+,(0,\infty)]$, $\psi_k \in C[R_+,R_-]$ *satisfy the conditions* (a), (b) *and* (c) *of Theorem 3.2. Suppose further that the maximal solution* $r(t) = r(t,t_0,u_0)$ *of* (3.4) *exists on* $[t_0,\infty)$. *Then*

$$m(t) \leq r(t), \; t \geq t_0.$$

The proof of this comparison result is similar to the proof of Theorem 3.2 with $m(t)$ playing the role of any solution $u(t)$ and $r(t)$ is the maximal solution of (3.5) defined by (3.7). We omit the details.

In order to study the stability of the trivial solution of (3.3), let us suppose that $f(t,0) \equiv 0$, $g(t,0) \equiv 0$, $I_k(0) = \psi_k(0) = 0$ and $\psi_k(u) < 0$ if $u > 0$.

THEOREM 3.4:   *Assume that*

(A_1) $V \in C^1[S(\rho), R_+]$, $\tau_k(x) = \gamma_k(V(x))$ *and for* $(t,x) \in R_+ \times S(\rho)$,

$$V'(x) = \frac{\partial V}{\partial x} f(t,x) \leq g(t,V(x)), \; t \neq \gamma_k(V(x));$$

(A_2) $x \in S(\rho)$ *implies* $x^+ = x + I_k(x) \in S(\rho)$ *and for* $x \in S(\rho)$,

$$V(x + I_k(x)) \geq V(x) + \psi_k(V(x));$$

$$\frac{\partial \gamma_k}{\partial u}[V(x + sI_k(x))]\left[\frac{\partial V}{\partial x}(x + sI_k(x))\right]I_k(x)$$

$$\leq \frac{\partial \gamma_k}{\partial u}[V(x) + s\psi_k(V(x))]\psi_k(V(x)), \; 0 \leq s \leq 1;$$

(A_3) *assumptions* (b) *and* (c) *of Theorem 3.2 hold and* $\gamma'_k(u)$ *is nondecreasing;*

(A_4) $b(|x|) \leq V(x) \leq a(|x|)$, $x \in S(\rho)$, *where* $a,b \in \mathcal{K} = \{\phi \in C[R_+,R_+] : \phi(0) = 0$ *and* $\phi(s)$ *is increasing*\}.

Then the stability properties of the trivial solution of (3.5) *imply the corresponding stability properties of the trivial solution of* (3.3).

PROOF: Let $0 < \epsilon < \rho$ and $t_0 \in R_+$ be given. Suppose that the trivial solution of (3.5) is equi-stable. Then, given $b(\epsilon) > 0$ and $t_0 \in R_+$, there exists a $\delta_1 = \delta_1(t_0,\epsilon) > 0$ such that

$$0 \leq u_0 < \delta_1 \text{ implies } u(t,t_0,u_0) < b(\epsilon), \; t \geq t_0$$

where $u(t) = u(t,t_0,u_0)$ is any solution of (3.5). Choose $\delta = \delta(t_0,\epsilon) > 0$ such that

$$a(\delta) < \delta_1$$

and let $u_0 = V(x_0)$. We claim that with this δ, the trivial solution of (3.5) is equi-stable. If this is not true, there exists a $t^* > t_0$ such that

$$\epsilon \leq |x(t^*)| \text{ and } |x(t)| < \epsilon, \; t_0 \leq t < t^*,$$

where $x(t)$ is a solution of (3.3).

If $x(t)$ does not meet any barrier σ_k in the interval $[t_0,t^*)$, then we will have $|x(t^*)| = \epsilon$ and there is nothing new in the standard proof for stability result. If, on the other hand, $x(t)$ has, just before t^*, hit the curve σ_k for some k, then

$$t_k < t^* \leq t_{k+1} \text{ and } |x(t^*)| \geq \epsilon.$$

As a result, since $|x(t_k)| = |x_k| < \epsilon < \rho$ and $|x_k^+| = |x(t_k^+)| = |x_k + I_k(x_k)| < \rho$ by assumption, it follows that there exists a t^0 with $t_k < t^0 \leq t^*$ satisfying $\epsilon \leq |x(t^0)| < \rho$. We set $m(t) = V(x(t))$ for $t_0 \leq t \leq t^0$ and wish to obtain the estimate

$$m(t) \leq r(t), \ t_0 \leq t \leq t^0,$$

where $r(t)$ is the maximal solution of (3.5). For this purpose, let us show that the conditions (ii) and (iii) of Theorem 3.1 hold so that $x(t)$ hit the surfaces S_k only once.

Because of (3.6) and $(A_1), (A_2), (A_3)$, we get the following:

$$\frac{\partial \tau_k(x)}{\partial x} f(t,x) = \left[\frac{\partial \gamma_k}{\partial u}(V(x))\right] V'(x) \leq \frac{\partial \gamma_k(V(x))}{\partial u} g(t, V(x)) < 1;$$

$$\left[\frac{\partial \tau_k}{\partial x}(x + s I_k(x))\right] I_k(x) = \left[\frac{\partial \gamma_k}{\partial u}(V(x + s I_k(x)))\right]\left[\frac{\partial V}{\partial x}(x + s I_k(x))\right] I_k(x)$$

$$\leq \left[\frac{\partial \gamma_k}{\partial u}(V(x) + s \psi_k(V(x)))\right] \psi_k(V(x)) < 0, \ 0 \leq s \leq 1;$$

$$\tau_{k+1}[x + I_k(x)] = \gamma_{k+1}[V(x + I_k(x))] \geq \gamma_{k+1}[V(x) + \psi_k(V(x))] > \gamma_k[V(x)];$$

$$m'(t) \leq g(t, m(t)), \ t \neq \gamma_k(m(t)),$$

$$m(t^+) \leq m(t) + \psi_k(m(t)), \ t = \gamma_k(m(t))$$

$$\text{and } m(t_0) \leq u_0.$$

Thus we find that all the hypotheses of (comparison) Theorem 3.3 are verified and therefore, the desired estimate follows, namely,

$$V(x(t)) \leq r(t, t_0, V(x_0)), \ t_0 \leq t \leq t^0.$$

Finally, employing (A_4) and (4.1), (4.2), (4.4), we obtain

$$b(\epsilon) \leq V(x(t^0, t_0, x_0)) \leq r(t^0, t_0, V(x_0)) < b(\epsilon)$$

in view of the fact $u_0 = V(x_0) \leq a(|x_0|) < a(\delta) < \delta_1$ which is a contradiction and thus we have proved that the trivial solution of (3.2) is equi-stable.

It is not difficult to construct the proofs of other stability properties utilizing the foregoing proof and the usual standard arguments. The proof of Theorem 3.4 is complete.

REFERENCES

1. Kaul, S., Lakshmikantham, V., and Leela, S., Comparison principle for impulsive differential equations with variable times and stability theory, *Nonlinear Analysis* (to appear).

2. Kaul, S., Lakshmikantham, V., and Leela, S., Extremal solutions, comparison principle and stability criteria for impulsive differential equations with variable times, *Nonlinear Analysis* (to appear).

3. Lakshmikantham, V., Leela, S., and Martynyuk, A.A., *Stability Analysis of Nonlinear Systems*, Marcel Dekker, New York 1989.

4. Lakshmikantham, V. and Leela, S., *Differential and Integral Inequalities*, Vol. I., Academic Press, New York 1969.

5. Lakshmikantham, V., Bainov, D.D., and Simeonov, P.S., *Theory of Impulsive Differential Equations*, World Scientific, Singapore 1989.

The Relationship between the Boundary Behavior of and the Comparison Principles Satisfied by Approximate Solutions of Elliptic Dirichlet Problems

Kirk E. Lancaster Department of Mathematics and Statistics, Wichita State University, Wichita, Kansas 67260-0033

1 INTRODUCTION

Suppose Ω is bounded domain in $\Re^n$ and Q is a quasilinear elliptic second order partial differential operator of the form

$$Qu(x) = \sum_{i,j=1}^{n} a^{ij}(x, Du(x))D_{ij}u(x) + b(x, u(x), Du(x)) \tag{1}$$

for $x \in \Omega$, where $a^{ij} \in C^0(\Omega \times \Re^n)$ and $b \in C^0(\Omega \times \Re \times \Re^n)$. For $\phi \in C^0(\partial\Omega)$, consider the Dirichlet problem

$$Qf = 0 \quad \text{in } \Omega \tag{2a}$$

$$f = \phi \quad \text{on } \partial\Omega. \tag{2b}$$

In some cases, the problem has a classical solution; that is, a function $f \in C^2(\Omega) \cap C^0(\overline{\Omega})$ which satisfies (2). For example, if $Q = \Delta$ is the Laplace operator and Ω is a Lipschitz domain, then (2) always has a classical solution. However, in many cases, a classical solution will not exist due, for example, to the behavior of b or geometric properties of Ω. In particular, when Q is not uniformly elliptic, a classical solution of (2) may not exist unless $\partial\Omega$ satisfies certain additional (geometric) conditions. We will consider the Dirichlet problems (2) in two-dimensional domains which fail to have classical solutions because of the geometry of $\partial\Omega$ and yet have smooth "approximate solutions" f which satisfy (2a) and are "generalized solutions" of (2) in some appropiate sense. Under different conditions on Q, Ω, and ϕ and, in certain cases assuming f satisfies special comparison principles near N, it can be proven that, at each point $N \in \partial\Omega$, the radial limits of f at N exist and the qualitative behavior of these radial limits can be determined.

In this note, three cases shall be considered in which the radial limits of f at N,

$$Rf(\theta, N) \equiv \lim_{r\to 0+} f(r\Theta + N), \qquad (3)$$

will turn out to exist for θ satisfying $r\Theta + N \in \Omega$ for all (or some) $r > 0$ sufficiently small when N is a point of $\partial\Omega$ at which f is discontinuous and $\Theta = (\cos(\theta), \sin(\theta))$. In two of these cases, the appropiate results will be stated without proof, while in the third the proof will be provided; we observe that these results have not appeared previously in the literature. Of particular interest is the relationship between, on the one hand, the operator Q and the particular notion of solution, including the comparison principle satisfied by solutions, and, on the other hand, the conditions on the domain and the boundary data which are required to establish the existence of radial limits of solutions. We notice that as the assumptions regarding the special type of comparison principle satisfied by the solution f (and the special structure of Q) are relaxed, increasingly more restrictive conditions are required on the domain Ω and the boundary data ϕ to obtain the existence of the radial limits. In particular, this suggests that the determination of the boundary behavior of smooth "approximate solutions" (or "candidate solutions") of elliptic Dirichlet problems with continuous boundary data in two-dimensional Lipschitz (or piecewise smooth) domains might require, or at least benefit from, the discovery of new comparison principles for smooth solutions of (2a) which might be similar to the general comparison principle.

1.1 Survey of the Literature

Since we will be considering domains in the plane, we will denote the coordinates of points in $\Re^2$ by (x, y) rather than by $x = (x_1, x_2)$. If Q is the minimal surface operator and Ω is not convex, classical solutions of (2) do not exist for some $\phi \in C^0(\partial\Omega)$; this was noted by Bernstein in 1912 and a number of authors, including Radó, Finn, Nitsche, Simon, and Parks, obtained examples of Lipschitz (or smooth) domains $\Omega \subset \Re^n$ ($n = 2$ or $n \geq 2$) and boundary functions $\phi \in C^0(\partial\Omega)$ for which no classical solution of (2) exists (e.g. [2], [12], [18], [5], [15], [21], [17]; see also [16], 353). For general classes of quasilinear elliptic equations, Serrin modified ideas of Finn and proved that when Q is singularly elliptic and $\partial\Omega$ does not satisfy appropiate boundary curvature conditions, classical solutions do not exist for some $\phi \in C^0(\partial\Omega)$ ([5], [19]; also [6], §14.4). In many cases, however, there exists a process for obtaining a candidate for a solution and these candidates, such as variational, Perron, and viscosity solutions, may be smooth in Ω and satisfy (2a). An important question in elliptic PDE theory is the boundary behavior of these "approximate solutions" when classical solutions do not exist.

Let us consider situations in which the behavior of "approximate solutions" at points of $\partial\Omega$ is known. Suppose Ω is a bounded Lipschitz domain in $\Re^2$, $N \in \partial\Omega$, Ω is locally convex at each point of $\partial\Omega$ in a deleted neighborhood of N, and Ω is not locally convex at N; notice

$$\Omega' = \{r\Theta + N : \alpha_N < \theta < \beta_N, 0 < r < r(\theta)\} \tag{4}$$

for some α_N, β_N with $\alpha_N \in (-\pi, \pi]$, $\pi < \beta_N - \alpha_N \leq 2\pi$, and $r(\cdot) = r_N(\cdot) > 0$, where $\Theta = (\cos(\theta), \sin(\theta))$, $\Omega' = \Omega \cap B$, and B is an open ball centered at N of sufficiently small radius. Let Q be the minimal surface operator and $\phi \in C^0(\partial\Omega)$. Then we know that the radial limits of f at N exist for all $\theta \in (\alpha_N, \beta_N)$. Further, if we define $Rf(\alpha_N, N) = Rf(\beta_N, N) = \phi(N)$, then $Rf(\cdot, N) \in C^0([\alpha_N, \beta_N])$ and, for some $\theta_0 = \theta_{0,N} \in (\alpha_N, \beta_N - \pi)$, $Rf(\cdot, N)$ is monotonic on $[\alpha_N, \theta_0]$, constant on $[\theta_0, \theta_0 + \pi]$, and monotonic on $[\theta_0 + \pi, \beta_N]$.

These results, together with generalizations to piecewise continuous ϕ, surfaces of prescribed mean curvature, and capillary surfaces, can be found in [3], [4], [8], [9], [10], and [11] (also [13]). The proofs of these results were obtained using the general maximum principle and, usually, parametric representations of the graphs of the variational solutions of (2). A special version of the maximum principle which we will use is given in the following

DEFINITION: We say that a solution f of $Qf = 0$ satisfies the **maximum principle** at N if and only if $u_0 \leq f \leq u_1$ in Ω' for every $\Omega' \subset \Omega$ and all $u_0, u_1 \in C^2(\Omega') \cap C^0(\overline{\Omega'})$ such that $\liminf (u_1 - f) \geq 0$ and $\liminf (f - u_0) \geq 0$ for every approach in Ω' to a point of $\partial\Omega' \setminus \{N\}$ and $Q(u_0) \geq 0 \geq Q(u_1)$ in Ω'.

Notice that if Q is uniformly elliptic and a solution $u \in C^2(\Omega)$ of $Qu = 0$ satisfies a growth condition near N or if (2a) is equivalent to the divergence of a bounded function of x, y, u, u_x, u_y and $u \in C^2(\Omega)$ satisfies (2a), then the "extended maximum principle" (e.g. [14]), otherwise known as the "Phragmén-Lindelöf principle" (e.g. [1]), or the general maximum principle ([22], Lemma 1) respectively implies u satisfies the maximum principle at N. To the best of our knowledge, other conditions under which the maximum principle at N holds are unknown, but it is conceivable that it might hold for special types of solutions, such as Perron or viscosity solutions.

2 MAIN RESULTS

Assume throughout that Ω is a bounded Lipschitz domain in $\Re^2$. We remark that some of these main results represent collaborative work with Brian White and some, including the behavior of Perron solutions, represent work of the author alone; these results are not necessarily stated in their greatest generality. In certain cases, we will assume a function f satisfies the condition $f \in C^0(\overline{\Omega'}\backslash\{N\})$ and

$$f = \phi \quad \text{on } \partial\Omega'\backslash\{N\} \tag{5}$$

for $N \in \partial\Omega$, $\phi \in C^0(\partial\Omega)$, and $\Omega' = \Omega \cap U$, where U is a neighborhood of N. We will usually assume Q has the form

$$Qu = a^{11}u_{xx} + 2a^{12}u_{xy} + a^{22}u_{yy}, \tag{6}$$

where $a^{ij} = a^{ij}(x, y, u_x, u_y)$, $a^{ij} \in C^0(\Omega \times \Re^2)$, $i, j = 1, 2$, and $a^{12} = a^{21}$.

2.1 Minimizers of Even Elliptic Parametric Functionals

Suppose $N \in \partial\Omega$. Then, since Ω is Lipschitz, there exist angles $\alpha_N^- \leq \alpha_N^+ < \beta_N^- \leq \beta_N^+ < \alpha_N^- + 2\pi$ such that the rays $\theta = \alpha_N^-$ and $\theta = \beta_N^+$ are "outer tangent rays" to $\partial\Omega$ at N and the rays $\theta = \alpha_N^+$ and $\theta = \beta_N^-$ are "inner tangent rays" to $\partial\Omega$ at N. Let $\mathcal{F} : \Re^3 \to \Re^+$ be smooth on $\Re^3 - \{(0,0,0)\}$, even, and elliptic such that $\mathcal{F}(\cdot)$ is positively homogeneous of degree 1. For $\phi : \partial\Omega \to \Re$ which is piecewise continuous, define $J : BV(\Omega) \to \Re$ by

$$J(u) = \int\int_\Omega \mathcal{F}(u_x(x,y), u_y(x,y), -1)dx\ dy + \int_{\partial\Omega} |u - \phi|g, \tag{7}$$

where $g(x,y) = \mathcal{F}(\vec{\eta}(x,y))$ for a.e. $(x,y) \in \partial\Omega$ and $\vec{\eta}$ is the outward unit normal to the boundary cylinder $\partial\Omega \times \Re$. Then we obtain

THEOREM 1: Let $\Omega \subset \Re^2$ be a Lipschitz domain and $\phi : \partial\Omega \to \Re$ be piecewise continuous on $\partial\Omega$. Suppose f minimizes J over $BV(\Omega)$. Then, for each $N \in \partial\Omega$,

(i) $Rf(\theta, N)$ exists when $\theta \in (\alpha_N^-, \beta_N^+)$ and $\theta \neq \alpha_N^+, \beta_N^-$.

(ii) $Rf(\cdot, N) \in C^0(\alpha_N^-, \alpha_N^+) \cup (\alpha_N^+, \beta_N^-) \cup (\beta_N^-, \beta_N^+)$.

(iii) The limits

$$Rf(\alpha_N^+, N) \equiv \lim_{\theta \to \alpha_N^+ +} Rf(\theta, N) \tag{8a}$$

and

$$Rf(\beta_N^-, N) \equiv \lim_{\theta \to \beta_N^- -} Rf(\theta, N) \tag{8b}$$

exist.

(iv) $Rf(\theta, N) = \phi^-(N)$ for $\alpha^- < \theta < \alpha^+$ and $Rf(\theta, N) = \phi^+(N)$ for $\beta^- < \theta < \beta^+$.
Here $\phi^-(N)$ and $\phi^+(N)$ are the left and right hand limits of ϕ at N and, when $\alpha^- < \theta < \alpha^+$ or $\beta^- < \theta < \beta^+$, $Rf(\theta, N)$ is defined to be the limit of $Rf(r\Theta + N)$ as $r \to 0$ for those $r > 0$ satisfying $r\Theta + N \in \Omega$ and $\Theta = (\cos(\theta), \sin(\theta))$.

Notice that $Rf(\cdot, N)$ is piecewise continuous on (α_N^-, β_N^+) and has jump discontinuities at $\theta = \alpha_N^+$ with a jump of magnitude $Rf(\alpha_N^+, N) - \phi^-(N)$ and at $\theta = \beta_N^-$ with a jump of magnitude $\phi^+(N) - Rf(\beta_N^-, N)$. Thus, from the perspective of the behavior of radial limits at N, the worst singularities of the graph of f at N occur in the directions $\theta = \alpha_N^+, \beta_N^-$ and these singularities lie on the wedge $\{(r\Theta + N, z) \in \Re^3 : r \geq 0, z \in \Re, \Theta = (\cos(\theta), \sin(\theta)), \theta = \alpha_N^+, \beta_N^-\}$. For $\theta \in [\alpha_N^+, \beta_N^-]$, $Rf(\theta, N)$ has the same qualitative behavior as that of nonparametric minimal surfaces (i.e. [9]); in particular, if $Rf(\alpha_N^+, N) = Rf(\beta_N^-, N)$, then on $[\alpha_N^+, \beta_N^-]$ this behavior is the same as that described in Theorem 3.3, [8] (except that $Rf(\alpha_N^+, N)$ and $Rf(\beta_N^-, N)$ need not equal either of $\phi^-(N)$ or $\phi^+(N)$.) We observe that the proof of Theorem 1 depends on the fact that our functional J arises from a parametric functional; while one might consider functionals J which do not arise in this way, it is not clear if the regularity necessary for the proof holds in this case.

2.2 Solutions Satisfying the Maximum Principle at N

Suppose next that Q has the same form as (6); let us denote by $\lambda = \lambda(x, y, p, q)$ and $\Lambda = \Lambda(x, y, p, q)$ the minimum and maximum eigenvalues of the coefficient matrix $(a^{ij}(x, y, p, q))$ of Q. Let $N \in \partial\Omega$ such that $\Omega' = \Omega \cap B$ can be represented as in (4) with B some open ball centered at N. Let us call the set $\Omega_N^C = \{(x_N + r\cos(\theta), y_N + r\sin(\theta)) : 0 < r < r(\theta), \beta_N - \pi < \theta < \alpha_N + \pi\}$ the **convex core** of Ω at N. We obtain the following

THEOREM 2: Let $\Omega \subset \Re^2$ be a Lipschitz domain with $N \in \partial\Omega$ such that Ω is locally convex at each point of $\partial\Omega \setminus \{N\}$ in some neighborhood of N, Ω is not locally convex at N, and $\partial\Omega \setminus \{N\}$ is C^1 near N. Suppose $\phi \in C^{0,1}(\partial\Omega)$, $f \in C^1(\overline{\Omega'} \setminus \{N\}) \cap C^2(\Omega')$ satisfies $Qf = 0$ in Ω', $f = \phi$ on $B \cap \partial\Omega \setminus \{N\}$, and the maximum principle at N, where $B = B(N, \epsilon)$ is the ϵ ball about N for some $\epsilon > 0$ and $\Omega' = \Omega \cap B$.

(A.) There exists $z_0 \in \Re$ such that, if we define $f(0,0) = z_0$, we have $f \in C^0(\overline{\Omega_N^C})$ and
$Rf(\theta, N) = z_0$ for $\beta_N - \pi \leq \theta \leq \alpha_N + \pi$. Further, $\lim_{n \to \infty} f(x_n, y_n)$ lies between
z_0 and $\phi(N)$ whenever $\{(x_n, y_n)\}$ is a sequence in Ω' which converges to N and
for which $\{f(x_n, y_n)\}$ has a limit.

(B.) Suppose additionally that $a^{ij} \in C^{1,\delta}(\overline{\Omega} \times \Re^2)$, $\Lambda = O((p^2 + q^2)\lambda)$ as $p^2 + q^2 \to \infty$

uniformly for $(x, y) \in \overline{\Omega}$, and, for some $\sigma \in [\frac{\pi}{2} - \beta_N, -\frac{\pi}{2} - \alpha_N]$,

$$\sin(\sigma)\frac{\partial a^{ij}}{\partial x}(x, y, p, q) = \cos(\sigma)\frac{\partial a^{ij}}{\partial y}(x, y, p, q) \tag{9}$$

for all $(x, y, p, q) \in \Omega \times \Re^2$. Assuming, for convenience, that $z_0 < \phi(N)$, let $T = \{(0, z) : z_0 < z < \phi(N)\}$ and $\Delta(z_1) = \{(s, z) : s > 0, s^2 + (z - z_1)^2 < \epsilon\}$ for $(0, z_1) \in T$ and $\epsilon > 0$ sufficiently small. Suppose that for each $(0, z_1) \in T$, $m \geq 0$ sufficiently large, and solution $g \in C^0(\overline{\Delta(z_1)}) \cap C^2(\Delta(z_1))$ of

$$\tilde{Q}_m g = 0 \quad \text{in } \Delta(z_1), \tag{10}$$

we have $g \in C^1(\overline{\Delta(z_1)})$ and $g_s(0, z) > 0$ for $(0, z) \in \partial\Delta(z_1)$ whenever $g > 0$ in $\Delta(z_1)$ and $g(0, z) = 0$ for $(0, z) \in \partial\Delta(z_1)$. Then
(a) $Rf(\theta, N)$ exists for $\theta \in [\alpha_N, \beta_N]$.
(b) $Rf(\cdot, N) \in C^0([\alpha_N, \beta_N])$.
(c) There exists $\theta_0 \in (\alpha_N, \beta_N - \pi)$ such that
$\quad Rf(\cdot, N)$ is decreasing on $[\alpha_N, \theta_0]$,
$\quad Rf(\cdot, N)$ is constant on $[\theta_0, \theta_0 + \pi]$,
$\quad Rf(\cdot, N)$ is increasing on $[\theta_0 + \pi, \beta_N]$.
Here we define $Rf(\alpha_N, N) = Rf(\beta_N, N) = \phi(N)$ and denote by $\tilde{Q}_m$ the operator given by

$$\tilde{Q}_m g(s, z) = \sum_{i,j=1}^{2} \tilde{a}_m^{ij}(s, z, g_s, g_z)\left(\frac{\partial}{\partial s}\right)^i \left(\frac{\partial}{\partial z}\right)^j g(s, z), \tag{11}$$

where cn denotes $\cos(\sigma)$, sn denotes $\sin(\sigma)$, and

$$\tilde{a}_m^{11}(s, z, p, q) = a^{11}\left(cn\ s, sn\ s, -cn\ \frac{p}{q} + sn\ \left(\frac{1}{q} + m\right), -sn\ \frac{p}{q} - cn\ \left(\frac{1}{q} + m\right)\right)$$

$$\tilde{a}_m^{12}(s, z, p, q) = -pqa^{11}\left(cn\ s, sn\ s, -cn\ \frac{p}{q} + sn\ \left(\frac{1}{q} + m\right), -sn\ \frac{p}{q} - cn\ \left(\frac{1}{q} + m\right)\right)$$

$$+ qa^{12}\left(cn\ s, sn\ s, -cn\ \frac{p}{q} + sn\ \left(\frac{1}{q} + m\right), -sn\ \frac{p}{q} - cn\ \left(\frac{1}{q} + m\right)\right),$$

$$\tilde{a}_m^{22}(s, z, p, q) = p^2 a^{11}\left(cn\ s, sn\ s, -cn\ \frac{p}{q} + sn\ \left(\frac{1}{q} + m\right), -sn\ \frac{p}{q} - cn\ \left(\frac{1}{q} + m\right)\right)$$

$$+ 2pa^{12}\left(cn\ s, sn\ s, -cn\ \frac{p}{q} + sn\ \left(\frac{1}{q} + m\right), -sn\ \frac{p}{q} - cn\ \left(\frac{1}{q} + m\right)\right)$$

$$+ a^{22}\left(cn\ s, sn\ s, -cn\ \frac{p}{q} + sn\ \left(\frac{1}{q} + m\right), -sn\ \frac{p}{q} - cn\ \left(\frac{1}{q} + m\right)\right).$$

REMARK: Let us set $s = \cos(\sigma)x + \sin(\sigma)y$ and $t = -\sin(\sigma)x + \cos(\sigma)y$ and write $z = f(x,y)$ as $z = \bar{f}(s,t)$. If $\bar{Q}$ is obtained from Q by rotation in the xy-plane such that $Qf = 0$ iff $\bar{Q}\bar{f} = 0$, then (9) implies the coefficients of $\bar{Q}$ are independent of t. Also, if $\bar{f}_t < m$ in the appropiate portion of Ω, then $z = \bar{f}(s,t) - mt$ and $\bar{Q}\bar{f}(s,t) = 0$ iff $t = g(s,z)$ and $\tilde{Q}_m g(s,z) = 0$; $\tilde{Q}_0$ is obtained from $\bar{Q}$ by rotation in the sz- plane. We note that equation (10) will often be degenerate on $T \cap \partial\Delta(z_1)$ and this possibility forces us to (essentially) assume the conclusions of the Hopf boundary point lemma in the hypotheses of (B.).

2.3 Perron Solutions

What can be said when our solutions are only assumed to satisfy the classical maximum principle? By translating our domain, we may set $N = (0,0)$. Suppose then that Q is as given by (6), Ω is symmetric with respect to the x-axis and locally convex at each point of $\partial\Omega \backslash \{N\}$, and the positive x direction points into Ω at N. Suppose also that $a^{ij} \in C^{1,\delta}(\overline{\Omega} \times \Re^2)$ for $i,j = 1,2$, Q is symmetric with respect to the x-axis, in the sense that $a^{ij}(x,-y,p,-q) = (-1)^{i+j}a^{ij}(x,y,p,q)$ for $i,j = 1,2$, and $\Lambda = O((p^2 + q^2)\lambda)$ as $p^2 + q^2 \to \infty$ uniformly for $(x,y) \in \overline{\Omega}$. We observe that these conditions imply the local solvability of arbitrary Dirichlet problems with continuous boundary data in discs (for Q) and the relative compactness of bounded families of solutions of (2a) in relatively compact subdomains of Ω and these are the ingredients required to obtain the existence of Perron solutions (which are smooth and satisfy (2a)). We will prove the following

THEOREM 3: Let $\Omega \subset \Re^2$ be a nonconvex domain and assume the suppositions above are correct.

(A.) There exists boundary data $\phi \in C^0(\partial\Omega)$ such that if $f \in C^0(\overline{\Omega} \backslash \{N\})$ is the upper Perron solution of the Dirichlet problem (2), then

(i) $Rf(\theta, N)$ exists for $\theta \in [-\frac{\pi}{2}, \frac{\pi}{2}]$.

(ii) For some $z_0 \in \Re$, $Rf(\theta, N) = z_0$ when $-\frac{\pi}{2} \le \theta \le \frac{\pi}{2}$.

(iii) If (x_n, y_n) is a sequence in $\overline{\Omega}$ converging to $(0,0)$ and if $z_0 \le \phi(0,0)$, then

$$z_0 \le \liminf_{n\to\infty} f(x_n,y_n) \le \limsup_{n\to\infty} f(x_n,y_n) \le \phi(0,0), \tag{12}$$

while if $z_0 \ge \phi(0,0)$, then

$$\phi(0,0) \le \liminf_{n\to\infty} f(x_n,y_n) \le \limsup_{n\to\infty} f(x_n,y_n) \le z_0. \tag{13}$$

(iv) If Q is singularly elliptic, then $z_0 \ne \phi(0,0)$ and $f \notin C^0(\overline{\Omega})$.

(B.) Suppose additionally that

$$a^{ij}(x,y,p,q) = a^{ij}(x,p,q), \tag{14}$$

$i,j = 1,2$. Assuming, for convenience, that $z_0 < \phi(N)$, let $T = \{(0,z) : z_0 < z < \phi(N)\}$ and $\Delta(z_1) = \{(x,z) : x > 0, x^2 + (z - z_1)^2 < \epsilon\}$ for $(0, z_1) \in T$ and $\epsilon > 0$ sufficiently small. Suppose that for each $(0, z_1) \in T$ and solution $g \in C^0(\overline{\Delta(z_1)}) \cap C^2(\Delta(z_1))$ of

$$\tilde{Q}g = 0 \quad \text{in} \quad \Delta(z_1), \tag{15}$$

we have $g \in C^1(\overline{\Delta(z_1)})$ and $g_x(0,z) > 0$ for $(0,z) \in \partial\Delta(z_1)$ whenever $g > 0$ in $\Delta(z_1)$ and $g(0,z) = 0$ for $(0,z) \in \partial\Delta(z_1)$. Then

(a) $Rf(\theta, N)$ exists for $\theta \in [\alpha_N, \beta_N]$.

(b) $Rf(\cdot, N) \in C^0([\alpha_N, \beta_N])$.

(c) There exists $\theta_0 \in (\alpha_N, \beta_N - \pi)$ such that
$Rf(\cdot, N)$ is decreasing on $[\alpha_N, \theta_0]$,
$Rf(\cdot, N)$ is constant on $[\theta_0, \theta_0 + \pi]$,
$Rf(\cdot, N)$ is increasing on $[\theta_0 + \pi, \beta_N]$.

Here we define $Rf(\alpha_N, N) = Rf(\beta_N, N) = \phi(N)$ and denote by $\tilde{Q}$ the operator

$$\tilde{Q}g(x,z) = \sum_{i,j=1}^{2} \tilde{a}^{ij}(x,z,g_x,g_z)(\frac{\partial}{\partial x})^i(\frac{\partial}{\partial z})^j g(x,z), \tag{16}$$

obtained by implicit differentiation of y as a function of x and z, where x, y and z are related by $Q(z+y) = 0$; that is, $z = f(x,y) - y$, $y = g(x,z)$, and the partial differential equation satisfied by g is $\tilde{Q}g = 0$ (the coefficients of $\tilde{Q}$ are those of $\tilde{Q}_1$ in (11) with $\sigma = 0$.)

   In the following chapter, we will prove Theorem 3. We will first prove (A.), making use of the maximum principle, the definition of Perron solutions, and the special properties of our domain and boundary data; these results, including sufficient conditions on the boundary data ϕ for (A.) to hold, are summarized in Theorems 4 and 5. We will then prove (B.) by considering the equation $z = f(x,y) - y$ (in appropiate subdomains of Ω) as $y = g(x,z)$ and using our assumptions to show that $Rf(\theta, N)$ exists and behaves as indicated. We note that equation (15) will usually not be uniformly elliptic in $\Delta(z_1)$, the usual results (i.e. regularity at the boundary and the Hopf boundary point lemma) do not follow immediately, even if we assume $|\nabla g(x,z)|$ is bounded, and we are therefore forced to assume in (B.) that these hold.

3 THE PERRON SOLUTION CASE

Let Ω be a bounded Lipschitz domain in $\Re^2$ with $N = (0,0) \in \partial\Omega$ such that Ω is locally convex at each point of $\partial\Omega\backslash\{N\}$ and Ω is not convex. Let us assume that Ω is symmetric with respect to the x-axis. Then we can write

$$\Omega = \{(r\cos(\theta), r\sin(\theta)) : -\beta < \theta < \beta,\ 0 < r < r(\theta)\} \tag{17}$$

with $\beta \in (\frac{\pi}{2}, \pi]$ and $0 < r(\theta) = r(-\theta)$ for $0 \le \theta < \beta$. Let us define $\partial^{\pm}\Omega = \{(x,y) \in \partial\Omega : \pm y \ge 0\}$, $\Omega^{\pm} = \{(x,y) \in \Omega : \pm y > 0\}$, and $\Omega_0 = \{(x,y) \in \Omega : x > 0\}$. Let $a, b, c \in C^0(\Omega \times \Re^2)$ such that $ac - b^2 > 0$ in $\Omega \times \Re^2$ and $a(x,y,p,q) = a(x,-y,p,-q)$, $b(x,y,p,q) = -b(x,-y,p,-q)$, and $c(x,y,p,q) = c(x,-y,p,-q)$ for all $(x,y) \in \Omega$ and $(p,q) \in \Re^2$.

Let $\phi \in C^0(\partial\Omega)$ and let Γ be the graph of ϕ; that is, $\Gamma = \{(x,y,\phi(x,y)) : (x,y) \in \partial\Omega\}$. We will assume that the following conditions are satisfied; in each of these conditions, "points" mean connected portions of Γ and may consist of one point or a connected curve.

(A1) $\phi(x,-y) = \phi(x,y)$ for $(x,y) \in \partial\Omega$.

(A2) $\phi(0,0) = \sup_{\partial\Omega} \phi$ and $\phi(0,0) > \phi(r(0),0)$.

(A3) No plane of the form $z = mx + z_1$ intersects Γ in more than four "points" when $z_1 \in (\inf_{\partial\Omega} \phi, \phi(0,0))$.

We will wish to consider certain other conditions on the boundary, such as:

(A4) No plane of the form $z = m(x - x_1) + z_1$ intersects Γ in more than two "points" when $m \ge 0$, $x_1 > 0$, and $(x_1,0,z_1)$ is a interior point of the convex hull of Γ.

(Notice that (A1), (A2), and (A4) imply that $\phi(r(0),0) = \inf_{\partial\Omega} \phi$.)

(A3') For some $\epsilon_0 > 0$, no plane of the form $z = m(x - x_1) + z_1$ intersects Γ in more than four "points" when (x_1,y_1,z_1) is an interior point of the convex hull of Γ for some y_1 and $|x_1| < \epsilon_0$.

(A3'') No plane of the form $z = m(x + x_1) + z_1$ intersects Γ in more than four "points" when (x_1,y_1,z_1) is an interior point of the convex hull of Γ for some y_1.

(A4') For some $\epsilon_1 > 0$, no plane of the form $z = m(y - y_1) + z_1$ intersects either Γ^+ or Γ^- in more than two "points" when (x_1,y_1,z_1) is an interior point of the convex hull of Γ for some x_1 and $|y_1| < \epsilon_1$, where $\Gamma^{\pm} = \{(x,y,z) \in \Gamma : \pm y \ge 0\}$.

(A4'') No plane of the form $z = m(y - y_1) + z_1$ intersects either Γ^+ or Γ^- in more than two "points" when (x_1,y_1,z_1) is an interior point of the convex hull of Γ for some x_1.

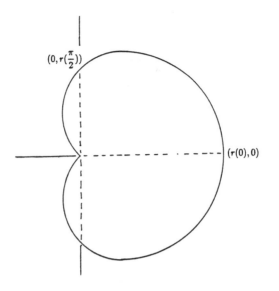

Figure 1

3.1 Some Lemmas

LEMMA 1: Suppose Ω' is an open subset of Ω, $a, b, c \in C^{1,\delta}(\Omega' \times \mathfrak{R}^2)$ for some $\delta \in (0,1)$, $u \in C^2(\Omega')$, and $Qu = 0$ in Ω'. Let $(x_0, y_0) \in \Omega'$ and define

$$v(x,y) = u(x_0,y_0) + u_x(x_0,y_0)(x - x_0) + u_y(x_0,y_0)(y - y_0). \qquad (18)$$

Then either $u \equiv v$ or there is an integer $n \geq 2$ such that the zeros of $u - v$ in a neighborhood of (x_0, y_0) lie on n C^1 curves $\gamma_1, \ldots, \gamma_n$ which divide a neighborhood of (x_0, y_0) into $2n$ disjoint open sectors $\omega_1, \ldots, \omega_{2n}$ with $u < v$ in $\omega_1 \cup \ldots \cup \omega_{2n-1}$ and $u > v$ in $\omega_2 \cup \ldots \cup \omega_{2n}$.

   If we set $w = u - v$, then $|\nabla w| = 0$ at (x_0, y_0). The lemma now follows from the following proposition. When the coefficients of Q are real analytic, Lemma 1 may be proven in a manner similar to that in [20], p. 380.

PROPOSITION: Suppose $\Omega' \subset \Omega$ is open, $a, b, c, d \in C^{1,\delta}(\Omega' \times \mathfrak{R}^2)$ for some $\delta \in (0,1)$, $w \in C^2(\Omega')$, and $Tw = 0$ in Ω', where

$$Tv = av_{xx} + 2bv_{xy} + cv_{yy} + d \qquad (19)$$

with $a = a(x,y,p,q)$, $b = b(x,y,p,q)$, $c = c(x,y,p,q)$, $d = d(x,y,p,q)$, $p = v_x(x,y)$, and $q = v_y(x,y)$. Suppose $ac - b^2 > 0$ in $\Omega' \times \mathfrak{R}^2$ and $d(x,y,0,0) = 0$ for $(x,y) \in \Omega'$. Let $(x_0, y_0) \in \Omega'$ and set $w_0 = w(x_0, y_0)$. Then either $w \equiv w_0$ or there is an integer

$n \geq 1$ such that the zeros of $w - w_0$ in a neighborhood of (x_0, y_0) lie on n smooth (C^1) curves $\gamma_1, \ldots, \gamma_n$ which divide a neighborhood of (x_0, y_0) into $2n$ disjoint open sectors, denoted in order around (x_0, y_0) by $\omega_1, \ldots, \omega_{2n}$, such that $w < w_0$ in $\omega_1 \cup \omega_3 \cup \ldots \cup \omega_{2n-1}$ and $w > w_0$ in $\omega_2 \cup \omega_4 \cup \ldots \cup \omega_{2n}$. Further, if $w \not\equiv w_0$, then $|\nabla w| \neq 0$ in a deleted neighborhood of (x_0, y_0) and $n = 1$ if and only if $|\nabla w| \neq 0$ at (x_0, y_0).

<u>Proof</u> : Suppose $w \not\equiv w_0$. If $|\nabla w| \neq 0$ at (x_0, y_0), the result follows with $n = 1$ from the implicit function theorem. Now we suppose that $|\nabla w| = 0$ at (x_0, y_0). Let us assume $ac - b^2 = 1$, since otherwise we may replace a, b, c, d by $\frac{a}{W}, \frac{b}{W}, \frac{c}{W}, \frac{d}{W}$ respectively, where $W^2 = ac - b^2$. Let $B = B(x_0, y_0)$ be an open disk centered at (x_0, y_0) satisfying $\overline{B} \subset \Omega'$. Then $w \in C^{2,\delta}(\overline{B})$. Define $M(x, y, p, q)$ and $N(x, y, p, q)$ in $C^{0,\delta}(\Omega' \times \Re^2)$ by

$$M = \frac{d(x,y,p,q) - d(x,y,0,q) + d(x,y,p,0)}{2p} \qquad \text{for } p \neq 0, \qquad (20a)$$

$$N = \frac{d(x,y,p,q) - d(x,y,p,0) + d(x,y,0,q)}{2q} \qquad \text{for } q \neq 0, \qquad (21a)$$

and

$$M(x,y,0,q) = \frac{1}{2}\left(\frac{\partial d}{\partial p}(x,y,0,q) + \frac{\partial d}{\partial p}(x,y,0,0)\right), \qquad (20b)$$

$$N(x,y,p,0) = \frac{1}{2}\left(\frac{\partial d}{\partial q}(x,y,p,0) + \frac{\partial d}{\partial q}(x,y,0,0)\right). \qquad (21b)$$

We define

$$Lv = a^0 v_{xx} + 2b^0 v_{xy} + c^0 v_{yy} + M^0 v_x + N^0 v_y, \qquad (22)$$

where $a^0(x,y) = a(x,y,w_x(x,y),w_y(x,y))$, $b^0(x,y) = b(x,y,w_x(x,y),w_y(x,y))$, $c^0(x,y) = c(x,y,w_x(x,y),w_y(x,y))$, $d^0(x,y) = d(x,y,w_x(x,y),w_y(x,y))$, $M^0(x,y) = M(x,y,w_x(x,y),w_y(x,y))$, and $N^0(x,y) = N(x,y,w_x(x,y),w_y(x,y))$. Notice that L is a linear, second order, uniformly elliptic partial differential operator and $Lw = 0$.

Let us introduce characteristic coordinates (e.g. [2]) for L. There is a diffeomorphism $(\xi, \eta) \rightarrow (x(\xi,\eta), y(\xi,\eta))$ of $\overline{D}$ onto $\overline{B}$ which maps $(0,0)$ to (x_0, y_0), where D is the open unit disk centered at $(0,0)$, such that $Lw = 0$ becomes

$$w_{\xi\xi} + w_{\eta\eta} = h(\xi,\eta)w_\xi + k(\xi,\eta)w_\eta, \qquad (23)$$

where

$$\begin{aligned} h = &\frac{x_\xi y_\eta}{J}\left(\tilde{b}\tilde{b}_\xi - \tilde{a}\tilde{c}_\xi + \tilde{b}_\eta\right) + \frac{y_\xi y_\eta}{J}\left(\tilde{a}\tilde{b}_\xi - \tilde{b}\tilde{a}_\xi - \tilde{a}_\eta\right) \\ &-\frac{x_\xi x_\eta}{J}\left(\tilde{c}\tilde{b}_\xi - \tilde{b}\tilde{c}_\xi + \tilde{c}_\eta\right) - \frac{y_\xi x_\eta}{J}\left(\tilde{b}\tilde{b}_\xi - \tilde{c}\tilde{a}_\xi - \tilde{b}_\eta\right) \\ &+ \tilde{M}(\xi,\eta)y_\eta - \tilde{N}(\xi,\eta)x_\eta, \end{aligned} \qquad (24)$$

$$k = \frac{x_\xi^2}{J}(\tilde{c}\tilde{b}_\xi - \tilde{b}\tilde{c}_\xi + \tilde{c}_\eta) + \frac{y_\xi y_\xi}{J}(\tilde{b}\tilde{b}_\xi - \tilde{c}\tilde{a}_\xi - \tilde{b}_\eta)$$

$$- \frac{x_\xi y_\xi}{J}(\tilde{b}\tilde{b}_\xi - \tilde{a}\tilde{c}_\xi + \tilde{b}_\eta) - \frac{y_\xi^2}{J}(\tilde{a}\tilde{b}_\xi - \tilde{b}\tilde{a}_\xi - \tilde{a}_\eta) \qquad (25)$$

$$- \tilde{M}(\xi,\eta)y_\xi + \tilde{N}(\xi,\eta)x_\xi,$$

$J = x_\xi y_\eta - x_\eta y_\xi$, $\tilde{a}(\xi,\eta) = a^0(x(\xi,\eta), y(\xi,\eta))$, $\tilde{b}(\xi,\eta) = b^0(x(\xi,\eta), y(\xi,\eta))$, $\tilde{c}(\xi,\eta) = c^0(x(\xi,\eta), y(\xi,\eta))$, $\tilde{M}(\xi,\eta) = M(x(\xi,\eta), y(\xi,\eta), \tilde{p}(\xi,\eta), \tilde{q}(\xi,\eta))$, $\tilde{N}(\xi,\eta) = N(x(\xi,\eta), y(\xi,\eta), \tilde{p}(\xi,\eta), \tilde{q}(\xi,\eta))$, $\tilde{p} = \frac{y_\eta}{J}w_\xi - \frac{y_\xi}{J}w_\eta$, and $\tilde{q} = -\frac{x_\eta}{J}w_\xi + \frac{x_\xi}{J}w_\eta$ (e.g. [10], (A7)-(A9)). Now $x_\eta = \tilde{b}x_\xi - \tilde{a}y_\xi$ and $y_\eta = \tilde{c}x_\xi - \tilde{b}y_\xi$ (e.g. [10], (A5)-(A6)), so

$$J = \tilde{a}y_\xi^2 - 2\tilde{b}x_\xi y_\xi + \tilde{c}x_\xi^2 \geq \lambda(x_\xi^2 + y_\xi^2) \qquad (26)$$

for some $\lambda > 0$. Thus $\frac{x_\xi^2}{J}, \ldots, \frac{y_\xi y_\eta}{J}$ are all bounded in $\overline{D}$. Hence h and k are bounded in $\overline{D}$.

Let us write $\zeta = \xi + i\eta$. Since $Lw = 0$, w satisfies (23). Thus there is an integer $m \geq 1$ such that $w(\zeta) = o(|\zeta|^m)$ and

$$\lim_{\zeta \to 0} \frac{w_\zeta(\zeta)}{\zeta^m} = Ke^{i\alpha} \qquad (27)$$

exists for some constants $K > 0$ and $\alpha \in [0, 2\pi)$ ([7], Lemma 3.6). This implies

$$w(\zeta) = Re\left(\frac{K}{m+1}e^{i\alpha}\zeta^{m+1}\right) + o(|\zeta|^{m+1}) \qquad (28)$$

or

$$w(re^{i\theta}) = \frac{K}{m+1}r^{m+1}\cos(\alpha + (m+1)\theta) + o(r^{m+1}) \qquad (29)$$

as $r = |\zeta| \to 0$. Thus there are $n = m+1 \geq 2$ 0-level curves of w near $(0,0)$ which pass through $(0,0)$ and these curves are tangent to the lines

$$\theta = \frac{(2k+1)\pi - 2\alpha}{2n}, \qquad k = 0, \ldots, n-1. \qquad (30)$$

Since $w(x,y) = w(\zeta(x,y))$ and $|\nabla w(x,y)| = 0$ if and only if $|w_\zeta(\zeta(x,y))| = 0$, the proposition follows. \qquad Q.E.D.

LEMMA 2: Suppose Ω' is an open subset of Ω. Let $(x_0, y_0) \in \Omega'$ and suppose $u \in C^2(\Omega') \cap C^0(\overline{\Omega'})$ is a solution of $Qu = 0$ in Ω' with $u_{xx}u_{yy} - u_{xy}^2 \neq 0$ at (x_0, y_0). Set

$$v(x,y) = u(x_0, y_0) + u_x(x_0, y_0)(x - x_0) + u_y(x_0, y_0)(y - y_0). \qquad (31)$$

Then either $u \equiv v$ or the zeros of $u - v$ in a neighborhood of (x_0, y_0) lie on two curves which intersect at (x_0, y_0) and divide a neighborhood of (x_0, y_0) into four disjoint open sectors $\omega_1, \omega_2, \omega_3, \omega_4$ such that $u < v$ in ω_1, ω_3 and $u > v$ in ω_2, ω_4.

The proof follows by establishing the representation $u - v = H(\overline{x}, \overline{y}) + o(r^2)$ as $r \to 0$, where $r^2 = (x - x_0)^2 + (y - y_0)^2$, $H(\overline{x}, \overline{y})$ is a homogeneous, harmonic polynomial of degree two, and $\overline{x}, \overline{y}$ are linear functions of $x - x_0$ and $y - y_0$, and using properties of harmonic functions to obtain the conclusions.

LEMMA 3: Suppose Ω' is a subdomain of Ω, $(x_0, y_0) \in \Omega'$, $u \in C^2(\Omega') \cap C^0(\overline{\Omega'})$, $Qu = 0$ in Ω', v is given by (13), and either $D^2 u(x_0, y_0) \neq 0$ or $a, b, c \in C^{1,\delta}(\Omega' \times \Re^2)$ for some $\delta \in (0, 1)$. Then there are $2n$, $n \geq 2$, points $p_1, \ldots, p_{2n}$ on $\partial\Omega'$ arranged in order around $\partial\Omega'$ such that $u - v < 0$ at $p_1, p_3, \ldots, p_{2n-1}$ and $u - v > 0$ at $p_2, p_4, \ldots, p_{2n}$.

The proof of this lemma is essentially the same as the proof of the lemma in §373 of [16]. Notice that the weak maximum principle for $w = u - v$ follows from the weak comparison principle, Theorem 10.1 of [6]. Notice also that if $D^2 u(x_0, y_0) \neq 0$, then $n = 2$ above.

LEMMA 4: Suppose Ω' is a subdomain of Ω and $u \in C^2(\Omega')$ with $Qu = 0$ in Ω'. Set $\sigma = \{(x, y) \in \Omega' : u_y(x, y) = 0\}$ and $\Sigma = \{(x, y) \in \Omega' : u_{xx}(x, y) = u_{xy}(x, y) = u_{yy}(x, y) = 0\}$. Then either $\Sigma = \Omega'$ or $\sigma \backslash \Sigma$ is dense in σ.

The proof follows using the Hopf boundary point lemma for weak solutions (e.g. [6], Thms. 3.4, 10.1).

DEFINITION: A function $u \in C^0(\Omega)$ is a **subsolution** (**supersolution**) of $Qu = 0$ in Ω iff for every ball $B \subset\subset \Omega$ and every solution $v \in C^0(\overline{B}) \cap C^2(B)$ of $Qv = 0$ in B with $u \leq v$ ($u \geq v$) on ∂B, we have $u \leq v$ ($u \geq v$) in B.

DEFINITION: Let $\psi \in L^\infty(\partial\Omega)$. We define the set of **subfunctions** (**superfunctions**) relative to ψ for $Qu = 0$, denoted $\mathbf{S}_\psi$ ($\mathbf{T}_\psi$), as those functions $u \in C^0(\overline{\Omega})$ which are subsolutions (supersolutions) of $Qu = 0$ in Ω with $u \leq \psi$ ($u \geq \psi$) on $\partial\Omega$.

Let us define U on $\overline{\Omega}$ by

$$U(x, y) = \inf_{v \in T_\phi} v(x, y) \tag{32}$$

for $(x, y) \in \overline{\Omega}$. Notice that U is the "upper Perron solution" for (2) and is well defined because of the weak maximum principle. Let us assume that we have the ingredients required to prove $U \in C^2(\Omega)$ and $QU = 0$ (e.g. [20], p. 375; [6]). Let $\phi \in C^0(\partial\Omega)$ satisfy conditions (A1), (A2), and (A3). We require some more information about elements of T_ϕ.

LEMMA 5: Let $u_1 \in T_\phi$. Then there exists $u_2 \in T_\phi \cap C^2(\Omega^+ \cup \Omega^-)$ with $u_2 \leq u_1$ such that

(1.) $u_2 = \phi$ on $\partial\Omega$,

(2.) $u_2(x,y) = u_2(x,-y)$ for $(x,y) \in \Omega$, and

(3.) $Qu_2 = 0$ in $\Omega^+ \cup \Omega^-$.

Suppose either $a, b, c \in C^{1,\delta}(\Omega \times \Re^2)$ for some $\delta \in (0,1)$ or (A3') holds. Then

(4.) $u_2(0,y_1) > u_2(0,y_2)$ if $0 \leq y_1 < y_2$.

Suppose further that condition (A4) holds. Then there exists $u_3 \in T_\phi$ with $u_3 \leq u_2$ in Ω such that u_3 satisfies (1.), (2.), (3.), (4.), and

(5.) $u_3(x_1,0) > u_3(x_2,0)$ if $0 \leq x_1 < x_2 \leq r(0)$.

The proof of Lemma 5 is based on the construction of successive decreasing superfunctions, each obtained from the previous one by solving a Dirichlet problem for (2a) in the convex domain Ω^+ or Ω_0. The properties (1.) - (5.) follow from Lemma 3, the properties of ϕ, and our construction.

Proof: Notice that ϕ cannot be linear, since ϕ cannot be constant because of (A2) and then ϕ cannot be linear because of (A3). Let $u_1 \in T_\phi$. In a manner analagous to pp. 23-24 and p. 103 of [6], we see that each of the following functions is in T_ϕ :

$$v_1(x,y) = \min\{u_1(x,y), u_1(x,-y), \phi(0,0)\} \tag{33}$$

$$v_2(x,y) = \begin{cases} v_1(x,y), & \text{for } (x,y) \in \overline{\Omega}\backslash\overline{\Omega_0} \\ w_1(x,y), & \text{for } (x,y) \in \overline{\Omega_0} \end{cases} \tag{34}$$

$$u_2(x,y) = \begin{cases} w_2(x,y), & \text{for } (x,y) \in \overline{\Omega^+} \\ w_2(x,-y), & \text{for } (x,y) \in \overline{\Omega^-} \end{cases} \tag{35}$$

$$v_3(x,y) = \begin{cases} u_2(x,y), & \text{for } (x,y) \in \overline{\Omega}\backslash\overline{\Omega_0} \\ w_3(x,y), & \text{for } (x,y) \in \overline{\Omega_0} \end{cases} \tag{36}$$

$$u_3(x,y) = \begin{cases} w_4(x,y), & \text{for } (x,y) \in \overline{\Omega^+} \\ w_4(x,-y), & \text{for } (x,y) \in \overline{\Omega^-}, \end{cases} \tag{37}$$

where w_1, w_2, w_3, and w_4 are solutions of $Qw = 0$ in the domains Ω_0, Ω^+, Ω_0, and Ω^+ respectively satisfying the boundary conditions $w_1 = v_1$ on $\partial\Omega_0 \cap \Omega$ and $w_1 = \psi$, where $\psi(x,y) = (x\phi(x,y)+(r(0)-x)v_1(x,y))/r(0)$, for $(x,y) \in \partial\Omega_0 \cap \partial\Omega$, $w_2 = v_2$ on $\partial\Omega^+ \cap \Omega$ and $w_2 = \phi$ on $\partial\Omega^+ \cap \partial\Omega$, $w_3 = u_2$ on $\partial\Omega_0$, and $w_4 = v_3$ on $\partial\Omega^+$ (see figure 2). Notice that w_1, w_2, w_3, and w_4 can be obtained using the Perron process in the convex domains Ω^+ and Ω_0 and they are each continuous on the closure of their domains. Also notice that $v_1(0,0) = \phi(0,0)$, $w_1(r(0),0) = \phi(r(0),0)$, and $u_1 \geq v_1 \geq v_2 \geq u_2 \geq v_3 \geq u_3$. Now $u_2 \in C^0(\overline{\Omega}) \cap C^2(\Omega^+ \cup \Omega^-)$ satisfies conclusions (1.), (2.) and (3.) of the lemma.

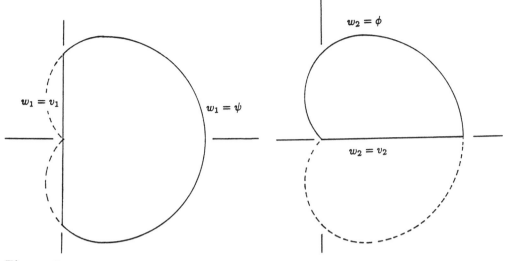

Figure 2

Observe that w_3 cannot be linear in Ω_0, since this would require $u_2(0, y)$ to equal $\phi(0,0)$ for all $(0, y) \in \overline{\Omega}$ and this would violate (A2) (because of conclusion (1.).)

Assuming for a moment that w_2 is linear on Ω^+, the fact that $w_2(0,0) = \phi(0,0) > \phi(0, r(\frac{\pi}{2})) = w_2(0, r(\frac{\pi}{2}))$ implies $u_2(0, y)$ $(= w_2(0, y))$ is a strictly decreasing function of y and so (4.) holds. Let us assume that w_2 is not linear. Suppose $y_1 \in (0, r(\frac{\pi}{2}))$ with $(u_2)_y(0, y_1) = 0$. When $a, b, c \in C^{1,\delta}(\Omega \times \Re^2)$, we will set $(x_0, y_0) = (0, y_1)$. Otherwise, since (A3$'$) holds, let us use Lemma 4 (with $\Omega' = \Omega^+$ and $u = w_2$) and select $(x_0, y_0) \in \Omega^+ \cap \sigma$ such that $-\epsilon_0 < x_0 \leq 0$ and $(w_2)_{xx}(w_2)_{yy} - (w_2)^2_{xy} \neq 0$ at (x_0, y_0), where ϵ_0 is as in (A3$'$); recall that $\sigma = \{(x, y) \in \Omega^+ : (w_2)_y(x, y) = 0\}$. Define

$$\tilde{v}(x, y) = w_2(x_0, y_0) + (w_2)_x(x_0, y_0)(x - x_0). \tag{38}$$

Since $(w_2 - \tilde{v})(x, y) = (w_2 - \tilde{v})(x, -y)$, either (A3) or (A3$'$) imply that $w_2 = \tilde{v}$ at no more than two points of $\partial\Omega^+ \cap \partial\Omega$. Let us define $D = \{(x, y) \in \Omega : u_2(x, y) < \tilde{v}(x, y)\}$. If $\tilde{v}(0, 0) \geq \phi(0, 0)$, then $x_0 < 0$ and $(w_2)_x(x_0, y_0) > 0$, since $w_2(x_0, y_0) < \phi(0, 0)$. Thus $\Omega_0 \subset D$, since $w_2 \leq \phi(0, 0)$, and $u_2 \neq \tilde{v}$ on $\partial\Omega^+ \cap \Omega$. Condition (A3$'$) together with the symmetry of $u_2 - \tilde{v}$ then imply $u_2 = \tilde{v}$ at no more than two points of $\partial\Omega^+$, in contradiction of Lemma 3 (with $\Omega' = \Omega^+$ and $u = u_2$). If $\tilde{v}(N) < \phi(N)$, then $N \notin \overline{D}$. Condition (A3$'$) and the symmetry of $u_2 - \tilde{v}$ implies that there are two points $(x_3, 0)$ and $(x_4, 0)$ with $0 < x_3 < x_4$ at which $u_2 = \tilde{v}$ and $\{(x, 0) : x_3 < x < x_4\} \subset D$; the other possibility, illustrated in figure 3a, clearly yields a contradiction. Hence there must be a component D_0 of D with $\partial D_0 \subset \Omega$ and $(x_0, y_0), (x_0, -y_0), (x_3, 0), (x_4, 0)$ are on ∂D_0 (see figure 3b). Since u_2 is a supersolution in Ω, $\tilde{v}$ is a solution, and $u_2 = \tilde{v}$ on $\partial D_0 \subset \Omega$, we see that $u_2 \geq \tilde{v}$ in D_0. However, since $u_2 < \tilde{v}$ in D_0, we have a contradiction. Thus $(u_2)_y \neq 0$ at $(0, y_1)$. Since $u_2(0, 0) = \phi(0, 0) \geq \phi(0, r(\frac{\pi}{2}))$,

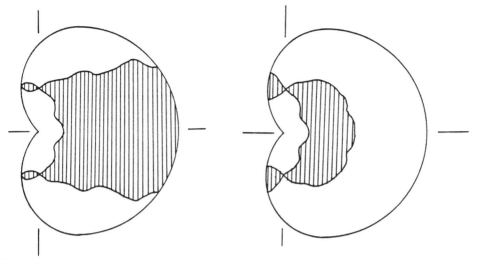

Figure 3

$\phi(0, r(\frac{\pi}{2})) = u_2(0, r(\frac{\pi}{2}))$, and $(u_2)_y(0, y) \neq 0$ when $0 < y < r(\frac{\pi}{2})$, it must be the case that $(u_2)_y(0, y) < 0$ for $0 < y < r(\frac{\pi}{2})$ and so (4.) holds. We notice that $\phi(0,0) > \phi(0, r(\frac{\pi}{2}))$, a fact which could have been obtained directly from (A3).

Let us now consider the last part of the lemma. Notice that $w_3(0, y) = u_2(0, y)$ and so $w_3(0, y)$ is strictly decreasing in y for $y \geq 0$. Further $(w_3)_y(x, 0) = 0$ for $x > 0$. Suppose (A4) holds and $(x_1, 0) \in \Omega$ is a point at which $(w_3)_x \geq 0$ and $(w_3)_{xx}(w_3)_{yy} - (w_3)_{xy}^2 < 0$. Let us define

$$\tilde{w}(x, y) = w_3(x_1, 0) + (w_3)_x(x_1, 0)(x - x_1). \tag{39}$$

Since $\tilde{w}(0, y)$ is constant, there is at most one $y_2 > 0$ such that $\tilde{w}(0, y_2) = w_3(0, y_2)$. Thus either $w_3(0, y) > \tilde{w}(0, y)$ for all $y \geq 0$ or $w_3(0, y) > \tilde{w}(0, y)$ for $0 \leq y < y_2$ and $w_3(0, y) < \tilde{w}(0, y)$ for $y > y_2$. In the first case, (A4) implies that there are at most two points of $\partial\Omega_0$ at which $w_3 = \tilde{w}$. In the second case, we see that $\phi(0, r(\frac{\pi}{2})) < \tilde{w}(0, y) < \phi(0, 0)$ for any y, since $\phi(0, r(\frac{\pi}{2})) = w_3(0, r(\frac{\pi}{2})) < w_3(0, y_2) < w_3(0, 0) = \phi(0, 0)$. Since ϕ and $\tilde{w}$ are continuous on $\partial\Omega$, there is at least one point $(x_5, y_5) \in \partial\Omega^+$ with $x_5 < 0$ at which $\tilde{w} = \phi$ and, from symmetry, $\tilde{w} = \phi$ at $(x_5, -y_5)$. (A4) then implies $\tilde{w} \neq \phi$ on $\partial\Omega$ except at $(x_5, \pm y_5)$. In particular, $w_3 = \phi < \tilde{w}$ on $\partial\Omega_0 \cap \partial\Omega$ and so $\tilde{w} \neq w_3$ on $\partial\Omega_0$ except at $(0, \pm y_2)$. If we apply Lemma 3 with $\Omega' = \Omega_0$ and $u = w_3$, we obtain a contradiction. Thus $(w_3)_x < 0$ at every point $(x, 0)$ at which $(w_3)_{xx}(w_3)_{yy} - (w_3)_{xy}^2 < 0$. Since $\sigma \backslash \Sigma$ is dense in σ (Lemma 4 with $\Omega' = \Omega_0$ and $u = w_3$) and $\{(x, 0) \in \Omega : x > 0\} \subset \sigma$, $(w_3)_x(x, 0) \leq 0$ for all $(x, 0) \in \Omega$. Now $u_3(x, 0) = w_3(x, 0)$ and so (5.) follows. Q.E.D.

LEMMA 6: Suppose $\phi \in C^0(\partial\Omega)$ satisfies (A1), (A2), (A3) and either (A3′) holds or $a, b, c \in C^{1,\delta}(\Omega \times \mathfrak{R}^2)$, for some $\delta \in (0,1)$. Let U be the upper Perron solution of (2). Then $\lim_{y \to 0} U(0,y)$ exists, $U(x,y) = U(x,-y)$ for $(x,y) \in \Omega$, and $U(0,y)$ is a decreasing function of $y > 0$. If $z_0 = \lim_{y \to 0} U(0,y)$ and $w \in C^0(\overline{\Omega_0}) \cap C^2(\Omega_0)$ is the solution of $Qw = 0$ in Ω_0 satisfying $w(N) = z_0$ and $w = U$ on $\partial\Omega_0\backslash\{N\}$, then $U \geq w$ in $\overline{\Omega_0}\backslash\{N\}$. If $\{(x_n, y_n)\}$ is any sequence in $\overline{\Omega}\backslash\{N\}$ converging to N, then

$$\liminf_{n \to \infty} U(x_n, y_n) \geq z_0. \tag{40}$$

Proof: Let us set $T_\phi^* = \{v \in T_\phi : v \text{ satisfies } (1.), (2.), (3.), \text{ and } (4.) \text{ of Lemma 5}\}$. According to Lemma 5,

$$U(x,y) = \inf_{v \in T_\phi^*} v(x,y) \tag{41}$$

for $(x,y) \in \overline{\Omega}$. Since $v(0,y)$ is decreasing in y for $y \geq 0$ and $v(x,y) = v(x,-y)$ for $(x,y) \in \overline{\Omega}$, we see that $U(0,y)$ is decreasing in $y > 0$ and $U(x,y) = U(x,-y)$ for $(x,y) \in \Omega$. Let $z_0 = \lim_{y \to 0} U(0,y)$ and let w be as indicated. If $v \in T_\phi$, then $v \geq U \geq w$ on $\partial\Omega_0\backslash\{N\}$. Since $v - w \in C^0(\overline{\Omega_0})$, $v - w \geq 0$ on $\partial\Omega_0$ and the maximum principle implies $v - w \geq 0$ in $\overline{\Omega_0}$. Therefore $U \geq w$ on $\overline{\Omega_0}\backslash\{N\}$. If $\psi \in C^0(\partial(\Omega^+\backslash\Omega_0))$ with $\psi = w$ on $\partial\Omega_0\cap\Omega^+$ and $\psi \leq \phi$ on $\partial\Omega_0\cap\partial\Omega^+$ and if $w_0 \in C^2(\Omega^+\backslash\Omega_0)\cap C^0(\overline{\Omega^+\backslash\Omega_0})$ is the solution of $Qw_0 = 0$ in $\Omega^+\backslash\Omega_0$ and $w_0 = \psi$ on $\partial(\Omega^+\backslash\Omega_0)$, then $w_0 \leq v$ on $\Omega^+\backslash\Omega_0$ for any $v \in T_\phi$. This implies $w_0 \leq U$ and the last part follows.   Q.E.D.

REMARK: If we were willing to assume that the solution f satisfies the maximum principle at N, Lemma 6 would immediately yield the following

PROPOSITION: Suppose either $a, b, c \in C^{1,\delta}(\Omega \times \mathfrak{R}^2)$, for some $\delta \in (0,1)$, or condition (A3′) of §1 holds. Let U be the upper Perron solution of the Dirichlet problem (2) and assume U satisfies the maximum principle at N. Then there exists $z_0 \in (\inf_{\partial\Omega} \phi, \phi(0,0)]$ such that

$$U(x,y) \to z_0 \quad \text{as} \quad (x,y) \to (0,0), \quad (x,y) \in \overline{\Omega_0}\backslash\{(0,0)\} \tag{42}$$

and if $\{(x_n, y_n)\}$ is any sequence in $\overline{\Omega}\backslash\{N\}$ converging to N, then

$$\liminf_{n \to \infty} U(x_n, y_n) \geq z_0. \tag{43}$$

One final lemma will be required for the proof of Theorem 3. This result is more general than is required for Theorem 3; the lemma as stated could be used in the proof of Theorem 2. Let $N = (x_N, y_N) \in \partial\Omega$ and $\theta_1, \theta_2 \in (\alpha_N^+, \beta_N^-)$ with $\theta_2 \geq \theta_1$. For some $R > 0$, we observe that

$$\{(r\cos(\theta) + x_N, r\sin(\theta) + y_N) : 0 < r < 2R, \theta_1 < \theta < \theta_2\} \subset \Omega. \tag{44}$$

Set $\Omega_N(\theta_1, \theta_2) = \{(r\cos(\theta) + x_N, r\sin(\theta) + y_N) : 0 < r < R, \theta_1 < \theta < \theta_2\}$.

LEMMA 7: Let $N \in \partial\Omega$ and let $\theta_1, \theta_2 \in [\alpha_N^+, \beta_N^-]$ with $0 < \theta_2 - \theta_1 < \pi$. Let $f \in C^2(\Omega)$ satisfy (2a). Suppose $z_1 = Rf(\theta_1, N)$ and $z_2 = Rf(\theta_2, N)$ both exist and $z_1 < z_2$ ($z_1 > z_2$). Let S_0 denote the graph of f over $\Omega_N(\theta_1, \theta_2)$ and let S be the closure (in $\Re^3$) of S_0. Assume that S is a C^1 manifold with boundary in a neighborhood of each point of $T = \{(x_N, y_N, z) : z \text{ lies strictly between } z_1 \text{ and } z_2\}$. Then:

(i) For each $z \in T$, there exists a unique angle $\theta(z) \in [\theta_1, \theta_2]$ such that the vertical half-plane H_z is the tangent cone to S at $\{x_N, y_N, z)\}$ and $\theta(\cdot)$ is a continuous, weakly monotonic function of $z \in T$, where $H_z = \{(r\cos(\theta(z)) + x_N, r\sin(\theta(z)) + y_N, t) : r \geq 0, t \in \Re\}$.

(ii) Suppose additionally that $a^{ij} \in C^1(\overline{\Omega} \times \Re^2)$, $\Lambda = O((p^2 + q^2)\lambda)$ as $p^2 + q^2 \to \infty$ uniformly for $(x, y) \in \overline{\Omega}$, and, for some $\sigma \in (\frac{\pi}{2} - \beta_N, -\frac{\pi}{2} - \alpha_N)$, (9) and the hypotheses of Theorem 2(ii) following (9) are satisfied. For each $\theta \in (\theta_1, \theta_2)$, $Rf(\theta, N)$ exists, $Rf(\cdot, N) \in C^0([\theta_1, \theta_2])$, $Rf(\cdot, N)$ maps $[\theta_1, \theta_2]$ onto $[z_1, z_2]$ ($[z_2, z_1]$ respectively), and for some $\theta_A, \theta_B \in [\theta_1, \theta_2]$ with $\theta_A < \theta_B$,

(a) $Rf(\theta, N) = z_1$ for $\theta_1 \leq \theta \leq \theta_A$.

(b) $Rf(\theta, N)$ is strictly increasing (respectively, decreasing) for $\theta_A \leq \theta \leq \theta_B$.

(c) $Rf(\theta, N) = z_2$ for $\theta_B \leq \theta \leq \theta_2$.

Proof: (i) We may assume $z_1 < z_2$. For each $z \in (z_1, z_2)$, there exists a unique angle $\theta(z) \in (\theta_1, \theta_2)$ such that the tangent cone to S at $(x_N, y_N, z) \in T$ is the half-plane H_z. Since S is a smooth manifold with boundary near each point of T, the tangent half-plane to S at $(x_N, y_N, z) \in T$ varies continuously with z and so $\theta(\cdot) \in C^0(T)$. Since S_0 is a graph, $\theta(\cdot)$ must be weakly monotonic.

(ii) We claim that $\theta(z_a) \neq \theta(z_b)$ if $z_a \neq z_b$, $z_a, z_b \in (z_1, z_2)$. Suppose otherwise and let $z_a, z_b \in (z_1, z_2)$ with $z_a < z_b$ and $\theta(z_a) = \theta(z_b)$. Let us rotate in the xy–plane about N so that σ becomes 0. Then the positive x axis points in the direction of σ. It is important to now observe that because of (9), in these rotated coordinates f is the solution of an equation of the form (6) in which the coefficients $a^{ij} = a^{ij}(x, y, p, q)$ are independent of y.

Let $\epsilon > 0$ satisfy $\epsilon < \min\{z_b - z_a, z_2 - z_1\}$ and let $S_j = S \cap B((x_N, y_N, z_j), \epsilon)$ for $j = a, b$. Then let us translate S_j and obtain

$$S'_j = \{(x - x_N, y - y_N, z - z_j) : (x, y, z) \in S_j\}, \quad j = a, b. \tag{45}$$

Since S is a graph, we see that $S'_1 \cap S'_2 \cap (\Omega \times \Re) = \emptyset$ and S'_2 lies to the right of S'_1 (i.e. closer to the half-space $\theta = \theta_2$). In a neighborhood of $(0, 0, 0)$, S'_j is the graph of a function g_j over (part of) $H = \{(x, 0, z) : x \geq 0, z \in \Re\}$, for $j = a, b$; that is,

$S_j \cap \{(x, y, z) : x \geq 0, \ y \in \Re, \ |(x, z)| < \delta\} = \{(x, g_j(x, z), z) : x \geq 0, |(x, z)| < \delta\}$, $j = a, b$, for some $\delta > 0$. Let us define $w = g_2 - g_1$ on $B^+ = \{(x, z) : x > 0, |(x, z)| < \delta\}$. Then $w > 0$ on B^+, $w(0, 0) = 0$, and $w_x(0, 0) = 0$. Implicit differentiation, together with the fact that the coefficients of Q are independent of y and z, shows that g_j is a solution of a second-order partial differential equation with the same form as (6), for $j = a, b$, and then the argument in the proof of Theorem 10.1 of [6] shows that w is the solution of a linear elliptic equation on B^+. Since $|\nabla w(0, 0)|$ is bounded, the linear equation is uniformly elliptic near $(0, 0)$. Our hypotheses (after (10)) imply $w_x(0, 0) > 0$, a contradiction. Thus our original assumption that $\theta(z_a) = \theta(z_b)$ must have been incorrect. Hence, we see that $\theta(z)$ is a strictly increasing function of z.

Since the tangent half-plane to S at $(x_N, y_N, z) \in T$ varies continuously with z, we have $\theta(\cdot) \in C^0((z_1, z_2))$. If we let $\theta_A = \lim_{z \to z_1^+} \theta(z)$ and $\theta_B = \lim_{z \to z_2^-} \theta(z)$, then we see that $\theta = \theta(z)$ defines z as a function of θ and $z(\cdot) \in C^0((\theta_A, \theta_B))$. Let us denote by Ω_A and Ω_B respectively the sets $\{(r \cos(\theta) + x_N, r \sin(\theta) + y_N) \in \Omega : r > 0, \theta_1 \leq \theta \leq \theta_A\}$ and $\{(r \cos(\theta) + x_N, r \sin(\theta) + y_N) \in \Omega : r > 0, \theta_B \leq \theta \leq \theta_2\}$.

We claim that $Rf(\theta, N)$ exists for $\theta_1 \leq \theta < \theta_2$, $Rf(\cdot, N) \in C^0([\theta_1, \theta_2])$, $Rf(\theta, N) = z_1$ for $\theta_1 \leq \theta \leq \theta_A$, $Rf(\theta, N)$ is strictly increasing for $\theta_A \leq \theta \leq \theta_B$, and $Rf(\theta, N) = z_2$ for $\theta_B \leq \theta \leq \theta_2$. Notice that for each $t \in (z_1, z_2)$, the intersection of the plane $z = t$ with S is a smooth curve $\gamma(t)$ which is tangent to the half-space H_t at (x_N, y_N, t). If $\sigma(t) = \{(x, y) : (x, y, t) \in \gamma(t)\}$, then $\sigma(t) \subseteq \overline{\Omega}$ and $\sigma(t)$ is tangent to the ray $\theta = \theta(t)$ at N. Let $\lambda(t) = \{(r \cos(\theta(t)) + x_N, r \sin(\theta(t)) + y_N) \in \Omega : r > 0\}$. For each $s \neq t$, $\sigma(s)$ has tangent direction $\theta(s) \neq \theta(t)$ and so, for some $\eta = \eta(s, t) > 0$, if $(x, y) \in \lambda(t)$ and $|(x - x_N, y - y_N)| < \eta$, then $f(x, y) < s$ if $t < s$ and $f(x, y) > s$ if $s < t$. Thus we see that $f(x, y)$ must converge to t as $(x, y) \in \lambda(t)$ converges to N. Hence $Rf(\theta(z), N)$ exists and equals z, for each $z \in (z_1, z_2)$. The same argument also shows that $f(x, y)$ converges to z_1 as $(x, y) \in \Omega_A$ approaches N and $f(x, y)$ converges to z_2 as $(x, y) \in \Omega_B$ approaches N. Finally, since $Rf(\theta, N) = z(\theta)$ for $\theta \in (\theta_A, \theta_B)$ and $z(\theta)$ is continuous in θ, we see that $Rf(\cdot, N) \in C^0([\theta_1 + \pi, \theta_2])$ and $Rf(\cdot, N)$ is strictly increasing on $[\theta_A, \theta_B]$. Q.E.D.

3.2 Some Additional Theorems

In this section, we will discuss some results for Perron solutions which have weaker hypotheses than Theorem 3; these results have some resemblance to Theorem 2(A.).

THEOREM 4: Suppose ϕ satisfies conditions (A3'), (A4), and (A4') of §1. Let U be the upper Perron solution of the Dirichlet problem (2). Then there exists $z_0 \in (\phi(r(0), 0), \phi(0, 0)]$ such that

$$U(x, y) \to z_0 \quad \text{as} \quad (x, y) \to (0, 0), \quad (x, y) \in \overline{\Omega_0} \backslash \{(0, 0)\} \tag{46}$$

and if $\{(x_n, y_n)\}$ is any sequence in $\overline{\Omega} \backslash \{N\}$ converging to N, then

$$\liminf_{n \to \infty} U(x_n, y_n) \geq z_0. \tag{47}$$

THEOREM 5: Suppose ϕ satisfies conditions (A3''), (A4), and (A4'') of §1. Let U be the upper Perron solution of the Dirichlet problem (2). Then
$U_x \leq 0$ in Ω, $U_y \leq 0$ in Ω^+, $U_y \geq 0$ in Ω^-,
$U(tx, ty)$ is a strictly decreasing function of $t > 0$ for each $(x, y) \in \overline{\Omega_0} \backslash \{N\}$
and there exists $z_0 \in (\phi(r(0), 0), \phi(0, 0)]$ such that (46) and (47) hold.

Proof of Theorem 4: We know that $U(0, y)$ is a decreasing function of $y > 0$ (Lemma 5). We will show that $U(x, y)$ is a decreasing function of x when $(x, y) \in \Omega$ and $|y| < \epsilon_1$. Let us assume this for a moment. If w is as in Lemma 6, we know that $U \geq w$. If $(x, y) \in \Omega^+ \cap \Omega_0$ with $|y| < \epsilon_1$, then $U(x, y) \leq U(0, y) \leq z_0$. Therefore $w \leq U \leq z_0$ in $\{(x, y) \in \Omega_0 : |y| < \epsilon_1\}$ and, since $\lim_{(x,y) \to (0,0)} w(x, y) = z_0$, (46) holds.
    Let $T_\phi^{**} = \{v \in T_\phi \cap C^2(\Omega^+ \cup \Omega^-) : v \text{ satisfies } (1.), (2.), (3.), (4.), (5.) \text{ of Lemma}$
$5\}$. Let $v \in T_\phi^{**}$ and suppose $(x_0, y_0) \in \Omega^+$ with $v_x(x_0, y_0) = 0$. From Lemma 4, we can find $(x_1, y_1) \in \Omega^+$ with $v_x(x_1, y_1) = 0$ and $v_{xx} v_{yy} - v_{xy}^2 < 0$ at (x_1, y_1). Define

$$\tilde{w}(x, y) = v(x_1, y_1) + v_y(x_1, y_1)(y - y_1) \tag{48}$$

and set $D = \{(x, y) \in \Omega : v(x, y) < \tilde{w}(x, y)\}$ and $E = \{(x, y) \in \Omega : v(x, y) > \tilde{w}(x, y)\}$. If $\tilde{w}(N) \geq \phi(N)$, then $\Omega^- \subset D$ and $v = \tilde{w}$ at no more than two points of $\partial \Omega^+$. However, this contradicts Lemma 3 (with $\Omega' = \Omega^+$ and $v = \tilde{w}$). If $\tilde{w}(N) < \phi(N)$ and we recall condition (5.) of Lemma 5, we see that $E \cap \{y = 0\}$ is an interval containing N and either $D \cap \{y = 0\} = \emptyset$ or $D \cap \{y = 0\}$ is an interval containing $(r(0), 0)$. In particular, $v = \tilde{w}$ at no more than one point of $\partial \Omega^+ \cap \Omega$ and together with (A4') we see $v = \tilde{w}$ at no more than three points of $\partial \Omega^+$, in contradiction of Lemma 3. Hence $v_x(x_0, y_0) \neq 0$. Since $v(0, 0) > v(r(0), 0)$, $v_x < 0$ in $\{(x, y) \in \Omega : |y| < \epsilon_1\}$. Thus $v(x, y)$, and so $U(x, y)$, is decreasing in x for $|y| < \epsilon_1$.   Q.E.D.

Proof of Theorem 5: From the proof of Theorem 4, we see that $U_x \leq 0$ in Ω when (A4'') holds. Further, the proof of (4.), Lemma 5 yields $U_y \leq 0$ in $\Omega^+ \backslash \Omega_0$ when (A3'') holds. It remains to show that $U_y \leq 0$ in $\Omega^+ \cap \Omega_0$.

Let $T_\phi^0 = \{v \in T_\phi \cap C^2(\Omega_0) : v$ satisfies (1.), (2.), (3.), (4.), (5.) of Lemma 5$\}$. The proof of Lemma 5 shows that

$$U(x,y) = \inf_{v \in T_\phi^0} v(x,y) \qquad (49)$$

for $(x,y) \in \overline{\Omega}$. Let $v \in T_\phi^0$ and suppose $(x_0, y_0) \in \Omega^+ \cap \Omega_0$ such that $v_y(x_0, y_0) = 0$. Let $(x_1, y_1) \in \Omega^+ \cap \Omega_0$ such that $v_y(x_1, y_1) = 0$ and $v_{xx}v_{yy} - v_{xy}^2 < 0$ at (x_1, y_1). Define

$$\tilde{v}(x,y) = v(x_1, y_1) + v_x(x_1, y_1)(x - x_1) \qquad (50)$$

and set $D = \{(x,y) \in \Omega : v(x,y) < \tilde{v}(x,y)\}$ and $E = \{(x,y) \in \Omega : v(x,y) > \tilde{v}(x,y)\}$. If $\tilde{v}(N) \geq \phi(N)$, then $\Omega \backslash \overline{\Omega_0} \subset D$ and so $v \neq \tilde{v}$ on $\partial\Omega_0 \cap \Omega$. Together with (A3'') and the symmetry of $v - \tilde{v}$, we see that there are two points $(x_3, 0)$ and $(x_4, 0)$ with $0 \leq x_3 < x_4$ at which $v = \tilde{v}$ and $\{(x,0) : x_3 < x < x_4\} \subset E$; also there is a component E_0 of E such that $\partial E_0 \subset \Omega_0 \cup \{N\}$. Since $Q(v - \tilde{v}) = 0$ in Ω_0 and $v = \tilde{v}$ on $\partial E_0 \subset \overline{\Omega_0}$, $v = \tilde{v}$ in E_0. Since $v > \tilde{v}$ in E_0, we have a contradiction.

Now suppose $\tilde{v}(N) < \phi(N)$. From (4.), Lemma 5, we see that $v(0,y) = \tilde{v}(0,y)$ for at most one $y \geq 0$. If such a y exists, the intermediate value theorem requires $v - \tilde{v} = 0$ at a point of $\partial\Omega^+ \backslash \overline{\Omega_0}$ or at $(0, r(\frac{\pi}{2}))$. Condition (A3'') then implies that there are two points $(x_3, 0)$ and $(x_4, 0)$ with $0 < x_3 < x_4$ at which $v = \tilde{v}$ and $\{(x,0) : x_3 < x < x_4\} \subset D$; further, there is a component D_0 of D with $\partial D_0 \subset \Omega$. Since v is a supersolution in Ω, $\tilde{v}$ is a solution, and $v = \tilde{v}$ on ∂D_0, $v \geq \tilde{v}$ in D_0, a contradiction. Thus we see that $v_y \neq 0$ in Ω^+. Since $v_y(0,y) < 0$ for $y > 0$, $v_y < 0$ in Ω^+. Hence $v(x,y)$, and so $U(x,y)$, is decreasing in y for $(x,y) \in \Omega^+$. Q.E.D.

3.3 Proof of Theorem 3

Let us pick boundary data $\phi \in C^0(\partial\Omega)$ which satisfies (A1)-(A4), (A3''), (A4''); for example, let (c,d) be any point on $\partial\Omega$ for which $d > 0$ and $c = \inf\{x : (x,y) \in \partial\Omega\}$ and define $\phi(x,y) = 0$ if $x > 0$ or $|y| > d$ and $\phi(x,y) = \epsilon(x - c)$ otherwise, for any positive ϵ. Let f denote the upper Perron solution of (2), so $f = U$. Using the results of Serrin ([19]), we see that if Q is singularly elliptic, then we may pick ϕ so that $z_0 < \phi(0,0)$ and $f \notin C^0(\overline{\Omega})$; for instance, by choosing ϵ large enough in the choice of ϕ above, f will be discontinuous at N.

Theorem 5 implies that for some $z_0 \leq \phi(0,0)$, $Rf(\theta, N)$ exists and equals z_0 for all $\theta \in [-\frac{\pi}{2}, \frac{\pi}{2}]$, conclusion (iii) of (A.) holds, and $f_y \leq 0$ in Ω. Let $O^\pm$ be the convex domains given by

$$O^\pm = \Omega^\pm \backslash \overline{\Omega_0}. \qquad (51)$$

We will prove that $Rf(\theta, N)$ exists for $\frac{\pi}{2} < \theta < \beta$ and has the desired properties; the corresponding results for $\alpha = -\beta < \theta < -\frac{\pi}{2}$ will follow by symmetry. Thus we consider the behavior of f over O^+.

Let us define $u(x, y) = f(x, y) - y$ for $(x, y) \in \overline{O^+} \backslash \{(0, 0)\}$. Clearly $Rf(\theta, N) = Ru(\theta, N)$ for $\frac{\pi}{2} \le \theta < \beta$ whenever either of these quantities exists. Since $f_y \le 0$, $u_y \le -1$ and the graph of u over O^+, denoted S_0, has a simple projection on the xz-plane. This means that there exists a function $g(x, z)$ over a domain in the xz-plane such that

$$z = u(x, y) \qquad \text{if and only if} \qquad y = g(x, z); \tag{52}$$

using implicit differentiation, we see that g is a solution of an elliptic partial differential equation, $\tilde{Q}g = 0$. Our hypotheses imply that at each point $(0, z)$, $z_0 < z < \phi(0, 0)$, g is C^1 and $g_x > 0$. Lemma 7 (with $\theta_1 = \frac{\pi}{2}$, $\theta_2 = \beta_N$, and $\sigma = 0$) then yields (B.) and our proof is complete.   Q.E.D.

REFERENCES

1. H. Bear and G. Hile: Behavior of Solutions of Elliptic Differential Inequalities near a Point of Discontinuous Boundary Data, Comm. Part. Diff. Equ. 8 (1983), 1175-1197.

2. S. Bernstein: Sur les equations du calcul des variations, Ann. Sci. Ec. Norm. Sup. 29 (1912), 431-485.

3. A. Elcrat and K. Lancaster: On the Behavior of a Nonparametric Minimal Surface in a Nonconvex Quadrilateral, Arch. Rat. Mech. Anal. 94 (1986), 209-226.

4. A. Elcrat and K. Lancaster: Boundary Behavior of Nonparametric Surfaces of Prescribed Mean Curvature near a Reentrant Corner, Trans. A.M.S. 297 (1987), 645-650.

5. R. Finn: Remarks relevant to Minimal Surfaces, and to Surfaces of Prescribed Mean Curvature, J. d'Anal. Math. 14 (1965), 139-160.

6. D. Gilbarg and N. Trudinger: Elliptic Partial Differential Equations of Second Order, second edition, Springer-Verlag, New York, 1983.

7. J. Jost: Harmonic Maps between Surfaces, Springer Lecture Notes in Math. 1062, 1980.

8. K. Lancaster: Boundary Behavior of a Nonparametric Minimal Surface in $\Re^3$ at a Nonconvex Point, Analysis 5 (1985), 61-69. Corrigendum: Analysis 6 (1986), 413.

9. K. Lancaster: Nonparametric Minimal Surfaces in $\Re^3$ whose Boundaries have a Jump Discontinuity, Inter. J. Math. Math. Sci. 14 (1988), 651-656.

10. K. Lancaster: Boundary Behavior near Reentrant Corners for Solutions of Certain Elliptic Equations, Rend. Cir. Math. Palermo 40 (1991), 189-214.

11. K. Lancaster and D. Siegel: Radial Limits of Capillary Surfaces (preprint).

12. J. Leray: Discussion d'un probleme de Dirichlet, J. Math. p. appl. 17 (1939), 89-104.

13. F. Lin: Uniqueness and Nonuniqueness of the Plateau Problem. Indiana U. Math. J. 36 (1987), 843-855.

14. N. Meyers and J. Serrin: The Exterior Dirichlet Problem for Second Order Elliptic Partial Differential Equations, J. Math. Mech. 9 (1960), 513-538.

15. J.C.C. Nitsche: On the Non-Solvability of Dirichlet's Problem for the Minimal Surface Equation, J. Math. Mech. 14 (1965), 779-788.

16. J.C.C. Nitsche: Lectures on Minimal Surfaces, vol. 1, Cambridge Univ. Press, Cambridge, 1989.

17. H. Parks: An Application of the Cauchy- Kovalevskii Theorem: the Minimal Surface Equation at Corners, 741-757, in Topics in Mathematical Analysis, World Sci. Press, Singapore, 1989.

18. T. Radó: Contributions to the Theory of Minimal Surfaces, Acta Litt. Sci. Univ. Szeged 6 (1932), 1-20.

19. J. Serrin: The Problem of Dirichlet for Quasilinear Elliptic Differential Equations with many Independent Variables, Philos. Trans. Roy. Soc. London Ser. A, 264 (1969), 413-496.

20. J. Serrin: The Dirichlet Problem for Surfaces of Constant Mean Curvature, Proc. London Math. Soc. 21 (1970), 361-384.

21. L. Simon: Boundary Behavior of Solutions of the Nonparametric Least Area Problem, Bull. Austral. Math. Soc. 26 (1982), 17-27.

22. J. Spruck: Infinite Boundary Value Problems for Surfaces of Constant Mean Curvature, Arch. Rat. Mech. Anal. 49 (1972), 1-31.

Numerical Solutions for Linear Integro–Differential Equations of Parabolic Type with Weakly Singular Kernels

YANPING LIN   Department of Mathematics, University of Alberta, Edmonton, Alberta T6G 2G1 Canada

1.   INTRODUCTION

In this paper we study numerical solutions by finite element for the following weakly singular parabolic integro-differential equation:

$$u_t + A(t)u(t) + \int_0^t (t-s)^{-\alpha} B(t,s)u(s)ds = f(t), \quad \text{in } \Omega \times J,$$

$$u = 0, \qquad \text{on } \partial\Omega \times J, \qquad (1.1)$$

$$u(\cdot,0) = v, \qquad \text{on } \Omega,$$

where $\Omega \subset R^d$ $(d \geq 1)$ is a bounded domain with smooth boundary $\partial\Omega$, $J = (0,T]$, $T > 0$, $0 < \alpha < 1$, f and v are known functions. $A(t)$ is a positive definite second order elliptic operator,

$$A(t) = -\sum_{i,j=1}^d \frac{\partial}{\partial x_i}\left(a_{ij}(x,t)\frac{\partial}{\partial x_j}\right) + a(x,t)I, \quad a(x,t) \geq 0,$$

and $B(t,\tau)$ is any second order operator,

$$B(t,\tau) = -\sum_{i,j=1}^d \frac{\partial}{\partial x_i}\left(b_{ij}(x,t,\tau)\frac{\partial}{\partial x_j}\right) + \sum_{i=1}^d b_i(x,t,\tau)\frac{\partial}{\partial x_i} + b(x,t,\tau)I,$$

261

with smooth coefficients in x, t, and τ.

When $\alpha = 0$ (smooth kernels), the questions of existence, uniqueness and continuous dependence of the solutions upon the data for the problem (1.1) have been studied extensively [10] and the references therein. Recently, numerical approximations have received a considerable attention. For example, Green-well Yanik and Fairweather [4] and Thomée and Zhang [13] considered finite element approximations and shown optimal L^2 error estimates for both smooth and nonsmooth data. Cannon and Lin [2], Lin [6] and Lin, Thomee and Wahlbin [7] used Ritz-Volterra projections, introduced in [2], studied both linear and nonlinear equations and obtained L^p ($2 \leq p < \infty$) optimal error estimates for both semi-discrete and Crank-Nicolson finite element approximations.

When $0 < \alpha < 1$, the work (to the author knowledge) by Renardy, Hrusa and Nohel [10, Chapter 4, Section 2] is the only published result in which the authors studied the existence and uniqueness of the solutions to a problem similar to (1.1) with small data. We want to mention the work by Lorenzi and Sinestriri [9] in which they studied the inverse problem with the unknown weakly singular kernels of the form $B(t, \tau) = (t-\tau)^{-\alpha} k(t-\tau)\Delta$, where Δ is Laplace operator in R^d and $k(\cdot)$ is an unknown smooth function and derive some regularities of the solutions.

Up to now it seems that very little attention has been given to the numerical approximation for the problem (1.1). However, Choi and MacCamy [3] have considered finite element method for the following integro-differential equation,

$$u_t = \int_0^t a(t-s)\Delta u(s)\,ds,$$

where $a(t) = t^{-\alpha}\exp(-t)$ ($0 < \alpha < 1$). This problem can arise from the modelling of viscoelasticity [10].

In [8] and [11] authors studied by finite difference methods for the following problem

$$u_t + \beta u u_x = \int_0^t (t-s)^{-1/2} u_{xx}(x, s)\,ds, \quad \beta = 0, 1.$$

Some error estimates and stability are demonstrated. The above equation can be considered as the intermediate state between parabolic $u_t = u_{xx}$ and hyperbolic $u_{tt} = u_{xx}$ for $\beta = 0$.

Let $\{S_h\}$ be a family of finite dimensional subspaces of $H_0^1(\Omega)$ with the following properties: for some $l \geq 2$,

$$\inf_{\chi \in S_h} (\|\chi - u\| + h\|\chi - u\|_1) \leq Ch^r\|u\|_r, \quad 1 \leq r \leq l, \quad u \in H^r(\Omega) \cap H_0^1(\Omega), \quad (1.2)$$

where C is a constant independent of u and h. $H^r(\Omega)$ is Hilbert spaces of order r with norms $\|\cdot\|_r$ and $H_0^1(\Omega)$ is the completion of $C_0^\infty(\Omega)$ under $\|\cdot\|_1$ norm.

The semi-discrete finite element approximation to the solution u of (1.1) is now defined by $u_h(t) : \bar{J} \rightarrow S_h$,

$$(u_{h,t}, \chi) + A(t; u_h, \chi) + \int_0^t (t-s)^{-\alpha} B(t, s; u_h(s), \chi)\,ds = (f, \chi), \quad \chi \in S_h,$$

$$u_h(0) = v_h,$$

where $v_h \in S_h$ is an appropriate approximation of v into S_h, $A(t; \cdot, \cdot)$ and $B(t, s; \cdot, \cdot)$ are the bilinear forms on $H_0^1(\Omega) \times H_0^1(\Omega)$, which are associated with the operators $A(t)$ and $B(t, s)$.

In next section a Ritz-Volterra type projection $W(t)$ is first introduced and its properties are given, and then L^2 error estimates for the semi-discrete finite element approximation (1.3) is stated. A special case for nonsmooth data will also be discussed there.

2. A RITZ-VOLTERRA PROJECTION W(t) AND ERROR ESTIMATES

In this section we define a Ritz-Volterra projection associated with our problem (1.1) and study its various error estimates, to the solution u, which will be used to derive L^2 error estimates for finite element approximations.

Following [2, 7], let u be the solution of (1.1). we now introduce the following Ritz-Volterra projection $W(t) : \bar{J} \to S_h$ defined by

$$A(t; u - W, \chi) + \int_0^t (t - s)^{-\alpha} B(t, s; u(s) - W(s), \chi) ds = 0, \quad \chi \in S_h. \qquad (2.1)$$

This $W(t)$ defined above is similar to that defined in [2, 7], but involved with a weakly singular kernel. Therefore, it is necessary to analyze the approximation of this new Ritz-Volterra projection to u in details.

LEMMA 2.1.   *There exists $C > 0$, independent of h and u, such that*

$$\|u(t) - W(t)\| \leq Ch^r \||u(t)\||_r, \quad t \in \bar{J}. \qquad (2.2)$$

$$\|u_t(t) - W_t(t)\|_1 \leq Ch^{r-1} (\||u(t)\||_r + \||u_t(t)\||_r) + Ct^{-\alpha}\|\eta(0)\|_1, \quad t > 0, \qquad (2.3)$$

$$\|\eta_t(t)\| \leq Ch^r (\||u(t)\||_r + \||u_t(t)\||_r) + Ct^{-\alpha}(h\|\eta(0)\|_1 + \|\eta(0)\|), \qquad (2.4)$$

Proof: The proof of this lemma follows from a modified arguments given in [2, 6, 7].

COROLLARY 2.1.   *There exist $C > 0$ such that for $\eta(t) = u(t) - W(t)$,*

$$\int_0^t (\|\eta(s)\| + h\|\eta(s)\|_1) \, ds \leq Ch^r \int_0^t \|u(s)\|_r \, ds, \quad t \in J, \qquad (2.5)$$

$$\int_0^t (\|\eta_t(s)\| + h\|\eta_t(s)\|_1) ds \leq Ch^r \left(\|v\|_r + \int_0^t \|u_t(s)\|_r \, ds\right), \quad t \in J. \qquad (2.6)$$

We now consider L^2 error estimates for the semi-discrete finite element approximation defined by (1.3).

THEOREM 2.1.   *Assume that u and u_h are the solutions of (1.1) and (1.3) respectively, and $v_h \in S_h$ is such that $\|v - v_h\| + h\|v - v_h\|_1 \leq Ch^r\|v\|_r$. Then there exists $C > 0$,* *independent of h and u, such that*

$$\|u(t) - u_h(t)\| \le Ch^r \left(\|v\|_r + \int_0^t \|u_t(s)\|_r \, ds \right), \quad t \in \bar{J}. \tag{2.7}$$

Proof: As usual, we write the error $e(t) = (u - W) + (W - u_h) = \eta + \theta$. It follow from Lemma 2.1 and Corollary 2.1 that

$$\|\eta(t)\| \le \|\eta(0)\| + \int_0^t \|\eta_t(s)\| \, ds \le Ch^r \left(\|v\|_r + \int_0^t \|u_t(s)\|_r \, ds \right). \tag{2.8}$$

It now remains to estimates θ in L^2. We see from (1.1) and (1.3) that θ satisfies

$$(\theta_t, \chi) + A(t; \theta, \chi) + \int_0^t (t - s)^{-\alpha} B(t, s; \theta(s), \chi) \, ds = -(\eta_t, \chi), \quad \chi \in S_h. \tag{2.9}$$

Letting $\chi = \theta \in S_h$ in (2.9), we obtain

$$\frac{1}{2} \frac{d}{dt} \|\theta\|^2 + A(t; \theta, \theta) \le \|\eta_t\| \, \|\theta\| + C \int_0^t (t - s)^{-\alpha} \|\theta(s)\|_1 \, \|\theta(t)\|_1 \, ds. \tag{2.10}$$

But, we see for $\epsilon > 0$ that

$$\int_0^t (t - s)^{-\alpha} \|\theta(s)\|_1 \, \|\theta(t)\|_1 ds \le \epsilon \int_0^t (t - s)^{-\alpha} \|\theta(t)\|_1^2 ds + C(\epsilon) \int_0^t (t - s)^{-\alpha} \|\theta(s)\|_1^2 \, ds$$

$$\le \epsilon T^{1-\alpha} (1 - \alpha)^{-1} \|\theta(t)\|_1^2 + C(\epsilon) \int_0^t (t - s)^{-\alpha} \|\theta(s)\|_1^2 ds.$$

It now follows by taking $\epsilon > 0$ small and fixed together with the positive definiteness of the operator $A(t)$ that

$$\frac{1}{2} \frac{d}{dt} \|\theta\|^2 + C \|\theta\|_1^2 \le \|\eta_t\| \, \|\theta\| + C \int_0^t (t - s)^{-\alpha} \|\theta(s)\|_1^2 \, ds, \tag{2.11}$$

and then, by integration from 0 to t and Lemma 2.1, that

$$\|\theta\|^2 + \int_0^t \|\theta\|_1^2 \, ds$$

$$\le C \left(\|\theta(0)\|^2 + \int_0^t \|\eta_t\| \, \|\theta\| \, ds + \int_0^t \int_0^\tau (\tau - s)^{-\alpha} \|\theta(s)\|_1^2 \, ds \, d\tau \right)$$

$$\le C \left(\|\theta(0)\|^2 + \int_0^t \|\eta_t\| \, \|\theta\| \, ds + \int_0^t (t - s)^{-\alpha} \left(\int_0^s \|\theta(\tau)\|_1^2 \, d\tau \right) ds \right). \tag{2.12}$$

Applying Lemma 2.1 to (2.12), it follows that

$$\|\theta(t)\|_1^2 + \int_0^t \|\theta(s)\|_1^2 \, ds \le C \left(\|\theta(0)\|^2 + \int_0^t \|\eta_t\| \, \|\theta\| \, ds \right.$$

$$+ \int_0^t (t-s)^{-\alpha} \int_0^s \|\eta_t(\tau)\| \, \|\theta(\tau)\| \, d\tau \, ds \Big)$$

$$\leq C\|\theta(0)\|^2 + C \int_0^t \|\eta_t\| \, \|\theta\| \, ds$$

$$\leq C\|\theta(0)\|^2 + \frac{1}{2} \sup_{0<s<t} \|\theta(s)\|^2 + C(\int_0^t \|\eta_t\| \, ds)^2.$$

Since this holds for all $t \in J$ we may conclude that

$$\|\theta(t)\| \leq \sup_{0<s<t} \|\theta(s)\| \leq C \left(\|\theta(0)\| + \int_0^t \|\eta_t\| \, ds \right). \tag{2.13}$$

Because $\theta(0) = (W(0) - v) + (v - v_h)$ we find by our assumptions and Lemma 2.1 and Corollary 2.1 that

$$\|\theta(t)\| \leq Ch^r \left(\|v\|_r + \int_0^t \|u_t(s)\|_r ds \right), \tag{2.14}$$

which together with (2.8) and triangle inequality complete the proof.   Q.E.D.

In our error estimates (2.7) we certainly need some regularity of the solution u of (1.1). In general, we can not expect to have the regularity for the solution u of (1.1) as we do for the solutions of equations with smooth kernels [2, 7]. However, in [9], the authors shown by using abstract semigroup theory that if the data is smooth and comparable, then one has that $u(t) \in C^2(\overline{\Omega})$ and $\Delta u_t(t) \in C(\overline{\Omega})$ for $t \in J$. Thus, our result (2.7) is optimal (with $r = 2$) for piecewise linear finite element approximations.

We shall now consider a semi-linear equation with non-smooth data. Namely, we study finite element solution for the following problem:

$$u_t + Au(t) - \int_0^t (t-s)^{-\alpha} f(t,s,u(s)) ds = 0, \quad \text{in } \Omega \times J, \tag{2.15}$$

$$u(\cdot, t) = 0, \quad \text{on } \partial\Omega \times J,$$

$$u(\cdot, 0) = v \in L^2(\Omega),$$

where A is a time independent positive definite elliptic operator and f is a given smooth function and is Lipschitz continuous in the last variable.

LeRoux and Thomee have studied (2.15) for $\alpha = 0$ [5] in which the authors derived optimal error estimates and also considered the time-discretizations. Here we only consider the semi-discrete case for $0 < \alpha < 1$ and obtain the same error bound as that given in [5]. The proof of this result is similar to that of [5]. For the completeness, we shall provide an example which shows that the convergence rate of higher than second is not possible in general.

THEOREM 2.2.   Assume that $v_h = P_h v$ and $\|v\| \leq R$. Then, there exists a constant $C = C(r, T, \alpha, R) > 0$ such that

$$\|u(t) - u_h(t)\| \leq C(r, T, \alpha, R) h^{2r} t^{-r}, \quad r < 1, \quad 0 < t \leq T, \tag{2.16}$$

where $P_h : L^2(\Omega) \to S_h$ be the L^2 projection. Notice that (2.9) is the same error estimate obtained by LeRoux and Thomee for $\alpha = 0$ in [5].

Proof: Let $E(t)$ be the semi-group generated by the operator $-A$, then we have for the exact solution $u(t)$ that

$$u(t) = E(t)v + \int_0^t E(t - s) \int_0^s (s - \tau)^{-\alpha} f(s, \tau, u(\tau)) \, d\tau \, ds$$

and the finite element solution $u_h(t)$ that

$$u_h(t) = E_h(t)P_h v + \int_0^t E_h(t - s) \int_0^s (s - \tau)^{-\alpha} P_h f(s, \tau, u_h(\tau)) \, d\tau \, ds,$$

where $E_h(t)$ is the semi-group generated by the operator $-A_h$ defined in section 2 (see [12]). We then obtain by writing the error $e(t) = u_h(t) - u(t)$ that

$$e(t) = F_h(t)v + \int_0^t F_h(t - s) \int_0^s (s - \tau) f(s, \tau, u(\tau)) \, d\tau \, ds$$

$$+ \int_0^t E_h(t - s) \int_0^t (s - \tau)^{-\alpha} P_h \left(f(s, \tau, u_h(\tau)) - f(s, \tau, u(\tau)) \right) \, d\tau \, ds.$$
(2.17)

where $F_h(t) = E_h(t)P_h - E(t)$ and satisfies for any $v \in L^2(\Omega)$ [12, 13] that

$$\|F_h(t)v\| \leq Ch^l t^{-l/2} \|v\|, \quad 0 \leq l \leq r, \quad r > 0.$$
(2.18)

By using (2.18) with $l = 2r$, the boundedness of v, and the stability of $E_h(t)$ and P_h [6, 22] together with the Lemma 2.1 and Corollary 2.1, we have

$$\|e(t)\| \leq Ch^{2r} t^{-r} + Ch^{2r} \int_0^t (t - s)^{-r} \int_0^s (s - \tau)^{-\alpha} \|f(s, \tau, u(\tau))\| \, d\tau \, ds$$

$$+ C \int_0^t \int_0^s (s - \tau)^{-\alpha} \|e(\tau)\| \, d\tau \, ds$$
(2.19)

$$\leq Ch^{2r} t^{-r} + C \int_0^t (t - s)^{-\alpha} \int_0^s \|e(\tau)\| \, d\tau.$$

Hence, (3.10) follows by applying lemma 1.1 to (2.19) with $0 < r < 1$.   Q.E.D.

Following [5], we consider, for $0 < \alpha < 1$, the system

$$u_t^1 - u_{xx}^1 = \int_0^t (t - s)^{-\alpha} f(u^2(s)) \, ds, \qquad\qquad (2.20)$$

$$u_t^2 - u_{xx}^2 = 0, \qquad\qquad 0 \leq x \leq 2\pi, \ t \geq 0, \qquad (2.21)$$

$$u^k(0, t) = u^k(2\pi, t), \qquad\qquad t \geq 0, \ k = 1, 2, \qquad (2.22)$$

$$u^k(x, t) = v^k(x), \qquad\qquad 0 \leq x \leq 2\pi, \ k = 1, 2. \qquad (2.23)$$

where $f(p) = 4p^2$ for $|p| \leq C$ and we imposed the periodic boundary condition instead of the standard Dirichlet boundary condition.

Let $h = 1/n$ and S_h be the span of $\{1, \cos x, \sin x, \cdots, \cos(n-1)x, \sin(n-1)x\}$. It is obvious that any smooth periodic function may be approximated to any higher order in S_h, that is, for any $r > 0$,

$$\|P_h v - v\| + h\|P_h v - v\|_1 \leq Cn^{-r}\|v^r\| \leq Ch^r\|v\|_r.$$

Now let $u_h(t) = (u_h^1(t), u_h^2(t))$ be the finite element solution in $S_h \times S_h$ for (2.20)-(2.23) with $u_h^k(0) = v_h^k = P_h v^k$, $k = 1, 2$. we then show that an estimate of the form

$$\|u_h^1(t) - u^1(t)\| + \|u^2(t) - u_h^2(t)\| \leq C(t, \alpha, R)h^r \quad \text{for} \quad \|v^1\| + \|v^2\| \leq R, \quad (2.24)$$

is not possible for any $t > 0$ if $r > 2$.

For this purpose, we choose $v^1 = 0$ and $v^2 = \cos nx$ which are bounded by $C = 2\pi$. Also we have $v_h^k = 0$ since v^2 is orthogonal to S_h, and hence, $u_h(t) = 0$ for $t \geq 0$. As a consequence, the error equals the exact solution of (2.20)–(2.23). But, we obtain by a simple calculation that

$$u^2(x, t) = \exp\{-n^2 t\} \cos nx$$

and

$$u^1(x, t) = \int_0^t g(s)\, ds + \int_0^t \exp\{-4n^2(t-s)\}g(s)\, ds\, \cos(2nx),$$

where

$$g(t) = 2\int_0^t (t-s)^{-\alpha} \exp\{-2n^2 s\}\, ds.$$

It can be checked easily that for any fixed $t > 0$,

$$\int_0^t g(s)\, ds = \frac{t^{1-\alpha}}{1-\alpha}n^{-2} + O\left(t^{1-\alpha}n^{-2}\exp\{-Cn^2 t\} + t^{-\alpha}n^{-4}\right) \quad \text{as } n \to \infty$$

and

$$\int_0^t \exp\{-4n^2(t-s)\}g(s)\, ds = O\left(t^{1-\alpha}n^{-2}\exp\{-Cn^2 t\} + t^{-\alpha}n^{-4}\right) \quad \text{as } n \to \infty$$

Thus, we have by $h = n^{-1}$ that for h sufficiently small,

$$\|u^1(t)\| + \|u^2(t)\| \geq C(t, \alpha, R)h^2.$$

This is a contradiction to (2.24) for $r > 2$.

REFERENCES

1. H. Brunner and P. J. van der Houwen, *The Numerical Solution of Volterra equations*, North-Holland (1986).

2. J. R. Cannon and Y. Lin, A priori L^2 error estimates for finite element methods for nonlinear diffusion equations with memory, *SIAM J. Numer. Anal.*, 27: 595–607 (1990).

3. U. Jin Choi and R. C. MacCamy, Fractional order Volterra equations, Volterra Intergodifferential Equations in Banach Spaces and Applications (ed. G. Da Prato and M. Iannelli), *Pitman Research Notes in Mathematics Series*, Vol. 190, 231–249 (1989).

4. C. E. Greenwell-Yanik and G. Fairweather, Finite element methods for parabolic and hyperbolic partial integro-differential equations, *Nonlinear Analysis*, 12: 785–809 (1988).

5. M. N. LeRoux and V. Thomee, Numerical solution of semilinear integro-differential equations of parabolic type with nonsmooth data, *SIAM J. Numer. Anal.*, 26: 1291–1309 (1989).

6. Y. Lin, Galerkin methods for nonlinear parabolic integro-differential equations with nonlinear boundary conditions, *SIAM J. Numer. Anal.*, 27: 608–621 (1990).

7. Y. Lin, V. Thomee, and L. Wahlbin, Ritz-Volterra projection onto finite element spaces and applications to integro-differential and related equations, *SIAM J. Numer. Anal.*, 28: 1047–1070 (1991).

8. J. C. Lopez-Marcos, A difference scheme for a nonlinear partial integrodifferential equation, *SIAM J. Numer. Anal.*, 27: 20–31 (1990).

9. A. Lorenzi and E. Sinestrari, An inverse problem in the theory of materials with memory, *Nonlinear Analysis*, 12, No. 12: 1317–1335 (1988).

10. M. Renardy, W. J. Hrusa and J. A. Nohel, Mathematical Problems in Viscoelasticity, *Longman Scientific & Technical*, England (1987).

11. J. M. Sanz-Serna, A numerical method for a partial integro-differential equation, *SIAM J. Numer. Math.*, 25: 319–327 (1988).

12. V. Thomee, Galerkin Finite Element Methods for Parabolic Problems, *Lecture Notes in Mathematics*, 1054, Springer-Verlag, 1984.

13. V. Thomee and N. Y. Zhang, Error estimates for semi-discrete finite element methods for parabolic integro-differential equations, *Math. Comp.*, 53: 121–139 (1989).

14. M. F. Wheeler, A priori L_2 error estimates for Galerkin approximation to parabolic partial differential equations, *SIAM J. Numer. Anal.*, 19: 723–759 (1973).

Impulsive Stabilization

XINZHI LIU   Department of Applied Mathematics, University of Waterloo, Waterloo, Ontario, Canada

ALLAN R. WILLMS   Department of Applied Mathematics, University of Waterloo, Waterloo, Ontario, Canada

1 INTRODUCTION

In this paper we shall investigate the problem of impulsively controlling a system of autonomous ordinary differential equations so as to keep solutions close to a given state, p, which need not be an equilibrium point of the system. Consider the following autonomous system

$$x' = f(x), \tag{1}$$

where $f \in C^1[D, \mathbb{R}^n]$, $D \subset \mathbb{R}^n$ is open. Let the space $X = \mathbb{R}^n$ be decomposed into the direct sum $X = Y \oplus Z$, where Y is an m-dimensional subspace of X, $1 \leq m < n$, and $Z = Y^\perp$. We call Y and Z the impulsive and non-impulsive subspaces respectively. Any vector $x \in X$, or vector function $f(x)$, may be expressed uniquely as $x = y + z$, or $f(x) = f_y(x) + f_z(x)$, where $y, f_y(x) \in Y$ and $z, f_z(x) \in Z$. We shall utilize this decomposition throughout this chapter and remark that the subscripts y and z on any vector in x shall denote its unique portion in Y and Z respectively, while the vectors y and z shall represent the unique portions of x.

Let U be the set of admissible controls u, where $u = \{(t_k, \Delta y_k)\}_{k=1}^\infty$ and

(i) $0 \leq t_1 < t_2 < \ldots < t_k < \ldots$, and $t_k \to \infty$ as $k \to \infty$,
(ii) $\Delta y_k \in Y$, $k = 1, 2, \ldots$, $|\Delta y_k| \leq A$ for some positive constant A.

Consider the impulsive control system associated with system (1)

$$
\begin{cases}
y' = f_y(x), & \\
z' = f_z(x), & t \neq t_k, \ k = 1, 2, \ldots \\
y(t_k^+) = y(t_k) + \Delta y_k, & k = 1, 2, \ldots \\
x(0) = x_0 .
\end{cases}
\tag{2}
$$

For a comprehensive treatment of impulsive differential equations see Lakshmikantham, Bainov and Simeonov (1989). Let $\phi(t, x)$ be a solution of (1). Then for each $u \in U$, $x(t) = x(t, x_0, u)$ is a solution of (2) given by

$$
x(t) = \phi(t - t_{k-1}, x_{k-1}^+), \qquad t \in (t_{k-1}, t_k], \ k = 1, 2, \ldots,
\tag{3}
$$

where $t_0 = 0$, $x_0^+ = x_0$, and

$$
x_k^+ = \phi(t_k - t_{k-1}, x_{k-1}^+) + \Delta y_k, \qquad k = 1, 2, \ldots .
\tag{4}
$$

Note from (3) and (4) that since f is C^1, ϕ is a continuous function of t, and since $\Delta y_k \in Y$, it follows that $z(t)$ is continuous for all $t \geq 0$, while $y(t)$ is continuous on each interval $(t_{k-1}, t_k]$, where $y(t) + z(t)$ is the decomposition of $x(t)$.

2 CRITERIA FOR STABILIZABILITY

We begin by stating the concept of impulsive stabilization of a point p. For $\alpha > 0$, let $B_\alpha(p) = \{x \in \mathbb{R}^n : |p - x| < \alpha\}$.

DEFINITION   A point $p \in D$ is said to be

(S_1) impulsively stabilizable if for any given $\epsilon > 0$, there exists a $\delta = \delta(\epsilon) > 0$ such that for each $x_0 \in B_\delta(p)$ there exists a $u \in U$ such that $x(t) \in B_\epsilon(p)$, for all $t \geq 0$, where $x(t) = x(t, x_0, u)$ is any solution of (2);

(S_2) asymptotically impulsively stabilizable if for any given $\epsilon > 0$, there exists a $\sigma = \sigma(\epsilon) > 0$ such that for each $x_0 \in B_\sigma(p)$ there exists a $u \in U$ such that $x(t) \in B_\epsilon(p)$, for all $t \geq 0$, and $\lim_{t \to \infty} x(t) = p$;

(S_3) impulsively unstabilizable if (S_1) fails to hold.

It should be noted that the point p in the above definition is, in general, not an equilibrium point of the system, in contrast to those found in the standard control theory (see for example Sontag, 1990). The type of stability defined above may be considered a special case of stability in terms of two measures, a concept expounded by Lakshmikantham and Liu (1989), and Liu (1990).

2.1 Necessary Conditions

It follows from the above definition that the vector field f must be tangent to the impulsive subspace at p for p to be impulsively stabilizable as indicated in the theorem below.

THEOREM 1   If $f(p) \notin Y$, then p is impulsively unstabilizable.

Proof: If $f(p) \notin Y$, then $f_z(p) = v \neq 0$. By continuity of f, there exists an $\epsilon > 0$ such that

$$|\text{Proj}_v f_z(x)| > 0, \qquad \forall\, x \in \overline{B_\epsilon(p)}, \tag{5}$$

where Proj_v denotes the orthogonal projection onto the one-dimensional subspace defined by $span\{v\}$. By continuity of the projection function and by the compactness of $\overline{B_\epsilon(p)}$, inequality (5) implies that there exists $m > 0$ such that

$$|\text{Proj}_v f_z(x)| \geq m > 0, \qquad \forall\, x \in \overline{B_\epsilon(p)}; \tag{6}$$

physically, m is the minimum speed in the positive v direction for all points in $\overline{B_\epsilon(p)}$. For any δ, $0 < \delta < \epsilon$, choose $x_0 = y_0 + z_0$ in $B_\delta(p)$. Since $z(t) = z(t, x_0, u)$ is continuous in t, (6) implies

$$|\text{Proj}_v(z(t) - z_0)| \geq mt, \qquad \forall\, t \geq 0, \text{ provided } x(t) \in \overline{B_\epsilon(p)}. \tag{7}$$

Since the projection is orthogonal, (7) implies

$$|z(t) - z_0| \geq mt, \qquad \forall\, t \geq 0, \text{ provided } x(t) \in \overline{B_\epsilon(p)}, \tag{8}$$

but $|z(t) - z_0| \leq |x(t) - x_0| \leq |x(t) - p| + |p - x_0|$, hence from (8) we have

$$|x(t) - p| \geq mt - |p - x_0| \geq mt - \delta, \qquad \forall\, t \geq 0,$$

so that $|x(t) - p| > \epsilon$ for t sufficiently large; consequently p is impulsively unstabilizable. ∎

2.2 Impulsively Invariant Sets

To motivate and help illustrate our subsequent theorem we shall embark on a short discussion of the problem of finding a control u that will create an invariant set of system (2).

Consider a system in $\mathbb{R}^3$ and a point p for which $f(p)$ is aligned with the x-axis. Let $Y = span\{(1,0,0)^T\}$ and $Z = span\{(0,1,0)^T, (0,0,1)^T\}$. Such a system meets the necessary condition of Theorem 1. Consider a closed curve C, lying in the plane through p parallel to Z such that p is in the interior of C. Generate a "cylinder", S, by constructing, through each point of C, a line segment of length 2ℓ parallel to Y such that its midpoint lies on C, (see Figure 1). The boundary of the cylinder is composed of the cylinder's wall (the line segments) and its two ends which are surfaces parallel to Z.

Our aim is to make the cylinder, S, invariant with the application of impulses in the x_1-direction. Consider a point x_0, starting within the the cylinder. The trajectory $\phi(t, x_0)$ will either stay within S or will reach the cylinder wall or the ends. Suppose $\phi(t^*, x_0) = A$ for some time t^*, where A is a point on one of the cylinder ends. It is then easy to see that an impulse of strength less than 2ℓ in the positive or negative x_1-direction as appropriate, will send the trajectory back into the interior of S. If however $\phi(t^*, x_0) = B$, where B is a point on the cylinder wall, then an impulse in Y can only carry the trajectory to some other

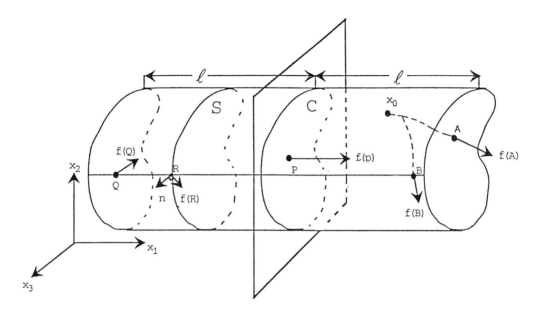

Figure 1: An invariant cylinder.

point on the cylinder wall along the line through B parallel to Y. If along this line segment there is a point Q where the vector field f is moving into the cylinder then an impulse to Q will keep $x(t)$ within S for at least a short time longer. If there is a point R along the line segment at which the velocity field is tangent to the cylinder wall and for which $\phi(t, R)$ lies either on or inside S for some positive time interval then an impulse to R would also keep $x(t)$ within S. If such points Q or R exist for each B on the cylinder wall then the cylinder S can be made invariant provided the sum of the time intervals between successive impulses is unbounded. We now formalize the preceding discussion on invariant sets.

THEOREM 2   Let $\Omega_z \subset Z$ be an open bounded region whose boundary is C^1, and let Ω_y be an open bounded region in Y. Define the "cylinder", Ω, and its wall, W, by $\Omega = \Omega_y \oplus \Omega_z$, and $W = \Omega_y \oplus \partial\Omega_z$. We assume $\Omega \subset D$. Let n be the unit outward normal to Ω defined on W, and define the set N as

$$N = \{w \in W : \quad f \cdot n|_w \leq 0\}^{O(W)},$$

where the superscript $O(W)$ denotes the interior of the set N with respect to W.

If $\mathrm{Proj}_z(N) = \partial\Omega_z$, then for any $x_0 \in \overline{\Omega}$, there exists a $u \in U$ such that $x(t, x_0, u) \in \overline{\Omega}$, for all $t \geq 0$.

Proof: See Liu and Willms (1993), or Willms (1993).

2.3 Sufficient Conditions

Sufficient conditions for impulsive stabilizability are given in the following theorem. Essentially, the conditions imposed assure that for any positive ϵ, there exists an impulsively

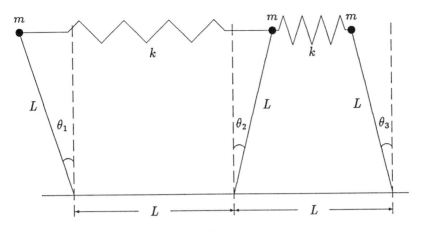

Figure 2: A system of inverted pendula.

invariant set contained within $B_\epsilon(p)$. The proof itself, although somewhat intuitive, is quite long and technical, for which reason the reader is referred to Liu and Willms (1993), or Willms (1993).

THEOREM 3   Let $p = p_y + p_z$ be a point in D and let $v \in C^1[Z, \mathbb{R}]$ be a positive definite function with respect to p_z. Define the sets

$$
\begin{aligned}
I &= \{x \in D \mid \nabla v(z) \cdot f_z(x) \le 0\}^O \\
I_\alpha &= \operatorname{Proj}_z (I \cap (B_\alpha^y(p_y) \oplus Z)) \cup \{p_z\} \\
J &= \{x \in D \mid \ \nabla v(z) \cdot f_z(x) < 0\} \\
J_\alpha &= \operatorname{Proj}_z (J \cap (B_\alpha^y(p_y) \oplus Z)) \cup \{p_z\}
\end{aligned}
$$

where the superscript O denotes the interior of the set, Proj_z denotes the orthogonal projection onto the Z subspace, and $B_\alpha^y(p_y)$ is the m-dimensional α-ball around p_y in the Y subspace.

(a) If I_α is a neighbourhood of p_z, for all $\alpha > 0$, then p is impulsively stabilizable.

(b) If J_α is a neighbourhood of p_z, for all $\alpha > 0$, then p is asymptotically impulsively stabilizable.

We remark that the openness of the set I in the above theorem is an essential requirement without which the theorem does not hold.

EXAMPLE 1   Consider a sequence of inverted pendula whose bobs are connected by springs as shown in Figure 2. The linear approximation of the governing equations for small oscillations of this system are given by

$$
\begin{aligned}
mL\ddot{\theta}_1 &= mg\theta_1 - kL(\theta_1 - \theta_2), \\
mL\ddot{\theta}_2 &= mg\theta_2 - kL(2\theta_2 - \theta_1 - \theta_3), \\
mL\ddot{\theta}_3 &= mg\theta_3 - kL(\theta_3 - \theta_2),
\end{aligned}
$$

where m is the mass of the bobs, g is the gravitational acceleration, L is the length of each pendulum and k is the spring constant. The above system reduces to the nondimensional first order system

$$x' = \begin{bmatrix} 0 & 1 & 0 & 0 & 0 & 0 \\ 1-\gamma & 0 & \gamma & 0 & 0 & 0 \\ 0 & 0 & 0 & 1 & 0 & 0 \\ \gamma & 0 & 1-2\gamma & 0 & \gamma & 0 \\ 0 & 0 & 0 & 0 & 0 & 1 \\ 0 & 0 & \gamma & 0 & 1-\gamma & 0 \end{bmatrix} x,$$

where $x = (\theta_1, \dot{\theta}_1, \theta_2, \dot{\theta}_2, \theta_3, \dot{\theta}_3)^T$, $\gamma = \dfrac{kL}{mg}$ and the differentiation is with respect to $\tau = \sqrt{\dfrac{g}{L}}\, t$. We wish to stabilize the oscillations about the unstable equilibrium point, $x = 0$, by applying impulses to x_2 and x_6, which corresponds to striking the outer bobs affecting an instantaneous change in velocity with no change in position. We shall assume that $2\gamma > 1$ which, in some sense, says that the spring forces dominate the gravitational forces for the centre bob.

In this situation we have $z = (x_1, x_3, x_4, x_5)^T$ and $y = (x_2, x_6)^T$. Define the Lyapunov function $v = \dfrac{1}{2}\left(x_1^2 + (2\gamma - 1)x_3^2 + x_4^2 + x_5^2\right)$, then

$$\begin{aligned}
\nabla v \cdot f_z &= (x_1, (2\gamma-1)x_3, x_4, x_5) \begin{pmatrix} x_2 \\ x_4 \\ \gamma x_1 \\ (1-2\gamma)x_3 \\ x_6 \end{pmatrix} \\
&= x_1 x_2 + \gamma x_1 x_4 + \gamma x_4 x_5 + x_5 x_6 \\
&= x_1\left(x_2 + \gamma x_4\right) + x_5\left(x_6 + \gamma x_4\right).
\end{aligned}$$

Consequently we have

$$\begin{aligned}
I \supset &\left[\{x : x_1 \geq 0, x_2 \leq -\gamma x_4\} \cup \{x : x_1 \leq 0, x_2 \geq -\gamma x_4\}\right] \\
&\cap \left[\{x : x_5 \geq 0, x_6 \leq -\gamma x_4\} \cup \{x : x_5 \leq 0, x_6 \geq -\gamma x_4\}\right].
\end{aligned}$$

For given α, if $|x_2|, |x_6| \leq \dfrac{\alpha}{\sqrt{2}}$ then $\|y\| < \alpha$, therefore

$$I_\alpha \supset \left\{x : |x_4| < \frac{\alpha}{\sqrt{2}\,\gamma}\right\},$$

showing that I_α is a neighbourhood of $p_z = 0$. It follows from Theorem 3 that $p_z = 0$ is impulsively stabilizable.

EXAMPLE 2   Suppose the owner of a resort on a small northern lake wishes to attract fishermen by increasing the population of two particular species of game fish in his lake. Upon looking into the matter he discovers that the cost of stocking his lake with these species is excessive while the cost of stocking his lake with the main prey species of these game fish is comparatively economical. The owner therefore wishes to determine how high he can keep the game species population by stocking the lake with the feeder fish. A model for this situation may be presented as below,

$$\dot{N}_1 = N_1 (b_1 - a_{11}N_1 - a_{12}N_2 - a_{13}N_3)$$
$$\dot{N}_2 = N_2 (b_2 - e_2 + a_{21}N_1 - a_{22}N_2 - a_{23}N_3)$$
$$\dot{N}_3 = N_3 (b_3 - e_3 + a_{31}N_1 - a_{32}N_2 - a_{33}N_3),$$

where N_1 is the feeder fish population, N_2 and N_3 are the game fish populations, and e_2, e_3 are the fishing efforts applied by the anglers (all constants in the above model are assumed to be positive). From Theorem 1, the candidate positive points, p, that could be stabilized are those satisfying

$$\begin{bmatrix} a_{22} & a_{23} \\ a_{32} & a_{33} \end{bmatrix} \begin{pmatrix} p_1 \\ p_2 \end{pmatrix} = \begin{pmatrix} b_2 - e_2 + a_{21}p_1 \\ b_3 - e_3 + a_{31}p_1 \end{pmatrix}. \tag{9}$$

Since the left hand side of (9) is positive it is clear that the fishing effort cannot exceed the sum of the game species' birth rate and the food availability. Furthermore, since the owner wishes to have p_2 and p_3 as large as possible, he will choose p_1 as large as possible, i.e. the maximum of what he can afford to stock or the maximum feeder fish population that the lake is capable of sustaining.

We assume that a point p is selected which satisfies (9), and choose a Lyapunov function

$$v = N_2 - p_2 - p_2 \ln \frac{N_2}{p_2} + N_3 - p_3 - p_3 \ln \frac{N_3}{p_3}.$$

Calculating the derivative we obtain

$$\begin{aligned} \dot{v} &= (N_2 - p_2)\left[a_{21}(N_1 - p_1) - a_{22}(N_2 - p_2) - a_{23}(N_3 - p_3)\right] \\ &\quad + (N_3 - p_3)\left[a_{31}(N_1 - p_1) - a_{32}(N_2 - p_2) - a_{33}(N_3 - p_3)\right] \\ &= -(N_2 - p_2, N_3 - p_3)\begin{bmatrix} a_{22} & a_{23} \\ a_{32} & a_{33} \end{bmatrix} \begin{pmatrix} N_2 - p_2 \\ N_3 - p_3 \end{pmatrix} \\ &\quad + (N_1 - p_1)\left[a_{21}(N_2 - p_2) + a_{31}(N_3 - p_3)\right]. \end{aligned} \tag{10}$$

If the determinant of the coefficient matrix

$$A = \begin{bmatrix} a_{22} & a_{23} \\ a_{32} & a_{33} \end{bmatrix},$$

is positive it immediately follows that $\dot{v} \leq 0$ when the second term on the right side of (10) is nonpositive, which, for fixed N_2, N_3, must occur when $N_1 - p_1$ is either positive or negative. It then follows that for any α, J_α is the entire non–impulsive subspace implying that p is globally asymptotically impulsively stabilizable. The same conclusion may be drawn if the determinant of A is zero and $\frac{a_{22}}{a_{23}} \neq \frac{a_{21}}{a_{31}}$.

The resort owner is thus assured of the possibility of maintaining a game fish population close to (p_2, p_3) by stocking his lake with a sufficient amount of feeder fish. It remains to determine an actual impulse program that will suit his needs and will be economically optimal. This is a question that resides in the realms of practical stability and optimal impulsive control (Liu, 1992).

3 CONCLUSION

Theorems 1 and 3 provide, respectively, necessary and sufficient conditions for a point p to be impulsively stabilizable, i.e. they establish the existence of an impulsive control program that will render the point stable. A constructive approach for actually determining an appropriate feedback control law could be generated from the notion of impulsively invariant sets described in Section 2.2 as follows. Once a Lyapunov function, v, is found that satisfies the conditions of Theorem 3, for any positive ϵ we choose a constant c such that the level set $v = c$ is contained within an ϵ–neighbourhood of p_z. The set $\{z : \|z\| \leq \epsilon, v(z) < c\}$ defines Ω_z, the non–impulsive portion of our invariant cylinder. We then choose an α such that $\Omega \subset B_\epsilon(p)$, where $\Omega = B_\alpha^y(p_y) \oplus \Omega_z$. Since Ω is an invariant cylinder, for all points $q \in \partial\Omega$ from which the trajectory induced by f leaves $\overline{\Omega}$ there is at least one impulse $\Delta y(q)$ such that $q + \Delta y(q)$ is either in the interior of Ω or at a point on $\partial\Omega$ from which the vector field f will keep the trajectory within $\overline{\Omega}$ for some positive time interval. For each such q we choose one of these $\Delta y(q)$ and define our feedback control law as $q \mapsto q + \Delta y(q)$. This law is implemented by observing the trajectory of an initial point within Ω and firing the appropriate impulse $\Delta y(q)$ each time the trajectory reaches a point q. In this manner the system is kept within a ϵ–neighbourhood of the desired point p.

We have considered the case where the impulses are confined to lie in a subspace of the underlying space. More restrictive or different situations could also be considered. In particular, the impulses may in addition be limited to a sector of the subspace, such as the positive sector. This is actually the case occurring in Example 2 where the impulses are only positive, (the owner stocks the feeder fish but does not remove them). In these situations the conditions for stabilizability will necessarily be more stringent.

REFERENCES

Lakshmikantham, V., Bainov, D.D., & Simeonov, P.S. (1989). *Theory of Impulsive Differential Equations*, World Scientific, Singapore.

Lakshmikantham, V., & Liu, X.Z. (1989). Stability criteria for impulsive differential equations in terms of two measures, *J. Math. Anal. Appl.*, *137*: 591.

Liu, X.Z. (1990). Stability theory for impulsive differential equations in terms of two measures, *Differential Equations Stability and Control* (S. Elaydi, ed.), Marcel Dekker, New York, p. 61.

Liu, X.Z. (1992). Practical stabilization of control systems with impulse effects, *J. Math. Anal. Appl.*, *166*: 563.

Liu, X.Z. & Willms, A.R. (1993). Impulsive stabilizability of autonomous systems, to appear *J. Math. Anal. Appl.*

Sontag, E.D. (1990). *Mathematical Control Theory*, Springer–Verlag, New York.

Willms, A.R. (1993). *Impulsive Control and Stabilization*, Master's Thesis, University of Waterloo, Waterloo, Ontario, Canada.

Comparison Methods and Stability Analysis
of Reaction Diffusion Systems

C.V. PAO, Department of Mathematics, North Carolina State University, Raleigh, North Carolina 27695-8205

1 INTRODUCTION

One of the widely used comparison methods for reaction diffusion systems is the method of upper and lower solutions. This method gives not only existence-comparison theorems for both parabolic and elliptic equations, suitable construction of upper and lower solutions can lead to various qualitative property of the system, including multiplicity of steady-state solutions, stability and instability of steady-state solutions, and blowing-up behavior of time-dependent solutions (cf. [1, 3-5, 7, 10, 11, 14]). It can also be used to develop computational algorithms for numerical solutions of the corresponding discrete reaction diffusion equations (e.g. see [6, 12, 13]). In this presentation we give an overview of the method for a class of finite coupled reaction diffusion systems and its application to the stability analysis and traveling wave solutions of some model problems arising from ecology.

The problems under consideration are the coupled system of parabolic equations

$$
\begin{aligned}
(u_i)_t - L_i u_i &= f_i(t, x, u_1, \ldots, u_N) && \text{in } D_T \\
B_i u_i &= g_i(t, x, u_1, \ldots, u_N) && \text{on } S_T \quad (i = 1, \ldots, N) \\
u_i(0, x) &= u_{i,0}(x) && \text{in } \Omega
\end{aligned}
\tag{1.1}
$$

and the corresponding elliptic system

$$-L_i u_i = f_i(x, u_1, \ldots, u_N) \qquad \text{in } \Omega$$

$$(i = 1, \ldots, N) \qquad (1.2)$$

$$B_i u_i = g_i(x, u_1, \ldots, u_N) \qquad \text{on } \partial\Omega$$

where Ω is a bounded domain in $\mathbb{R}^n$ with boundary $\partial\Omega$, $D_T = (0, T] \times \Omega$, $S_T = (0, T] \times \partial\Omega$ for any $T < \infty$, and L_i and B_i are linear operators given in the form

$$L_i u_i \equiv \sum_{j,\ell=1}^{n} a_{j,\ell}^{(i)}(x) \partial^2 u_i / \partial x_j \partial x_\ell + \sum_{j=1}^{n} b_j^{(i)}(x) \partial u / \partial x_j$$

$$B_i u_i \equiv \alpha_i(x) \partial u / \partial \nu + \beta_i(x) u$$

with $\partial/\partial\nu$ denoting the outward normal derivative on $\partial\Omega$. It is assumed that for each $i = 1, \ldots, N$, L_i is a uniformly elliptic operator in $\overline{\Omega}$ with smooth coefficients, α_i and β_i are smooth nonnegative functions on $\partial\Omega$ with $\alpha_i + \beta_i > 0$, and $u_{i,0}$ is a Hölder continuous function satisfying the boundary condition at $t = 0$. The functions f_i and g_i are assumed Hölder continuous in (t, x) and continuously differentiable in $\mathbf{u} \equiv (u_1, \ldots, u_N)$.

To describe a monotone iteration process we write $\mathbf{u}$ in the form

$$\mathbf{u} \equiv (u_i, [\mathbf{u}]_{a_i}, [\mathbf{u}]_{b_i}) \text{ or } \mathbf{u} \equiv (u_i, [\mathbf{u}]_{c_i}, [\mathbf{u}]_{d_i})$$

where a_i, b_i, c_i and d_i are nonnegative integers satisfying

$$a_i + b_i = c_i + d_i = N - 1 \quad (i = 1, \ldots, N)$$

Then problems (1.1) and (1.2) may be expressed in the forms

$$(u_i)_t - L_i u_i = f_i(t, x, u_i, [\mathbf{u}]_{a_i}, [\mathbf{u}]_{b_i}) \quad \text{in } D_T$$

$$B_i u_i = g_i(t, x, u_i, [\mathbf{u}]_{c_i}, [\mathbf{u}]_{d_i}) \quad \text{on } S_T \qquad (1.3)$$

$$u_i(0, x) = u_{i,0}(x) \qquad \text{in } \Omega$$

and

$$-L_i u_i = f_i(x, u_i, [\mathbf{u}]_{a_i}, [\mathbf{u}]_{b_i}) \qquad \text{in } \Omega$$

$$B_i u_i = g_i(x, u_i, [\mathbf{u}]_{c_i}, [\mathbf{u}]_{d_i}) \qquad \text{on } \partial\Omega \qquad (1.4)$$

respectively. The systems (1.3) and (1.4) will be used for the construction of monotone convergence sequences when the functions $\mathbf{f} \equiv (f_1, \ldots, f_N), \mathbf{g} \equiv (g_1, \ldots, g_N)$ possess some quasimonotone properties with respect to $\mathbf{u} \equiv (u_1, \ldots, u_N)$. Here a vector function $\mathbf{f} \equiv (f_1, \ldots, f_N)$ is said to possess a quasimonotone property if for each $i = 1, \ldots, N$, there exist nonnegative integers a_i, b_i with $a_i + b_i = N - 1$ such that $f_i(\cdot, u_i, [\mathbf{u}]_{a_i}, [\mathbf{u}]_{b_i})$ is monotone nondecreasing in $[\mathbf{u}]_{a_i}$ and monotone nonincreasing in $[\mathbf{u}]_{b_i}$. When $a_i = 0$ or $b_i = 0, \mathbf{f}$ is said to be quasimonotone nonincreasing and quasimonotone nondecreasing, respectively. Similar definition holds for the vector function $\mathbf{g} \equiv (g_1, \ldots, g_N)$ with respect to the components $[\mathbf{u}]_{c_i}$ and $[\mathbf{u}]_{d_i}$. In the following section we present a monotone iterative scheme for the construction of solutions to (1.3), (1.4) when $\mathbf{f}$ and $\mathbf{g}$ are quasimonotone functions. We also give existence-comparison theorems for the systems (1.1) and (1.2) when $\mathbf{f}$ and $\mathbf{g}$ are not necessarily quasimonotone.

2 EXISTENCE-COMPARISON THEOREMS

When $\mathbf{f} \equiv (f_1, \ldots, f_N)$ and $\mathbf{g} \equiv (g_1, \ldots, g_N)$ are quasimonotone with respect to their respective components $[\mathbf{u}]_{a_i}, [\mathbf{u}]_{b_i}$ and $[\mathbf{u}]_{c_i}, [\mathbf{u}]_{d_i}$, where $a_i + b_i = c_i + d_i = N - 1$, we can construct two monotone convergent sequences from a suitable iteration process using upper and lower solutions as initial iterations.

Definition 2.1. A pair of smooth functions $\tilde{\mathbf{u}} \equiv (\tilde{u}_1, \ldots, \tilde{u}_N), \tilde{\mathbf{u}} \equiv (\hat{u}_1, \ldots, \hat{u}_N)$ are called coupled upper and lower solutions of (1.3) if $\tilde{\mathbf{u}} \geq \hat{\mathbf{u}}$ and if

$$(\tilde{u}_i)_t - L_i \tilde{u}_i \geq f_i(t, x, \tilde{u}_i, [\tilde{\mathbf{u}}]_{a_i}, [\hat{\mathbf{u}}]_{b_i})$$

$$\text{in } D_T$$

$$(\hat{u}_i)_t - L_i \hat{u}_i \leq f_i(t, x, \hat{u}_i, [\hat{\mathbf{u}}]_{a_i}, [\tilde{\mathbf{u}}]_{b_i})$$

$$B_i \tilde{u}_i \geq g_i(t, x, \tilde{u}_i, [\tilde{\mathbf{u}}]_{c_i}, [\hat{\mathbf{u}}]_{d_i}) \tag{2.1}$$

$$\text{on } S_T \qquad (i = 1, \ldots, N)$$

$$B_i \hat{u}_i \leq g_i(t, x, \hat{u}_i, [\hat{\mathbf{u}}]_{c_i}, [\tilde{\mathbf{u}}]_{d_i})$$

$$\tilde{u}_i(0, x) \geq u_{i,0}(x) \geq \hat{u}_{i,0}(x)$$

$$\text{in } \Omega$$

Let $\tilde{\mathbf{u}}, \hat{\mathbf{u}}$ be a pair of coupled upper and lower solutions of (1.3). Define

$$\langle \hat{\mathbf{u}}, \tilde{\mathbf{u}} \rangle \equiv \{\mathbf{u} \in \mathcal{C}(\overline{D}_T); \hat{\mathbf{u}} \leq \mathbf{u} \leq \tilde{\mathbf{u}}\}$$

$$\overline{c}_i(t, x) \equiv \max\{-\frac{\partial f_i}{\partial u_i}(t, x, \mathbf{u}); \mathbf{u} \in \langle \hat{\mathbf{u}}, \tilde{\mathbf{u}} \rangle\} \tag{2.2}$$

$$\overline{b}_i(t, x) \equiv \max\{-\frac{\partial g_i}{\partial u_i}(t, x, \mathbf{u}); \mathbf{u} \in \langle \hat{\mathbf{u}}, \tilde{\mathbf{u}} \rangle\}$$

where $\mathcal{C}(\overline{D}_T)$ denotes the set of continuous functions $\mathbf{u}$ in $\overline{D}_T \equiv [0, T] \times \overline{\Omega}$. Then by letting

$$F_i(t, x, u_i, [\mathbf{u}]_{a_i}, [\mathbf{u}]_{b_i}) = \overline{c}_i u_i + f_i(t, x, u_i, [\mathbf{u}]_{a_i}, [\mathbf{u}]_{b_i})$$

$$G_i(t, x, u_i, [\mathbf{u}]_{c_i}, [\mathbf{u}]_{d_i}) = \overline{b}_i u_i + g_i(t, x, u_i, [\mathbf{u}]_{c_i}, [\mathbf{u}]_{d_i})$$

$$\mathcal{L}_i u_i = (u_i)_t - L_i u_i + \overline{c}_i u_i$$

$$\mathcal{B}_i u_i = B_i u_i + \overline{b}_i u_i \qquad (i = 1, \ldots, N) \tag{2.3}$$

we may write (1.3) in the equivalent form

$$\mathcal{L}_i u_i = F_i(t, x, u_i, [\mathbf{u}]_{a_i}, [\mathbf{u}]_{b_i}) \qquad \text{in } D_T$$

$$\mathcal{B}_i u_i = G_i(t, x, u_i, [\mathbf{u}]_{c_i}, [\mathbf{u}]_{d_i}) \qquad \text{on } S_T \tag{2.4}$$

$$u_i(0, x) = u_{i,0}(x) \qquad \text{in } \Omega$$

Our construction of monotone sequences is through the iteration process

$$\mathcal{L}_i \overline{u}_i^{(k)} = F_i(t, x, \overline{u}_i^{(k-1)}, [\overline{\mathbf{u}}^{(k-1)}]_{a_i}, [\underline{\mathbf{u}}^{(k-1)}]_{b_i})$$

$$\mathcal{B}_i \overline{u}_i^{(k)} = G_i(t, x, \overline{u}_i^{(k-1)}, [\overline{\mathbf{u}}^{(k-1)}]_{c_i}, [\underline{\mathbf{u}}^{(k-1)}]_{d_i})$$

$$\mathcal{L}_i \underline{u}_i^{(k)} = F_i(t, x, \underline{u}_i^{(k-1)}, [\underline{\mathbf{u}}^{(k-1)}]_{a_i}, [\overline{\mathbf{u}}^{(k-1)}]_{b_i}) \qquad (i = 1, \ldots, N) \qquad (2.5)$$

$$\mathcal{B}_i \underline{u}_i^{(k)} = G_i(t, x, \underline{u}_i^{(k-1)}, [\underline{\mathbf{u}}^{(k-1)}]_{c_i}, [\overline{\mathbf{u}}^{(k-1)}]_{d_i})$$

$$\overline{u}_i^{(k)}(0, x) = \underline{u}_i^{(k)}(0, x) = u_{i,0}(x)$$

for $k = 1, 2, \ldots$, where $\overline{u}_i^{(0)} = \tilde{u}_i$ and $\underline{u}_i^{(0)} = \hat{u}_i$. The above iteration process leads to the following existence-comparison result.

THEOREM 2.1. *Let $\tilde{\mathbf{u}}, \hat{\mathbf{u}}$ be a pair of coupled upper and lower solutions of (1.3), and let $\mathbf{f}, \mathbf{g}$ be quasimonotone C^1-functions in $\langle \hat{\mathbf{u}}, \tilde{\mathbf{u}} \rangle$. Then the sequences $\{\overline{\mathbf{u}}^{(k)}\} \equiv \{\overline{u}_1^{(k)}, \ldots, \overline{u}_N^{(k)}\}, \{\underline{\mathbf{u}}^{(k)}\} \equiv \{\underline{u}_1^{(k)}, \ldots, \underline{u}_N^{(k)}\}$ given by (2.5) with $\overline{\mathbf{u}}^{(0)} = \tilde{\mathbf{u}}$ and $\underline{\mathbf{u}}^{(0)} = \hat{\mathbf{u}}$ converge monotonically to a unique solution $\mathbf{u}^*$ of (1.3). Moreover,*

$$\hat{\mathbf{u}} \leq \underline{\mathbf{u}}^{(k)} \leq \underline{\mathbf{u}}^{(k+1)} \leq \mathbf{u}^* \leq \overline{\mathbf{u}}^{(k+1)} \leq \overline{\mathbf{u}}^{(k)} \leq \tilde{\mathbf{u}} \quad in \ \overline{D}_T \qquad (2.6)$$

A proof of the above theorem can be found in Chapter 9 of [10]. In this theorem the main requirements are the existence of a pair of coupled upper and lower solutions and the quasimonotone property of $\mathbf{f}$ and $\mathbf{g}$. Since $\tilde{\mathbf{u}} = \mathbf{u}$ and $\hat{\mathbf{u}} = \mathbf{u}$ are coupled upper and lower solutions when $\mathbf{u}$ is a solution, the first requirement can be fulfilled unless the problem has no solution in $\overline{D}_T$. On the other hand, when one or both of the functions $\mathbf{f}$ and $\mathbf{g}$ are not quasimonotone there is also an existence-comparison theorem in relation to a generalized coupled upper and lower solutions which are defined in the following

Definition 2.2. A pair of smooth functions $\tilde{\mathbf{u}} \equiv (\tilde{u}_1, \ldots, \tilde{u}_N)$, $\hat{\mathbf{u}} \equiv (\hat{u}_1, \ldots, \hat{u}_N)$ are called generalized upper and lower solutions of (1.1) if $\tilde{\mathbf{u}} \geq \hat{\mathbf{u}}$ and if

$$(\tilde{u}_i)_t - L_i \tilde{u}_i \geq f_i(t, x, \mathbf{v})$$
$$\text{for } \mathbf{v} \in \langle \hat{\mathbf{u}}, \tilde{\mathbf{u}} \rangle \text{ with } v_i = \tilde{u}_i$$
$$B_i \tilde{u}_i \geq g_i(t, x, \mathbf{v})$$
$$(\hat{u}_i)_t - L_i \hat{u}_i \leq f_i(t, x, \mathbf{v}) \qquad (2.7)$$
$$\text{for } \mathbf{v} \in \langle \hat{\mathbf{u}}, \tilde{\mathbf{u}} \rangle \text{ with } v_i = \hat{u}_i$$
$$B_i \hat{u}_i \leq g_i(t, x, \mathbf{v})$$
$$\tilde{u}_i(0, x) \geq u_{i,0}(x) \geq \hat{u}_i(0, x) \qquad (i = 1, \ldots, N)$$

It is seen from the above definition that when $\mathbf{f}$ and $\mathbf{g}$ are quasimonotone in $\langle \hat{\mathbf{u}}, \tilde{\mathbf{u}} \rangle$ this definition coincides with that in Definition 2.1. The following theorem gives an existence-comparison result for nonquasimonotone $\mathbf{f}, \mathbf{g}$.

THEOREM 2.2. *Let $\tilde{\mathbf{u}}, \hat{\mathbf{u}}$ be a pair of generalized upper and lower solutions of (1.1), and let $\mathbf{f}, \mathbf{g}$ be C^1-functions in $\langle \hat{\mathbf{u}}, \tilde{\mathbf{u}} \rangle$. Then there exists a unique solution $\mathbf{u}^*$ to (1.1) and $\mathbf{u}^* \in \langle \hat{\mathbf{u}}, \tilde{\mathbf{u}} \rangle$.*

Proof. A proof of the theorem for the case $\mathbf{g} \equiv \mathbf{g}(t, x)$, independent of $\mathbf{u}$, can be found in Chapter 8 of [10]. The proof for the general case $\mathbf{g} \equiv \mathbf{g}(t, x, \mathbf{u})$ follows from a similar argument and is omitted.

We next give some existence-comparison results for the elliptic system (1.2) for quasimonotone as well as nonquasimonotone functions $\mathbf{f}$ and $\mathbf{g}$.

Definition 2.3. Two smooth functions $\tilde{\mathbf{u}} \equiv (\tilde{u}_1, \ldots, \tilde{u}_N), \hat{\mathbf{u}} \equiv (\hat{u}_1, \ldots, \hat{u}_N)$ are called coupled upper and lower solutions of (1.4) if $\tilde{\mathbf{u}} \geq \hat{\mathbf{u}}$ and if

$$
\begin{aligned}
-L_i \tilde{u}_i &\geq f_i(x, \tilde{u}_i, [\tilde{\mathbf{u}}]_{a_i}, [\hat{\mathbf{u}}]_{b_i}) \\
-L_i \hat{u}_i &\leq f_i(x, \hat{u}_i, [\hat{\mathbf{u}}]_{a_i}, [\tilde{\mathbf{u}}]_{b_i}) \\
B_i \tilde{u}_i &\geq g_i(x, \tilde{u}_i, [\tilde{\mathbf{u}}]_{c_i}, [\hat{\mathbf{u}}]_{d_i}) \\
B_i \hat{u}_i &\leq g_i(x, \hat{u}_i, [\hat{\mathbf{u}}_{c_i}, [\tilde{\mathbf{u}}]_{d_i})
\end{aligned}
\qquad
\begin{aligned}
&\text{in } \Omega \\
&\text{on } \partial\Omega
\end{aligned}
\tag{2.8}
$$

Define $\langle \hat{\mathbf{u}}, \tilde{\mathbf{u}} \rangle$, $\overline{c}_i(x)$ and $\overline{b}_i(x)$ with respect to $\tilde{\mathbf{u}}(x)$ and $\hat{\mathbf{u}}(x)$ as that in (2.2) and F_i, G_i as that in (2.3). Without loss of generality we may assume that $\overline{c}_i(x) \geq 0$ and $\overline{b}_i(x) \geq 0$. Then problem (1.2) may be written as

$$
\begin{aligned}
-L_i u_i + \overline{c}_i u_i &= F_i(x, u_i, [\mathbf{u}]_{a_i}, [\mathbf{u}]_{b_i}) \\
B_i u_i + \overline{b}_i u_i &= G_i(x, u_i, [\mathbf{u}]_{c_i}, [\mathbf{u}]_{d_i})
\end{aligned}
\tag{2.9}
$$

Using the initial iterations $\overline{\mathbf{u}}^{(0)} = \tilde{\mathbf{u}}$ and $\underline{\mathbf{u}}^{(0)} = \hat{\mathbf{u}}$ we construct two sequences $\{\overline{\mathbf{u}}^{(k)}\} \equiv \{\overline{u}_1^{(k)}, \ldots, \overline{u}_N^{(k)}\}, \{\underline{\mathbf{u}}^{(k)}\} \equiv \{\underline{u}_1^{(k)}, \ldots, \underline{u}_N^{(k)}\}$ from the iteration process

$$
\begin{aligned}
-L_i \overline{u}_i^{(k)} + \overline{c}_i \overline{u}_i^{(k)} &= F_i(x, \overline{u}_i^{(k-1)}, [\overline{\mathbf{u}}^{(k-1)}]_{a_i}, [\underline{\mathbf{u}}^{(k-1)}]_{b_i}) \\
B_i \overline{u}_i^{(k)} + \overline{b}_i \overline{u}_i^{(k)} &= G_i(x, \overline{u}_i^{(k-1)}, [\overline{\mathbf{u}}^{(k-1)}]_{c_i}, [\underline{\mathbf{u}}^{(k-1)}]_{d_i}) \\
-L_i \underline{u}_i^{(k)} + \overline{c}_i \underline{u}_i^{(k)} &= F_i(x, \underline{u}_i^{(k-1)}, [\underline{\mathbf{u}}^{(k-1)}]_{a_i}, [\overline{\mathbf{u}}^{(k-1)}]_{b_i}) \\
B_i \underline{u}_i^{(k)} + \overline{b}_i \underline{u}_i^{(k)} &= G_i(x, \underline{u}_i^{(k-1)}, [\underline{\mathbf{u}}^{(k-1)}]_{c_i}, [\overline{\mathbf{u}}^{(k-1)}]_{d_i})
\end{aligned}
\tag{2.10}
$$

Just as in the case of parabolic system these sequences possess the monotone property

$$\hat{\mathbf{u}} \leq \underline{\mathbf{u}}^{(k)} \leq \underline{\mathbf{u}}^{(k+1)} \leq \overline{\mathbf{u}}^{(k+1)} \leq \overline{\mathbf{u}}^{(k)} \leq \tilde{\mathbf{u}} \quad \text{in } \overline{\Omega}, \quad k = 1, 2, \ldots.$$

Therefore the limits

$$\lim \overline{\mathbf{u}}^{(k)}(x) = \overline{\mathbf{u}}(x) \quad \text{and} \quad \lim \underline{\mathbf{u}}^{(k)}(x) = \underline{\mathbf{u}}(x) \quad \text{as } k \to \infty$$

exist and satisfy the relation $\underline{\mathbf{u}}^{(k)} \leq \underline{\mathbf{u}} \leq \overline{\mathbf{u}} \leq \overline{\mathbf{u}}^{(k)}$ for every k . It can be shown by letting $k \to \infty$ in (2.10) that $\overline{\mathbf{u}} \equiv (\overline{u}_1, \ldots, \overline{u}_N)$ and $\underline{\mathbf{u}} \equiv (\underline{u}_1, \ldots, \underline{u}_N)$ satisfy the relation

$$-L_i\overline{u}_i = f_i(x, \overline{u}_i, [\overline{\mathbf{u}}]_{a_i}, [\underline{\mathbf{u}}]_{b_i})$$

$$B_i\overline{u}_i = g_i(x, \overline{u}_i, [\overline{\mathbf{u}}]_{c_i}, [\underline{\mathbf{u}}]_{d_i})$$

$$-L_i\underline{u}_i = f_i(x, \underline{u}_i, [\underline{\mathbf{u}}]_{a_i}, [\overline{\mathbf{u}}]_{b_i}) \tag{2.11}$$

$$B_i\underline{u}_i = g_i(x, \underline{u}_i, [\underline{\mathbf{u}}]_{c_i}, [\overline{\mathbf{u}}]_{d_i})$$

Unlike the parabolic system (1.3) the limits $\overline{\mathbf{u}}, \underline{\mathbf{u}}$ are, in general, not solutions of (1.4). A counter example showing that neither $\overline{\mathbf{u}}$ nor $\underline{\mathbf{u}}$ is a solution is given in Chapter 8 of [10]. However, if $\mathbf{f}$ and $\mathbf{g}$ are both quasimonotone nondecreasing in $\langle \hat{\mathbf{u}}, \tilde{\mathbf{u}} \rangle$, then $\overline{\mathbf{u}}$ and $\underline{\mathbf{u}}$ are solutions. In fact, they are the respective maximal and minimal solutions of (1.4) in the sense that if $\mathbf{u}^*$ is any other solution of (1.4) in $\langle \hat{\mathbf{u}}, \tilde{\mathbf{u}} \rangle$ then $\underline{\mathbf{u}} \leq \mathbf{u}^* \leq \overline{\mathbf{u}}$. This is given in the following theorem (cf. [10]).

THEOREM 2.3. *Let* $\tilde{\mathbf{u}} \equiv (\tilde{u}_1, \ldots, \tilde{u}_N)$, $\hat{\mathbf{u}} \equiv (\hat{u}_1, \ldots, \hat{u}_N)$ *be a pair of ordered upper and lower solutions of (1.4), and let* $\mathbf{f}, \mathbf{g}$ *be quasimonotone nondecreasing* C^1*-functions in* $\langle \hat{\mathbf{u}}, \tilde{\mathbf{u}} \rangle$. *Then the sequences* $\{\overline{\mathbf{u}}^{(k)}\} \equiv \{\overline{u}_1, \ldots, \overline{u}_N\}$ *and* $\{\underline{\mathbf{u}}^{(k)}\} \equiv \{\underline{u}_1^{(k)}, \ldots, \underline{u}_N^{(k)}\}$ *given by (2.10) with* $\overline{\mathbf{u}}^{(0)} = \tilde{\mathbf{u}}$ *and* $\underline{\mathbf{u}}^{(0)} = \hat{\mathbf{u}}$ *converge monotonically to a maximal solution* $\overline{\mathbf{u}}$ *and a minimal solution* $\underline{\mathbf{u}}$ *of (1.4), respectively. Moreover,*

$$\hat{\mathbf{u}} \leq \underline{\mathbf{u}}^{(k)} \leq \underline{\mathbf{u}}^{(k+1)} \leq \underline{\mathbf{u}} \leq \overline{\mathbf{u}} \leq \overline{\mathbf{u}}^{(k+1)} \leq \overline{\mathbf{u}}^{(k)} \leq \tilde{\mathbf{u}} \quad \text{in } \Omega$$

$$k = 1, 2, \ldots \tag{2.12}$$

When $\mathbf{f}$ or $\mathbf{g}$ is not quasimonotone it is possible to show a similar existence-comparison theorem for (1.4) as that in Theorem 2.2. Here a generalized upper and lower solutions

$\tilde{\mathbf{u}} \equiv (\tilde{u}_1, \ldots, \tilde{u}_N), \hat{\mathbf{u}} \equiv (\hat{u}_1, \ldots, \hat{u}_N)$ *are required to satisfy* $\tilde{\mathbf{u}} \geq \hat{\mathbf{u}}$ *and the relation*

$$
\begin{aligned}
-L_i \tilde{u}_i &\geq f_i(x, \mathbf{v}) \\
&\qquad \text{for } \mathbf{v} \in \langle \hat{\mathbf{u}}, \tilde{\mathbf{u}} \rangle \text{ with } v_i = \tilde{u}_i \\
B_i \tilde{u}_i &\geq g_i(x, \mathbf{v}) \\
-L_i \hat{u}_i &\leq f_i(x, \mathbf{v}) \qquad\qquad\qquad (i = 1, \ldots, N) \qquad (2.13) \\
&\qquad \text{for } \mathbf{v} \in \langle \hat{\mathbf{u}}, \tilde{\mathbf{u}} \rangle \text{ with } v_i = \hat{u}_i \\
B_i \hat{u}_i &\leq g_i(x, \mathbf{v})
\end{aligned}
$$

The above relation is reduced to that in (2.8) when $\mathbf{f}$ *and* $\mathbf{g}$ *are quasimonotone functions. Using the Schauder fixed point theorem we can show the following existence-comparison result (see Chapter 9 of [10]).*

THEOREM 2.4. *Let* $\tilde{\mathbf{u}}, \hat{\mathbf{u}}$ *be generalized upper and lower solutions of (1.2), and let* $\mathbf{f}, \mathbf{g}$ *be* C^1*-functions in* $\langle \hat{\mathbf{u}}, \tilde{\mathbf{u}} \rangle$. *Then there exists at least one solution* $\mathbf{u}^*$ *to (1.2) and* $\mathbf{u}^* \in \langle \hat{\mathbf{u}}, \tilde{\mathbf{u}} \rangle$.

3. ELLIPTIC SYSTEMS IN UNBOUNDED DOMAINS

The method of upper and lower solutions for coupled systems of parabolic and elliptic equations can be extended to various types of reaction diffusion systems including integroparabolic and integroelliptic equations and systems with nonlocal boundary conditions (cf. [10]). It can also be extended to discretized reaction diffusion systems for numerical solutions (cf. [6, 12, 13]). In this section we give some existence-comparison results for finite elliptic systems in unbounded domains, including elliptic systems in $\mathbb{R}^n$. This type of problem has received considerable attention in recent years and our existence-comparison results in $\mathbb{R}^n$ can be used to study traveling wave solutions (cf. [8,9,11]). In the following discussion we limit our attention to the exterior of a bounded domain and to the whole space $\mathbb{R}^n$.

Let Ω_e be the exterior of a bounded domain, and let $\partial\Omega$ be the (inner) boundary surface of Ω_e. Consider the coupled system of elliptic equations

$$
\begin{aligned}
-L_i u_i &= f_i(x, u_1, \ldots, u_N) && \text{in } \Omega_e \\
&&& \qquad (i = 1, \ldots, N) \qquad (3.1) \\
B_i u_i &= g_i(x, u_1, \ldots, u_N) && \text{on } \partial\Omega
\end{aligned}
$$

As for elliptic systems in bounded domains we have the following definition of generalized upper and lower solutions.

Definition 3.1. A pair of smooth functions $\tilde{\mathbf{u}} \equiv (\tilde{u}_1, \ldots, \tilde{u}_N)$, $\hat{\mathbf{u}} \equiv (\hat{u}_1, \ldots, \hat{u}_N)$ are called generalized upper and lower solutions of (3.1) if $\tilde{\mathbf{u}} \geq \hat{\mathbf{u}}$ and if

$$
\begin{aligned}
-L_i\tilde{u}_i &\geq f_i(x, \mathbf{v}) \\
&\qquad \text{for } \mathbf{v} \in \langle \hat{\mathbf{u}}, \tilde{\mathbf{u}} \rangle \text{ with } v_i = \tilde{u}_i \\
B_i\tilde{u}_i &\geq g_i(x, \mathbf{v}) \\
-L_i\hat{u}_i &\leq f_i(x, \mathbf{v}) \\
&\qquad \text{for } \mathbf{v} \in \langle \hat{\mathbf{u}}, \tilde{\mathbf{u}} \rangle \text{ with } v_i = \hat{u}_i \\
B_i\hat{u}_i &\leq g_i(x, \mathbf{v})
\end{aligned}
\qquad (i = 1, \ldots, N) \qquad (3.2)
$$

Let $\{B_m\}$ be an increasing sequence of balls in $\mathbb{R}^n$ containing $\partial\Omega$, and let S_m be the surface of the ball B_m. Define $\Omega_m = \Omega_e \cap B_m$ and consider the boundary-value problem

$$
\begin{aligned}
-L_iu_i &= f_i(x, u_1, \ldots, u_N) \quad \text{in } \Omega_m \\
B_iu_i &= g_i(x, u_1, \ldots, u_N) \quad \text{on } \partial\Omega
\end{aligned}
\qquad (i = 1, \ldots, N) \qquad (3.3)
$$

With the additional boundary condition

$$
u_i(x) = u_i^{(0)}(x) \qquad \text{on } S_m, \quad (i = 1, \ldots, N) \qquad (3.4)
$$

where $u_i^{(0)}(x)$ is either an upper solution or a lower solution, Problem (3.3) , (3.4) becomes a system of elliptic equations in the bounded domain Ω_m. In view of (3.2) the restrictions of $\tilde{\mathbf{u}}, \hat{\mathbf{u}}$ to $\overline{\Omega}_m$ are also generalized upper and lower solutions of (3.3), (3.4). Hence by Theorem 2.4 this system possesses at least one solution $\mathbf{u}^{(m)}$ and $\hat{\mathbf{u}} \leq \mathbf{u}^{(m)} \leq \tilde{\mathbf{u}}$ in $\overline{\Omega}_m$. Moreover, if $\mathbf{u}^{(m+\ell)}$ is a solution of (3.3) where Ω_m is replaced by $\Omega_{m+\ell}$, $\ell = 1, 2, \ldots$, then the restriction of $\mathbf{u}^{(m+\ell)}$ to $\overline{\Omega}_m$ satisfies (3.3) in $\overline{\Omega}_m$. This implies that Problem (3.3) has a sequence of solutions $\{\mathbf{u}^{(m+\ell)}\}$ such that

$$
\hat{\mathbf{u}} \leq \mathbf{u}^{(m+\ell)} \leq \tilde{\mathbf{u}} \quad \text{in } \overline{\Omega}_m, \quad \ell = 0, 1, 2, \ldots
$$

It can be shown that $\mathbf{u}^{(m+\ell)} \in C^{2+\alpha}(\overline{\Omega}_m)$ and

$$
\|\mathbf{u}^{(m+\ell)}\|_{C^{2+\alpha}(\overline{\Omega}_m)} \leq K
$$

where $\alpha \in (0,1)$ and K is a constant independent of ℓ. It follows by the Arzela-Ascoli theorem and a diagonal selection process that the sequence $\{\mathbf{u}^{(m+\ell)}\}$ contains a subsequence which converges to a solution of (3.1). This leads to the following conclusion (see [11] for more details).

THEOREM 3.1. *Let $\tilde{\mathbf{u}}, \hat{\mathbf{u}}$ be a pair of generalized upper and lower solutions of (3.1), and let $\mathbf{f}, \mathbf{g}$ be C^1-functions in $\langle \hat{\mathbf{u}}, \tilde{\mathbf{u}} \rangle$. Then there exists at least one solution $\mathbf{u}^*$ to (3.1) and $\mathbf{u}^* \in \langle \hat{\mathbf{u}}, \tilde{\mathbf{u}} \rangle$.*

If $\mathbf{f}$ and $\mathbf{g}$ are quasimonotone nondecreasing then a solution of (3.1) can be constructed from the linear iteration process

$$-L_i u_i^{(m)} + \bar{c}_i u_i^{(m)} = \bar{c}_i u_i^{(m-1)} + f_i(x, U^{(m-1)}) \qquad in \ \Omega_m$$

$$B_i u_i^{(m)} + \bar{b}_i u_i^{(m)} = \bar{b}_i u_i^{(m-1)} + g_i(x, U^{(m-1)}) \qquad on \ \partial\Omega \quad (i = 1, \ldots, N) \qquad (3.5)$$

$$u_i^{(m)} = u_i^{(0)} \qquad on \ S_m$$

where $\bar{c}_i$ and $\bar{b}_i$ are the nonnegative functions in $\overline{\Omega}_e$ similar to that in (2.2), $u_i^{(0)}$ is either $\tilde{u}_i$ or $\hat{u}_i$ and $U^{(m)} \equiv (U_1^{(m)} \ldots, U_N^{(m)})$ is the extention of $\mathbf{u}^{(m)}$ given by

$$U_i^{(m)}(x) \equiv \begin{cases} u_i^{(m)}(x) & when \ x \in \overline{\Omega}_m \\ u_i^{(0)}(x) & when \ x \in \overline{\Omega}_e \backslash \overline{\Omega}_m \end{cases} \qquad (i = 1, \ldots, N) \qquad (3.6)$$

Denote the solution of (3.5) and its extention in (3.6) by $\overline{\mathbf{u}}^{(m)} \equiv (\overline{u}_1^{(m)}, \ldots, \overline{u}_N^{(m)})$ and $\overline{U}^{(m)} \equiv (\overline{U}_1^{(m)}, \ldots, \overline{U}_N^{(m)})$ when $\mathbf{u}^{(0)} = \tilde{\mathbf{u}}$, and by $\underline{\mathbf{u}}^{(m)} \equiv (\underline{u}_1^{(m)}, \ldots, \underline{u}_N^{(m)})$ and $\underline{U}^{(m)} \equiv (\underline{U}_1^{(m)}, \ldots, \underline{U}_N^{(m)})$ when $\mathbf{u}^{(0)} = \hat{\mathbf{u}}$. It can be shown by the same reasoning as for boundary-value problems in bounded domains that the sequences $\{\overline{U}^{(m)}\}, \{\underline{U}^{(m)}\}$ possess the monotone property

$$\hat{\mathbf{u}} \leq \underline{U}^{(m)} \leq \underline{U}^{(m+1)} \leq \overline{U}^{(m+1)} \leq \overline{U}^{(m)} \leq \tilde{\mathbf{u}} \quad in \ \overline{\Omega}_e$$

This monotone property yields the following conclusion (see [11]).

THEOREM 3.2. *Let $\tilde{\mathbf{u}}, \hat{\mathbf{u}}$ be a pair of ordered upper and lower solutions of (3.1), and let $\mathbf{f}, \mathbf{g}$ be quasimonotone nondecreasing C^1-functions in $\langle \hat{\mathbf{u}}, \tilde{\mathbf{u}} \rangle$. Then there exist a maximal* solution $\bar{\mathbf{u}}$ and a minimal solution $\underline{\mathbf{u}}$ to (3.1) such that $\hat{\mathbf{u}} \le \underline{\mathbf{u}} \le \bar{\mathbf{u}} \le \tilde{\mathbf{u}}$ in $\overline{\Omega}_e$. Moreover the sequences $\{\overline{U}^{(m)}\}$, $\{\underline{U}^{(m)}\}$ converge monotonically to $\bar{\mathbf{u}}$ and $\underline{\mathbf{u}}$, respectively.

When $\Omega_e = \mathbb{R}^n$ Problem (3.1) is reduced to the equation

$$-L_i u_i = f_i(x, u_1, \ldots, u_N) \quad \text{in } \mathbb{R}^n \quad (i = 1, \ldots, N) \tag{3.7}$$

(without boundary condition). Here the differential inequalities in (3.2) are reduced to

$$\begin{aligned}
-L_i \tilde{u}_i &\ge f_i(x, \mathbf{v}) \quad \text{for } \mathbf{v} \in \langle \hat{\mathbf{u}}, \tilde{\mathbf{u}} \rangle \text{ with } v_i = \tilde{u}_i \\
-L_i \hat{u}_i &\le f_i(x, \mathbf{v}) \quad \text{for } \mathbf{v} \in \langle \hat{\mathbf{u}}, \tilde{\mathbf{u}} \rangle \text{ with } v_i = \hat{u}_i
\end{aligned} \tag{3.8}$$

By the same reasoning as that for the exterior problem (3.1) we have the following conclusion.

THEOREM 3.3. Let $\tilde{\mathbf{u}}, \hat{\mathbf{u}}$ be generalized upper and lower solutions of (3.7), and let $\mathbf{f}$ be a C^1-function in $\langle \hat{\mathbf{u}}, \tilde{\mathbf{u}} \rangle$. Then there exists at least one solution $\mathbf{u}^*$ to (3.7) and $\mathbf{u}^* \in \langle \hat{\mathbf{u}}, \tilde{\mathbf{u}} \rangle$.

If $\mathbf{f}$ is quasimonotone nondecreasing in $\langle \hat{\mathbf{u}}, \tilde{\mathbf{u}} \rangle$ then we can construct a monotone sequence from the linear iteration process

$$\begin{aligned}
- L_i u_i^{(m)} + \bar{c}_i u_i^{(m)} &= \bar{c}_i u_i^{(m-1)} + f_i(x, U^{(m-1)}) \quad \text{in } B_m \\
u_i^{(m)} &= u_i^{(0)} \qquad\qquad\qquad\qquad\qquad \text{on } S_m
\end{aligned} \tag{3.9}$$

where $U^{(m)} \equiv (U_1^{(m)}, \ldots, U_N^{(m)})$ is defined as that in (3.6) and $u_i^{(0)}$ is either $\tilde{u}_i$ or $\hat{u}_i$. Denote again the sequence obtained from (3.9) and its extention by $\bar{\mathbf{u}}^{(m)}$ and $\overline{U}^{(m)}$, respectively, when $u_i^{(0)} = \tilde{u}_i$, and by $\underline{\mathbf{u}}^{(m)}$ and $\underline{U}^{(m)}$, respectively, when $u_i^{(0)} = \hat{u}_i$. Then we have (cf. [11])

THEOREM 3.4. Let $\tilde{\mathbf{u}}, \hat{\mathbf{u}}$ be a pair of ordered upper and lower solutions of (3.7), and let $\mathbf{f}$ be a quasimonotone nondecreasing C^1-function in $\langle \hat{\mathbf{u}}, \tilde{\mathbf{u}} \rangle$. Then there exist a maximal solution $\bar{\mathbf{u}}$ and a minimal solution $\underline{\mathbf{u}}$ to (3.7) such that $\hat{\mathbf{u}} \le \underline{\mathbf{u}} \le \bar{\mathbf{u}} \le \tilde{\mathbf{u}}$ in $\mathbb{R}^n$. Moreover, the sequences $\{\overline{U}^{(m)}\}, \{\underline{U}^{(m)}\}$ converge monotonically to $\bar{\mathbf{u}}$ and $\underline{\mathbf{u}}$, respectively.

4. APPLICATIONS

The existence-comparison theorems given in Sections 2 and 3 can be used to study various qualitative properties of the solution of reaction diffusion systems, including existence and nonexistence of positive steady-state solutions, stability or instability of a steady-state solution, and blowing-up behavior of time-dependent solutions. Theorems 3.3 and 3.4 can also be used to investigate traveling wave solutions for reaction-diffusion systems. In this section we discuss this kind of properties for two model problems arising from ecology. Our first problem is the Volterra-Lotka coorporating model which is given by the following coupled system of two equations:

$$
\begin{aligned}
u_t - D_1 \nabla^2 u &= u(a_1 - b_1 u + c_1 v) & & (t > 0, x \in \Omega) \\
v_t - D_2 \nabla^2 v &= v(a_2 + b_2 u - c_2 v) & & \\
B_1 u = B_2 v &= 0 & & (t > 0, x \in \partial\Omega) \\
u(0, x) = u_0(x), v(0, x) &= v_0(x) & & (x \in \Omega)
\end{aligned}
\tag{4.1}
$$

where u and v represent the two coorporating population species in a bounded habitat Ω with initial population densities $u_0 \geq 0, v_0 \geq 0$. The constants D_i, a_i, b_i and $c_i, i = 1, 2$, are all positive and the functions $\mathbf{f}, \mathbf{g}$ are defined by

$$
\mathbf{f}(u, v) \equiv (u(a_1 - b_1 u + c_1 v), v(a_2 + b_2 u - c_2 v)), \quad \mathbf{g}(u, v) \equiv (0, 0)
$$

The corresponding steady-state problem is given by

$$
\begin{aligned}
-D_1 \nabla^2 u &= u(a_1 - b_1 u + c_1 v) & & (x \in \Omega) \\
-D_2 \nabla^2 v &= v(a_2 + b_2 u - c_2 v) & & \\
B_1 u = B_2 v &= 0 & & (x \in \partial\Omega)
\end{aligned}
\tag{4.2}
$$

It is obvious that $\mathbf{f}$ is a quasimonotone nonincreasing C^1-function for all $u \geq 0, v \geq 0$, and $(0, 0)$ is a solution of (4.2) for any a_i, b_i and c_i. Our interest here is to determine whether and when Problem (4.2) has a positive solution and when the time-dependent solution of (4.1) converges to a positive solution.

The answer to the above problem is related to the smallest eigenvalue λ_i of the eigen-value problem

$$D_i \nabla^2 \Phi_i + \lambda_i \Phi_i = 0 \text{ in } \Omega, \quad B_i \Phi_i = 0 \text{ on } \partial\Omega \tag{4.3}$$

where $i = 1, 2$. It is well known that when $\beta_i(x) \not\equiv 0$, λ_i is real and positive, and its corresponding eigenfunction Φ_i is positive ($\lambda_i = 0$ and $\Phi_i = $ constant when $\beta_i(x) \equiv 0$). In terms of λ_i and the various physical constants in (4.2) we have the following conclusion concerning the existence and nonexistence of a positive solution. (Proofs of the following three theorems can be found in Chapter 12 of [10]).

THEOREM 4.1. *If $a_i > \lambda_i, i = 1, 2$, then Problem (4.2) has no positive solution when $b_2/b_1 > c_2/c_1$, and it has a positive max-min solution $(\overline{u}_s, \underline{v}_s)$ and a positive min-max solution $(\underline{u}_s, \overline{v}_s)$ when $b_2/b_1 < c_2/c_1$. On the other hand, if $a_i \leq \lambda_i$ and $b_2/b_1 \leq c_2/c_1$ then Problem (4.2) has only the trivial solution $(0, 0)$.*

In addition to the existence and nonexistence of a positive solution to (4.2) the same condition given in Theorem 4.1 also determines the asymptotic behavior of the time-dependent solution of (4.1), including the blowing-up behavior of the solution. Specifically, we have the following results.

THEOREM 4.2. *Let $a_i > \lambda_i, i = 1, 2$. (i) If $b_2/b_1 > c_2/c_1$ then for any nontrivial $(u_0, v_0) \geq (0, 0)$ there exists a finite T^* such that a unique solution (u, v) to (4.1) exists in $[0, T^*) \times \overline{\Omega}$ and blows-up in $\overline{\Omega}$ as $t \to T^*$. (ii) If $b_2/b_1 < c_2/c_1$ then a unique global solution (u, v) exists and converges to $(\overline{u}_s, \underline{v}_s)$ as $t \to \infty$ for one class of (u_0, v_0) and converges to $(\underline{u}_s, \overline{v}_s)$ for another class of (u_0, v_0). Moreover (u, v) converges to (u_s, v_s) for any nontrivial $(u_0, v_0) \geq (0, 0)$ if $(\overline{u}_s, \underline{v}_s) = (\underline{u}_s, \overline{v}_s) \equiv (u_s, v_s)$.*

THEOREM 4.3. *If $a_i < \lambda_i, i = 1, 2$, and $b_2/b_1 \leq c_2/c_1$ then for any $(u_0, v_0) \geq (0, 0)$ a unique global solution (u, v) to (4.1) exists and converges to $(0, 0)$ as $t \to \infty$.*

The conclusions in Theorems 4.2 and 4.3 states that when the natural growth rates a_1, a_2 overcome the effect of diffusion (which is measured by λ_1, λ_2) the two species u, v coexist and reach to a positive steady sate as $t \to \infty$ if the product $b_1 c_2$ of the self-regulating rates dominates the product b_2c_1 of the coorporating rates. On the other hand, if the domination of the above product rates is reversed then both species will grow to explosion in finite time, no matter how small the initial population of the two species may be. However, if the natural growth rates a_1, a_2 are dominated by the effect of diffusion then both species will be extinct as $t \to \infty$.

We next study the Volterra-Lotka competition model

$$u_t - \nabla^2 u = u(a_1 - b_1 u - c_1 v)$$
$$v_t - \nabla^2 v = v(a_2 - b_2 u - c_2 v)$$
$$\text{in } \mathbb{R}^n \qquad (4.4)$$

where a_i, b_i and $c_i, i = 1, 2$, are positive constants. Although the asymptotic behavior of the solution for this system can be investigated by the method of upper and lower solutions for parabolic systems our interest here is to determine when the system possesses a traveling wave solution. By a traveling wave solution of (4.4) we mean a bounded solution in the form $(u(\omega \cdot x + ct), v(\omega \cdot x + ct))$ for some constant $c > 0$ and a unit vector $\omega \equiv (\omega_1, \ldots, \omega_n)$ in $\mathbb{R}^n$ (that is, $\omega_1^2 + \ldots + \omega_n^2 = 1$). It is easy to see by the change of variable $\xi = \omega \cdot x + ct$ that Problem (4.4) is reduced to the form

$$-u_{\xi\xi} + cu_\xi = u(a_1 - b_1 u - c_1 v)$$
$$-v_{\xi\xi} + cv_\xi = v(a_2 - b_2 u - c_2 v)$$
$$(\xi \in \mathbb{R}^1) \qquad (4.5)$$

This implies that every bounded solution of (4.5) is a traveling wave solution of (4.4). Since Problem (4.5) is a special case of (3.7) in $\mathbb{R}^1$, the existence of a positive solution is ensured if there exist a pair of positive upper and lower solutions. For the system (4.5) the differential inequalities for upper and lower solutions, denoted by $\tilde{\mathbf{u}} \equiv (\tilde{u}, \tilde{v})$ and $\hat{\mathbf{u}} \equiv (\hat{u}, \hat{v})$, are given by

$$-\tilde{u}_{\xi\xi} + c\tilde{u}_\xi \geq \tilde{u}(a_1 - b_1\tilde{u} - c_1\hat{v})$$
$$-\tilde{v}_{\xi\xi} + c\tilde{v}_\xi \geq \tilde{v}(a_2 - b_2\hat{u} - c_2\tilde{v})$$
$$-\hat{u}_{\xi\xi} + c\hat{u}_\xi \leq \hat{u}(a_1 - b_1\hat{u} - c_1\tilde{v})$$
$$-\hat{v}_{\xi\xi} + c\hat{v}_\xi \leq \hat{v}(a_2 - b_2\tilde{u} - c_2\hat{v})$$
$$(4.6)$$

To construct a pair of positive upper and lower solutions we use the solution U of the scalar equation

$$-U_{\xi\xi} + cU_\xi = \sigma U(1 - U) \qquad (\xi \in \mathbb{R}^1) \qquad (4.7)$$

where σ is a positive constant. It is known that if $c \geq (2\sigma)^{1/2}$ then a positive solution $U(\xi)$ to (4.7) exists and satisfies the relation

$$0 \leq U(\xi) \leq 1, \quad \lim_{\xi \to -\infty} U(\xi) = 0, \quad \lim_{\xi \to \infty} U(\xi) = 1$$

(cf. [2]). Define

$$\sigma^* \equiv \min\{a_2(a_1/a_2 - c_1/c_2), \ a_1(a_2/a_1 - b_2/b_1)\}$$

Then $\sigma^* > 0$ if

$$c_1/c_2 < a_1/a_2 < b_1/b_2 \tag{4.8}$$

It can be verified by direct calculation that if $c > (2\sigma^*)^{1/2}$ then for sufficiently small constants $\delta_1 > 0, \delta_2 > 0$, the pair

$$(\tilde{u}, \tilde{v}) = (a_1/b_1, a_2/c_2), (\hat{u}, \hat{v}) = (\delta_1 U, \delta_2 U)$$

satisfy the inequalities in (4.6) and $(\tilde{u}, \tilde{v}) \geq (\hat{u}, \hat{v})$. As a consequence of Theorem 4.3 we have the following result.

THEOREM 4.4. If (4.8) holds then for any $c > (2\sigma^*)^{1/2}$ Problem (4.4) has a positive traveling wave solution $(u(\omega \cdot x + ct), v(\omega \cdot x + ct))$ which satisfies the relation

$$\delta_1 U(\omega \cdot x + ct) \leq u(\omega \cdot x + ct) \leq a_1/b_1$$
$$\delta_2 U(\omega \cdot x + ct) \leq v(\omega \cdot x + ct) \leq a_2/c_2$$

where $U(\omega \cdot x + ct)$ is the solution of (4.7) with $\xi \equiv \omega \cdot x + ct$.

References

1. H. Amman, *Fixed point equations and nonlinear eigenvluae problems in ordered Banach spaces*, SIAM Review, **18** (1976), 620-709.

2. D. G. Aronson and H. F. Weinberger, *Nonlinear diffusion in population genetics, combustion and nerve propagation*, Lecture Notes in Mathematics, **446**, 5-49, Springer-Verlag, New York (1975).

3. C. Cosner and A. C. Lazer, *Stable coexistence states in the Volterra-Lotka competition model with diffusion*, SIAM J. Appl. Math., **44** (1984), 1112-1132.

4. J. Hernandez, *Some existence and stability results for solutions of reaction diffusion systems with nonlinear boundary conditions*, Nonlinear Differential Equations: Invariance, Stability and Bifurcation, Academic Press, New York (1981), 161-173.

5. G. S. Ladde, V. Lakshmikantham and A. S. Vatsala, *Monotone Iterative Techniques for Nonlinear Differential Equations*, Pittman, Boston (1985).

6. A. C. Lazer, A. W. Leung and A. M. Diego, *Monotone scheme for finite difference equations concerning steady-state prey-predator interactions*, J. Comput. Appl. Math., **8** (1982), 242-252.

7. A. W. Leung, *Equilibria and stabilities for competing species reaction-diffusions with Dirichlet boundary data*, J. Math. Anal. Appl., **73** (1980), 204-218.

8. W. M. Ni, *On the elliptic equation $\Delta u + K(x)u^{(u+2)/(n+1)} = 0$, its generalization, and applications in geometry*, Indiana Univ. Math. J., **31** (1982), 493-529.

9. E. S. Noussair and C. A. Swanson, *Positive solutions of seminlinear elliptic equations in unbounded domains*, J. Differential Equations, **57** (1985), 349-372.

10. C. V. Pao, *Nonlinear Parabolic and Elliptic Equations* , Plenum Press, New York (1992).

11. C. V. Pao, *Nonlinear elliptic systems in unbounded domains*, J. Nonlinear Anal., TMA (to appear).

12. C. V. Pao, *Numerical methods for semilinear parabolic equations*, SIAM J. Numer. Anal., **24** (1987), 24-35.

13. C. V. Pao, *Numerical solutions for some coupled system of nonlinear boundary-value problems*, Numer. Math., **51** (1987), 381-394.

14. W. Walter, *On the existence and nonexistence in the large solutions of parabolic differential equations with a nonlinear boundary condition*, SIAM J. Math. Anal., **6** (1975), 85-90.

Some Applications of the Maximum Principle to a Free Stekloff Eigenvalue Problem and to Spatial Gradient Decay Estimates

G.A. PHILIPPIN, Département de mathématiques et de statistique, Université Laval, Sainte-Foy, Québec, Canada, G1K 7P4

1   Introduction

The scope of this paper it to review a number of possible applications of the maximum principle. The second section of the paper deals with a free boundary Stekloff eigenvalue problem. The third section is concerned with gradient estimates for harmonic functions defined in a convex region $\Omega \subset \mathbb{R}^2$ in terms of boundary data. Our estimates turn out to be valid in more general cases, such as for classical solutions of the minimal surface equation. Similar methods of investigation are then applied in the last section of the paper to obtain pointwise gradient decay estimates for solution of the Laplace equation.

2   A free Stekloff eigenvalue problem

In this section we consider the Stekloff eigenvalue problem defined in a simply connected domain Ω with Lipschitz boundary $\partial\Omega$

$$(2.1) \qquad \qquad \triangle u = 0 \text{ in } \Omega,$$

$$(2.2) \qquad \qquad \frac{\partial u}{\partial n} = pu \text{ on } \partial\Omega.$$

In (2.2), $\frac{\partial u}{\partial n}$ is the exterior normal derivative of u on $\partial\Omega$. Let us denote by

$$0 = p_1 < p_2 \leq p_3 \leq \cdots$$

the eigenvalues for which the Stekloff problem (2.1), (2.2) has nontrivial solutions

$$u_1(= \text{ const.}), u_2, u_3, \cdots$$

Let us make problem (2.1), (2.2) overdetermined by imposing the further boundary condition

$$(2.3) \qquad \qquad |\nabla u| = \text{ const. on } \partial\Omega.$$

We observe that u_2 satisfies (2.1), (2.2), (2.3) if Ω is a disc of radius p_2. It is natural to ask whether discs are the only domains for which problem (2.1), (2.2), (2.3) is solvable. The next theorem provides a positive answer to this question :

Theorem 1. Let u_2 satisfy (2.1), (2.2), and (2.3) with $p = p_2$. Then Ω must be a disc of radius p_2.

For the proof of Theorem 1 we make use of the following

Lemma 1. u_2 has no critical point in Ω.

Several proofs of Lemma 1 are available in the literature. See for instance [1,7].

Let us define

(2.4) $$H := u_x + i u_y,$$

with $u := u_2$. H is an analytic non vanishing function in Ω in view of Lemma 1. This implies that

(2.5) $$\log H = \log |H| + i \arg(H) = \log |\nabla u| + i \arg(H)$$

is analytic in Ω. We are then led to the following boundary value problem for $\log |\nabla u|$:

(2.6) $$\triangle(\log |\nabla u|) = 0 \text{ in } \Omega \ , \ \log |\nabla u| = \ \text{const. on } \partial\Omega \ ,$$

so that we must have

(2.7) $$|\nabla u| = \text{const. in } \Omega.$$

From (2.7) and (2.1) we conclude then that

(2.8) $$0 = \triangle(|\nabla u|^2) = u_{xx}^2 + 2u_{xy}^2 + u_{yy}^2 \ ,$$

i.e. all second order derivatives of u vanish. This shows that u is a linear function, i.e.

(2.9) $$u = x$$

in appropriate coordinates. We now make use of condition (2.2) rewritten as

(2.10) $$n_1 - p_2 x = 0 \text{ on } \partial\Omega \ ,$$

where (n_1, n_2) is the exterior unit normal vector. Differentiating identity (2.10) along the boundary leads to

(2.11) $$\left(n_2 \frac{\partial}{\partial x} - n_1 \frac{\partial}{\partial y} \right) (n_1 - p_2 x) = 0 \ ,$$

i.e.

(2.12) $$n_2, n_{1,1} - n_1 n_{1,2} - n_2 p_2 = (n_{1,1} + n_{2,2})n_2 - n_2 p_2 = 0 \text{ on } \partial\Omega.$$

Equation (2.12) may be rewritten under the form

$$(2.13) \qquad\qquad n_2(k - p_2) = 0 \text{ on } \partial\Omega ,$$

where k stands for the curvature of $\partial\Omega$. Equation (2.13) implies that $\partial\Omega$ consists of vertical line segments and arcs of circles. The only possible configuration for Ω is the union of two semicircles symmetric about the x and y axes possibly separated by a rectangular region of height h. But identity (2.9) cannot hold if $h > 0$ as it is easily checked from the Rayleigh principle

$$(2.14) \qquad\qquad p_2 \leq R[\varphi] := \frac{\int_\Omega |\nabla\varphi|^2 dx}{\oint_{\partial\Omega} \varphi^2 ds} ,$$

valid for every function φ piecewise continuously differentiable in Ω satisfying the orthogonality condition

$$(2.15) \qquad\qquad \oint_{\partial\Omega} \varphi ds = 0 ,$$

with equality in (2.14) if and only if $\varphi = u_2$. In fact if $h > 0$ we have

$$(2.16) \qquad\qquad R[y] < R[x] ,$$

so that x cannot be second eigenfunction of (2.1), (2.2). This shows that we must have $h = 0$, i.e. Ω is a disc.

The above result remains valid under less severe conditions than (2.3). In fact we have the next result.

Theorem 2. Let u_2 satisfy (2.1), (2.2) and any one of the following conditions

$$(2.17) \qquad\qquad |\nabla u_2| \exp\left(\lambda \frac{\partial u_2}{\partial x}\right) = \text{ const. on } \partial\Omega ,$$

or

$$(2.18) \qquad\qquad |\nabla u_2| \exp\left(\lambda \left[x \frac{\partial u_2}{\partial x} + y \frac{\partial u_2}{\partial y} - u_2\right]\right) = \text{ const. on } \partial\Omega ,$$

for some unknown $\lambda \in \mathbb{R}$. Then Ω must be a disc of radius p_2. (i.e. λ must actually be zero and condition (2.17) or (2.18) reduces to (2.3)).

The proof of Theorem 2 makes use of the following

Lemma 2. Let the function ϕ be defined as

$$(2.19) \qquad\qquad \phi(x) := |\nabla u|^2 e^v ,$$

with $\triangle u = \triangle v = 0$ in $\Omega \subset \mathbb{R}^2$, $|\nabla u|^2 \neq 0$ in Ω. If ϕ takes its maximum or minimum at an interior point P of Ω , then ϕ must be identically constant in Ω.

For the proof of Lemma 2 we observe that the function ϕ satisfies the differential equation

$$(2.20) \qquad \triangle\phi - \phi^{-1}|\nabla\phi|^2 = 0 \text{ in } \Omega.$$

The conclusion of Lemma 2 is now a simple consequence of Hopf's first maximum principle [3, 9].

To establish the first part of Theorem 2 we select ϕ according to (2.19) with

$$(2.21) \qquad u := u_2 , v := 2\lambda u_x.$$

It then follows from Lemma 2 and (2.17) that

$$(2.22) \qquad \phi = \text{ const. in } \Omega,$$

i.e.

$$(2.23) \qquad |\nabla u|^2 - ce^{-2\lambda u_x} = 0 \text{ in } \Omega,$$

with $u := u_2$ and $c = $ const. Differentiating (2.23) with respect to x and y we obtain a system of two linear equations for $u_{xx}(= -u_{yy})$ and u_{xy} :

$$(2.24) \qquad \begin{cases} u_{xx}(u_x + c\lambda e^{-2\lambda u_x}) + u_{xy}u_y = 0, \\ -u_{xx}u_y + u_{xy}(u_x + c\lambda e^{-2\lambda u_x}) = 0, \end{cases}$$

valid in Ω. It follows then that at each point of Ω either

$$(2.25) \qquad u_{xx} = u_{xy} = u_{yy} = 0,$$

or

$$(2.26) \qquad (u_x + c\lambda e^{-2\lambda u_x})^2 + u_y^2 = 0$$

holds. It can actually be seen that either (2.25) or (2.26) must hold throughout Ω. The first case (2.25) has already been investigated. In the second case (2.26) we must have

$$(2.27) \qquad u_y = u_x + c\lambda e^{-2\lambda u_x} = 0 \text{ in } \Omega ,$$

from which we obtain

$$(2.28) \qquad u(x, y) = c_1 + c_2 x ,$$

since u is harmonic. In both cases we observe that u is a linear function so that we have again (2.9) in appropriate coordinates. We can now continue the proof as before. The second part of Theorem 2 is established using the same techniques. Further results on the free boundary problem (2.1), (2.2), (2.3) are given in [2].

3 Estimates for $|\nabla u|$ in terms of boundary data

In this section we estimate $|\nabla u|$ for classical solutions of the elliptic equation

$$(3.1) \qquad \nabla \cdot (g(|\nabla u|^2)\nabla u) = 0 \text{ in } \Omega$$

in terms of boundary data, where Ω in a smooth convex plane domain and g is a positive C^1 function satisfying the ellipticity condition

$$(3.2) \qquad g(s) + 2sg'(s) \geq 0 \quad \forall s > 0.$$

We want to establish the following results

Theorem 3. Let u be a classical solution of problem (3.1), (3.2) that satisfies the Dirichlet condition

$$(3.3) \qquad u(s) = \varphi(s) \quad \forall s \in \partial\Omega ,$$

where φ is a given C^2 function. Then we have the estimate

$$(3.4) \qquad |\nabla u|^2 \leq \max_{\partial\Omega}\{(\varphi')^2 + k^{-2}(\varphi'')^2\} .$$

If (3.3) is replaced by the Neumann condition

$$(3.5) \qquad \frac{\partial u}{\partial n} = \psi(x) \text{ on } \partial\Omega ,$$

where ψ is a given C^1 function, then we have the estimate

$$(3.6) \qquad |\nabla u|^2 \leq \max_{\partial\Omega}\{k^{-2}(\psi')^2 + \psi^2\} .$$

In (3.4) and in (3.6), k is the curvature of $\partial\Omega$.

We remark that these estimates are independant of g. The proof of Theorem 3 is based on the fact that $|\nabla u|^2$ takes its maximum value on $\partial\Omega$. Let us consider the particular case corresponding to $g \equiv 1$. Since we have

$$(3.7) \qquad \triangle(|\nabla u|^2) = u_{xx}^2 + 2u_{xy}^2 + u_{yy}^2 \geq 0 \text{ in } \Omega ,$$

the quantity $\phi := |\nabla u|^2$ takes its maximum at some point $P \in \partial\Omega$. At P we must have

$$(3.8) \qquad \frac{1}{2}\frac{\partial\phi}{\partial n} = u_n u_{nn} + u_s u_{sn} - ku_s^2 \geq 0 .$$

The second normal derivative u_{nn} may be computed from the differential equation assumed to hold at P, rewritten in normal coordinates as follows

$$(3.9) \qquad \triangle u = u_{nn} + ku_n + u_{ss} = 0 ,$$

where k is the curvature of $\partial\Omega$. Combining (3.8) and (3.9) we obtain

$$(3.10) \qquad -k|\nabla u|^2 - u_n u_{ss} + u_s u_{sn} \geq 0 \text{ at } P .$$

Moreover we have

(3.11) $$\frac{1}{2}\frac{\partial \phi}{\partial s} = u_n u_{ns} + u_s u_{ss} = 0 \text{ at } P \ .$$

In the Dirichlet case we want to elimintate u_{ns} from (3.10) and (3.11) since u_{ns} cannot be expressed in terms of boundary data. This will be achieved when we compute the following combination

(3.12) $$(3.10)\, u_n^2 - (3.11) u_n u_s \geq 0 \text{ at } P \ .$$

Note that we connot drop a factor u_n in the above inequality since the sign of u_n is unknown at P. Inequality (3.12) reduces to

(3.13) $$k u_n^2 \leq -u_n u_{ss} \text{ at } P \ .$$

Since $\partial\Omega$ convex we may square inequality (3.13). This leads to

(3.14) $$u_n^2 \leq \frac{u_{ss}^2}{k^2} \text{ at } P \ ,$$

from which we conclude

(3.15) $$|\nabla u|^2 \leq (u_s^2 + u_n^2)_P \leq \max_{\partial\Omega}\{(\varphi')^2 + k^{-2}(\varphi'')^2\} \ .$$

In the Neumann case we want to eliminate u_{ss} from (3.10) and (3.11) since u_{ss} cannot be expressed in terms of boundary data. We are then led to the inequality

(3.16) $$(3.10)\, u_s^2 + (3.11)\, u_n u_s \geq 0 \text{ at } P \ ,$$

i.e.

(3.17) $$k u_s^2 \leq u_s u_{ns} \text{ at } P \ .$$

Since Ω is convex we may square inequality (3.17). This leads to

(3.18) $$u_s^2 \leq \frac{u_{ns}^2}{k^2} \text{ at } P \ ,$$

from which we conclude

(3.19) $$|\nabla u|^2 \leq (u_s^2 + u_n^2)_P \leq \max_{\partial\Omega}\{k^{-2}(\psi')^2 + \psi^2\} \ .$$

We refer the reader to [6,8] for further details.

4   Spatial decay estimates for $|\nabla u|$

In this section we derive spatial decay estimates for the gradient of a harmonic function in a semi infinite strip $\Omega := \{(x,y) \in \mathbb{R}^2 | x > 0, 0 < y < H\}$. We consider the following Dirichlet problem

(4.1)
$$\begin{cases} \triangle u = 0 \text{ in } \Omega \ , \\[2mm] u(x,0) = u(x,H) = 0 \ \ x \geq 0 \ , \\[2mm] u(0,y) = f(y) \ , \ 0 \leq y \leq H \ , \end{cases}$$

where $f(y)$ $(\not\equiv 0)$ is a given C^2 function satisfying $f(0) = f(H) = 0$.

It is well known that $|\nabla u|$ decays like $e^{-\frac{\pi}{H}x}$ as $x \longrightarrow \infty$ for bounded solutions of (4.1). This fact and Lemma 2 imply that the function

$$(4.2) \qquad \phi := e^{2\lambda x}|\nabla u|^2 \ , \ \lambda < \frac{\pi}{H} \ ,$$

takes its maximum value either on the sides $y = 0$, $y = H$, or $x = 0$. We are then led to the decay estimate

$$(4.3) \qquad |\nabla u| \le Ke^{-\frac{\pi}{H}x} \text{ in } \Omega \ ,$$

with $K := \sqrt{\phi_{\max}}$. We may estimate K using the techniques described in Section 3. We have three cases to investigate.

(a) Suppose that ϕ takes its maximum value at some point P on the horizontal side $y = H$. Then we must have

$$(4.4) \qquad \frac{1}{2}\phi_y(P) = (u_x u_{xy} + u_y u_{yy})e^{2\lambda x} \text{ at } P \ .$$

With

$$(4.5) \qquad u = u_x = u_{xx} = -u_{yy} = 0 \text{ at } P \ ,$$

we obtain

$$(4.6) \qquad \phi_y(P) = 0 \ ,$$

in contradiction to Hopf's second maximum principle [4,9]. The same argument applies on the side $y = 0$, so that ϕ cannot take its maximum value on any horizontal side of Ω.

(b) Suppose that ϕ takes its maximum value at some point P on the vertical side $x = 0$. In this case we need obviously an upper bound for $u_x^2(P)$. We proceed as in Section 3, assuming $u_x(P) \ne 0$. At P we must have

$$(4.7) \qquad \frac{1}{2}\phi_x = u_x u_{xx} + u_y u_{yx} + \lambda|\nabla u|^2 < 0 \ , \ 0 < \lambda < \frac{\pi}{H} \ ,$$

with strict inequality in view of Hopf's second maximum principle [4,9]. Moreover we have at P

$$(4.8) \qquad \frac{1}{2}\phi_y = u_x u_{xy} + u_y u_{yy} = 0.$$

The quantity $u_{xy}(P)$ cannot be expressed in terms of boundary data, so that we must get rid of it by forming an appropriate combination of (in)equalities (4.7) and (4.8). We are then led to the inequality

$$(4.9) \qquad (4.7)u_x^2 - (4.8)u_x u_y < 0 \text{ at } P \ ,$$

i.e.

(4.10) $$\lambda u_x^2 < u_x u_{yy} \ , \ 0 < \lambda < \frac{\pi}{H}.$$

We obtain the desired upper bound for $u_x^2(P)$ by squaring both sides of inequality (4.10) and dividing through $\lambda^2 u_x^2(P) > 0$. We obtain

(4.11) $$u_x^2(P) \le \frac{u_{yy}^2(P)}{\lambda^2} = \frac{(f''(P))^2}{\lambda^2}.$$

(c) Suppose that ϕ takes its maximum value at a corner P of Ω. Then we must have

(4.12) $$\phi_{\max} = \max\left((f'(0))^2, (f'(H))^2\right).$$

The above discussion leads to the following estimate

(4.13) $$\phi_{\max} \le \max_{\partial\Omega}\left\{(f')^2 + \lambda^{-2}(f'')^2\right\} \le \max_{\partial\Omega}\left\{(f')^2 + \frac{H^2}{\pi^2}(f'')^2\right\}.$$

Similar results are available for various boundary conditions, in $\mathbb{R}^3$ and for more general differential equations including the minimal surface equation. We refer the reader to [5] for further informations.

References

1. G. Alessandrini, R. Magnanini, Elliptic equations in divergence form, geometric critical points of solutions, and Stekloff eigenfunctions, preprint.

2. G. Alessandrini, R. Maganini, Symmetry and non symmetry for the overdetermined Stekloff eigenvalue problem, ZAMP, (to appear).

3. E. Hopf, Elementare Bemerkung über die Lösung partieller Differentialgleichungen zweiter Ordnung von elliptischen Typus, Sitzungsber. Preuss. Akad. Wiss., **19**, (1927), pp. 147-152.

4. E. Hopf, A remark on linear elliptic differential equations of the second order, Proc. Amer. Math. Soc. **3**, (1952), pp. 791-793.

5. C.O. Horgan, L.E. Payne, G.A. Philippin, Pointwise gradient decay estimates for solutions of the Laplace and minimal surface equations, J. of Diff. and Int. Equations. (to appear).

6. L.E. Payne. Isoperimetric inequalities, Maximum principles and their applications, unpublished lecture notes, University of New Castle, (1972).

7. L.E. Payne, G.A. Philippin, Some overdetermined boundary value problems for harmonic functions, ZAMP, **42**, (1991), pp. 864-873.

8. L.E. Payne, G.A. Philippin, On gradient maximum principles for quasilinear elliptic equations, J. of Nonlinear Analysis, (to appear).

9. M.H. Protter, H.F. Weinberger, Maximum principles in differential equations, Springer Verlag, (1984).

Comparison Methods in Control Theory

EMILIO O. ROXIN   Department of Mathematics, University of Rhode Island, Kingston, Rhode Island 02881

ABSTRACT   The purpose of this paper is to present the role of comparison methods in the theory of dynamical systems, particularly in the area of control theory. While many well known results fall within the area of such comparison methods, it is by thinking about comparison methods in a more general way, that it becomes evident that these methods play a very important role and hold the key for the solution of many problems.

1   COMPARISON METHODS IN AN ABSTRACT SETTING

Consider a mathematical problem which we will call P. Assume that this problem has a unique solution, which we will call x. The problem P is, of course, to "find" (maybe to calculate) x. If this happens to be too difficult, we can do two things:

   a) We may try to find an "approximate" solution. This should be a x' such that using some norm $|\cdot|$, $|x - x'|$ is "small." To find x' is not enough, if we are not able to estimate how small this difference is. But in most cases the situation is even worse: we usually just substitute the problem P by another problem P' which approximates P. We select P' such that we can solve it, and take x' to be the solution of P'. This alone is of course not enough to insure that $|x - x'|$ is small.

   b) We can try to embed the problem P in a family of problems of similar type, and also the solution x in a family of possible solutions (this last is usually much easier). We then have to define some order, such that between two problems P, P', and also between two possible solutions x, x', the relations $P < P'$ and $x < x'$ may hold. The idea is then to prove that if x is the solution of P, and x' is the solution of P', and if $P < P'$, then also $x < x'$. This is the abstract setting of a typical comparison results.

Since the solution x of a problem is usually a real number, a vector or a function $x(t)$, the meaning of the order $x < x'$ is usually obvious. In most cases the difficult part is to find a suitable definition of the order $P < P'$ in such a way that $P < P'$ implies $x < x'$.

2 LYAPUNOV STABILITY

Within the context of the above abstract formulation, the Lyapunov's "second method" of proving stability of a dynamical system can be presented as follows.

Let

$$\dot{x} = f(t, x), \quad \text{with } x \text{ in } R^n, \; f(t, 0) = 0 \text{ for all } t \tag{1}$$

be the differential equation defining a dynamical system. Assume that the conditions for existence and uniqueness of a solution of the initial value problem are satisfied. We wish to study the stability of the solution

$$x(t) = 0. \tag{2}$$

For that purpose we define the scalar comparison equation

$$\dot{y} = g(t, y), \quad \text{where } y \text{ in } R, \tag{3}$$

having the property that for any solution of (3), y tends to zero as $t \to \infty$.

If we can find a (differentiable) scalar function $V(t, x)$ such that

$$\partial V / \partial t + (\partial V / \partial x) \, f(t, x) \le g(t, V(t, x)), \tag{4}$$

then it will follow that along any solution $x(t)$ of (1),

$$dV / dt \le g(t, V). \tag{5}$$

Let the solution $x(t)$ of (1) satisfying $x(0) = x_0$, and compare $V(t, x(t))$ with $y(t)$ defined by (3) and $y(0) = y_0 = V(0, x_0)$. Assuming some additional properties of $V(t, x)$ and $g(t, y)$ (like being positive definite), it then follows that

$$dV(t, x(t)) / dt \le dy / dt \tag{6}$$

wherever $V(t, x(t)) = y(t)$ (these are one-dimensional) and therefore $V(t, x(t)) \le y(t)$ for all $t > 0$. Finally, as $t \to \infty$, $y(t) \to 0$ implies $V(t, x(t)) \to 0$. According to the assumptions made on $V(t, x)$, this then gives the desired proof of asymptotic stability.

Therefore the method for proving stability using Lyapunov functions is really based on a comparison between the solutions of equations (1) and (3).

3 DIFFERENTIAL GAMES

The setting of a deterministic two-players zero-sum differential game is as follows. The evolution of a system is given by a differential equation

$$\dot{x} = f(t, x, u, v), \; x(t) \in R^n, \; u(t) \in U \subset R^p, \; v(t) \in V \subset R^q, \tag{7}$$

where U, V are given sets, and $u(t)$ and $v(t)$ are the control functions of the players. Suppose this system starts at an initial state $x_0 = x(t_0)$. Given is also a "payoff" function

$$J = g(T, x(T)) \tag{8}$$

which "player u" tries to maximize and "player v" tries to minimize. Here T is the end-time of the game, which may be given in the formulation of the problem or determined by further conditions.

The well known theory (see Isaacs, Friedman (1971)) leads to an

$$\text{"upper Value" } V^+(t_0, x_0) = \lim V^\delta, \text{ and a}$$
$$\text{"lower Value" } V^-(t_0, x_0) = \lim V_\delta, \text{ for } \delta \to 0,$$

which represent the best player u, respectively v, can force to achieve. The corresponding strategies are to constantly play the control u, respectively v, so as to achieve

$$H^+(p, t, x) = \min_v \max_u [p(t) \cdot f(t, x(t), u(t), v(t))]$$
$$(\text{"}v\text{" choses first, "}u\text{" second})$$
$$H^-(p, t, x) = \max_u \min_v [p(t) \cdot f(t, x(t), u(t), v(t))]$$
$$(\text{"}u\text{" choses first, "}v\text{" second}).$$

Here $H(p, t, x, u, v)$ is the "Hamiltonian" and $p = p(t)$ is the "adjoint variable" satisfying, together with $x(t)$, the canonical system (see Isaacs, Friedman (1971)). The "Isaacs condition"

$$H^+ = H^- = H^*$$

leads then to the "Value" $V(t_0, x_0)$ of the game:

$$V^+(t_0, x_0) = V^-(t_0, x_0) = V(t_0, x_0).$$

All these "Values" are, of course, functions of the starting point of the game. The Value function $V(t, x)$ satisfies the partial differential equation

$$\partial V/\partial t + H^*(\nabla_x V, t, x) = 0. \tag{9}$$

The practical computation of the solution of differential games becomes extremely cumbersome, hence easier to get estimations are highly desired. A very powerful result in this respect is the following.

Theorem of Friedman (See Friedman (1973).) Assume, with the above notation, that the differential game as a "Value" $V = V^+ = V^-$. If there is a function $U(t, x)$ satisfying

$$\partial U/\partial t + H^+(\nabla_x U, t, x) \leq 0 \tag{10}$$

and the boundary condition

$$U(t, x) \geq g(t, x) \tag{11}$$

on the manifold determining the end conditions of the game, then

$$V(t_0, x_0) = V^+(t_0, x_0) \leq U(t_0, x_0). \tag{12}$$

It is clear that (10) is a comparison equation for (9), hence we are in some sense comparing the solutions of two related differential games, and one gives an upper estimate of the other.

4   CONTROL SYSTEMS

The subject of control systems is in between dynamical systems (no control) and differential games (two opposing controls). Hence we also expect to find that comparison methods will play an important role here. A typical control system is given by a differential system

$$\dot{x} = f(x, u), \ x(t) \in R^n, \ u(t) \in U \subset R, \tag{13}$$

where for simplicity we consider an autonomous system; $u(t)$ is the control function which should be selected in the interval $[t_0, T]$ in such a way that with given initial conditions

$$x(t_0) = x_0, \tag{14}$$

the "objective functional"

$$J = g(T, x(T)) + \int_{t_0}^{T} f^0(x(\tau), u(\tau)) \, d\tau \tag{15}$$

should be maximized. Here the end-time T can be prescribed or left open subject to additional conditions.

As in the case of differential games, an adjoint variable $p(t)$ and the Hamiltonian $H = f^0(x, u) + p \cdot f(x, u)$ are defined, and according to the maximum principle the optimal control $u^*(t)$ is determined by maximizing

$$\max_{u \in U} H(p, x, u) = H(p, x, u^*) = H^*(p, x) \tag{16}$$

at (almost) every t in $[t_0, T]$.

The "Value" function $V(t, x)$ defined by

$$V(t_0, x_0) = \max \ J = J^* \quad \text{(corresponding to } u^*(t)) \tag{17}$$

satisfies the "Hamilton–Jacobi equation"

$$\partial V / \partial t + H^*(\nabla_x V, x) = 0. \tag{18}$$

This is similar to (is a particular case of) a differential game as seen above, and Friedman's theorem can therefore also be applied, leading to an estimate function $U(t, x)$ such that

$$V(t, x) \leq U(t, x). \tag{19}$$

5   ESTIMATING THE ATTAINABLE SET

The attainable set of the above control system, at time t, starting at (t_0, x_0), is the set of points $x(t)$ belonging to "admissible" solutions $x(\cdot)$ which, with "admissible" controls $u(\cdot)$, satisfy (13) and (14).

This attainability set $A(t, t_0, x_0)$ shows which states x are attainable at a time t, by using all admissible controls. Notice that this set does not relate to any criterion of optimality. In many cases (for example, for linear systems) this set is convex and compact, hence determined by a support function

$$\varphi(p) = \max[p \cdot x \mid x \in A(T, t_0, x_0)], \tag{20}$$

p being an outward normal to the boundary of the attainable set. This support function can be found by maximizing the scalar product

$$J = p \cdot x(T), \quad \text{for the end-time } T. \tag{21}$$

As seen, the solution to this optimization problem can be estimated by comparison methods. We can, for example, compare the given control system (13) with a simpler one, easier to solve. Another problem related to sets of attainability are the "invariant sets." For control systems, such "invariance" can be taken in either a strong or a weak sense (see Roxin (1965, 1967)). Of particular interest is the stability of invariant sets (Roxin (1966 a-b-c)). As shown in the mentioned papers, different types of stability of invariant sets can be ascertained by Lyapunov functions, again an application of comparison methods.

Finally, any optimization related to a control system can be converted into a question about attainability sets. This is done by defining as additional coordinate the value of the quantity to be maximized, and then asking for the support function of the attainable set in the direction of this new coordinate.

6 INFINITE HORIZON PROBLEMS

Optimal control problems with infinite horizon are those whose performance index, to be optimized, involves an infinite interval of time, typically

$$J = \int_0^\infty f^0(t, x, u) \, dt. \tag{22}$$

In such a case the definition of optimality should be specified further (see Stern and Roxin–Stern). We may require that this integral should be minimal among the possible convergent integrals. More reasonable, in some sense, is to require that for "all sufficiently large end-time T", the integral $J(T)$ is optimal. Or, that for "arbitrarily large values of T", $J(T)$ is optimal. Such criteria of optimality introduce the possibility that for two controls u_1 and u_2, the corresponding J_1 and J_2 are not comparable (neither $J_1 \leq J_2$ nor $J_1 \geq J_2$). In this case, the abstract formulation given at the very beginning of this paper should be adapted to a partial order, both for the "problems" and for the "solutions."

REFERENCES

Friedman, A. (1971). *Differential Games*, Wiley, New York.

Friedman, A. (1973). *Lecture Notes on Differential Games*, Springer, New York.

Roxin, E. O. (1965). Stability in general control systems, *J. Diff. Eqs.*, **1**, 115–150.

Roxin, E. O. (1966a). On stability in control systems, *SIAM J. Control*, **3**, 357–372.

Roxin, E. O. (1966b). On asymptotic stability in control systems, *Rendiconti Circolo Mat. Palermo II*, **15**, 193–297.

Roxin, E. O. (1966c). On finite stability in control systems, *Rendiconti Circolo Mat. Palermo II*, **15**, 273–282.

Roxin, E. O. (1967). Problems about the set of attainability, *Lecture Notes Summer Course 1966 CIME*, Edizione Cremonese, 241–369.

Roxin E. O. and Stern, L. E. (1982). "Some recent developments in the infinite time optimal control problem," in Lakshmikantham, V. (ed.), Proceedings Internat. Conference Nonlinear Phenomena in Math. Sci., Academic Press, New York, pp. 859–868.

Stern, L. E. (1980). *The Infinite Horizon Optimal Control Problem*, Ph.D. Thesis, University of Rhode Island, Kingston, RI.

The Self-Destruction of the Perfect Democracy

RUDOLF STARKERMANN   Grabemattweg 14, CH-5443 Nd'Rohrdorf, Switzerland

ABSTRACT

The attribute **perfect** to qualify a democracy is justified if this form of a society allows free exchange of information among its constituent parts, i.e., if there is no danger of being persecuted if attitudes differ from the dominant or of any doctrine of the community. The presented mathematical model illustrates some developments and its results when such free exchange exists. The main point of interest is the growing number of constituents, which form a democratic system. It is distinguished between bilateral and multilateral information exchange within the social partners. The essay demonstrates how the volition of the constituents, with which they strive toward their goals, has to shrink, and how the speed of their action decreases, if a stable democracy grows cancerously in the number of constituent elements and of information exchange among them. The democracy grows toward self-destruction.

1   INTRODUCTION

The mathematical model investigates certain developments and its results of a **perfect** democracy if the constituents, who strive toward their own goal each, exchange freely their attitudes among each other and, therefore, seem to lose energy and time with respect to their own goal striving process. In the following, the individual constituent part is called the social **unit**. Each unit pursues its **own goal**, but it has either bilateral or multilateral communication with the other units of the system. Each unit has its volition with which it strives toward the goal, and the whole system has a certain speed, or velocity, of acting. A unit can be a single person, or a group of people with their homogeneous goal, or even a larger agglomeration of constituents with their goal they persue. The goal has to be seen as the self-realization of the unit. It becomes obvious that it is a **natural law** that volition and velocity decline with the increasing number of unites which participate in the system.

The essay is a parallel investigation to a similar topic, but of different parameters taken into account. The parallel-paper, "The Self-Destruction of the True Democracy", will be published by IEEE Service Center, 445, Hoes Lane, P.O.Box 1331, Piscataway. NJ 08855-1331, USA. This IEEE-paper will be presented at the 1993-IEEE/SMC Conference on "Systems, Man and Cybernetics" in Le Touquet, France, October, 1993.

Firstly, the mathematical model of the **unit** will be shortly explained. Secondly, the **structure** of the growing democracy is given, and thirdly, the parameters **volition** and **velocity** of action will be illustrated as a function of the agglomeration of units. The **functions** will be the volitions

and the velocities. And the parameters will be:
a) bilateral and multilateral information exchange,
b) slow, medium and fast information exchange.

Although the freedom to express everyone's attitude *ad libitum* seems to be the ideal form of living in togetherness, there is the unavoidable danger of instability of a social system. This danger requires a reduction of the units' volitions with which they strive toward their goal. If instability occurs, a system becomes unable to approach the goal, and it goes asunder in totality.

In order to built a model of several units to form a social society, the unit has to be formed first.

2 THE SOCIAL UNIT

Fig. 1 depicts the mathematical structure of the unit on the basis of a control loop. The loop has self-control of its proximity with respect to its goal, and the loop fights the effect of exogeneous, not goal related, disturbances. The closed loop is merely known in the technical aera as self-controlling action based on **natural laws**. But as natural laws are universal, the parallelism is given as a synonym **"technical construct - social construct"**. Fig. 1 illustrates the model and lists the nomenclature.

Technical Construct of a control loop of a multiply controlled process, or plant:

u_{11}	Variable to be controlled, reference signal
u_{12}	Disturbance for which there is a built-in correction
u_{13}	Disturbance for which there is no built-in correction, e.g., an undesired change of the reference signal
Σ	Summing point of signals
V_{12}, V_{21}	Exchanged information from and to a second controlled variable x_2; crossed feed back
A_{12}, A_{21}	Planned information exchange between units (e.g., for autonomization)
S_{12}, S_{21}	Mutual physical dependency of the controlled variables of a plant of dual control
ε_1	Error signal
G_1	Proportional factor of a P-controller
T	Time constant of natural delay of acting
m_1	Exponent; it determines the grade of delay
S_{11}	Unit's inherent information transfer within a technical plant (it is generally time dependent)
x_1	Moment to moment goal achievement
C_1	Transfer factor of x_1 to y_1
y_1	Information available for assessing unit's functioning
R_1	Feedback factor (generally taken as -1)
UC	Inherent information block, the process or plant to be controlled

The Social Construct of the individual P_1

u_{11}	Prime individual's goal, identical to self-realization
u_{12}	Exogenous disturbance of which the effect is controlled and fought, e.g., noise
u_{13}	Uncontrollable intake of, e.g., drugs, leading to drug addiction
Σ	Accumulation of information messages
V_{12}, V_{21}	Transmitting to and receiving information from a second individual P_2 by, e.g., observation

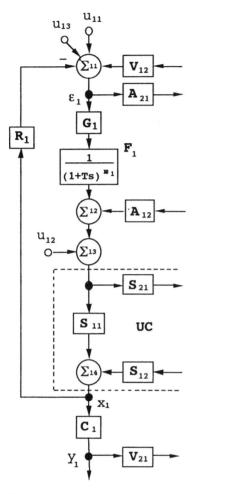

A_{12}, A_{21} Conscious support directed to or from a second individual P_2 for adaptation, or for mutual criticism of inability to adapt, with the result of help or harm

S_{12}, S_{21} Unconscious, inherent information exchange between units P_1 and P_2

ε_1 **Difference** between the intended goal of self-realization and, e.g., drug intake (u_{13}) along with the received information (V_{12}) and the individual's degree of perception of goal achievement (R_1)

G_1 Volition, or will-power, to achieve the goal

T & m_1 Individual's reaction time or retention of action

S_{11} Individual's unconscious transfer of information in relation to itself

x_1 Individual's moment-to-moment achieved self

C_1 Relationship between individual's self-achievement and observable behaviour

y_1 Observable behaviour of the individual

R_1 Individual's degree of perception of goal achievement

UC Totality of unconscious information as held by the prime individual, exchanged with another and held by another individual

u_{11}, u_{12}, u_{13} can change as a function of time; but herein they are considered as constant. - Disturbances can enter a unit anywhere where it is "open" toward the environment.

Fig. 1: Structural diagram (model) of the social unit P_1.

Explanations for Fig. 1.

The single partner (individual or person), the unit, has a goal (u_{11}). This is the individual's **self-realization**. Whatever a living being does, it wants to realize itself. Self-realization is survival. The individual becomes constantly and at random disturbed by the environment or from inside itself (u_{12}, u_{13}), and the individual has self-control of what it does (via $-x_1 R_1$, the negative feedback signal) - and it has interaction with other individuals. This interaction is two-fold. There is conscious interaction (V_{12}, V_{21}, A_{12}, A_{21}) and unconscious interaction (S_{12}, S_{21}), simultaneously. Furthermore, an individual has a delay or a retention of action, (T, m_1), and it has a will-power or volition (G_1) to reach its goal. Additionally, it fights the influence of disturbances, disturbances which are not goal related (u_{12}). Goal-related disturbances (u_{13}), cannot be corrected, they become part of the goal, as supporting, (+); or as damaging, (-). All these characteristics, together as a package, form an automatic control loop, or in other words, a self-controlling feedback device. This is essentially and in short the **social unit**.

Concerning interaction, a parallelism is set in the unevoidable **unconscious mutual** interaction of living beings on one hand, and the inherent mutual interaction of physical variables in the technical realm on the other hand.

As a technical example of inherent mutual interaction of three variables: pressure, temperature and relative humidity in an air vessel. It is not possible to change one variable without influencing the other two. Mutual dependencies exist as a natural law, and they can be comprehended mathematically, if the physics of the vessel's content is known.

And what about the conscious interaction? It can be seen in what is imposed as actions on other beings, be this by observation, or by any means of the media, or by direct physical actions. And technically: also what we impose on physical systems as multiple control with optimization, with autonomization, adaptation, parameter estimation, etc.

Fig. 1 illustrates these facts in the form of a diagram which can easily be modelled in the computer (Macintosh, Simulab).

In order to have the potential of instability available -- because a unit can become unstable -- the number of first order differential equations in series (as the most simple approach to model the dynamics) must be three at least. - An unstable system is no longer able to strive toward its goal. The same is true for a system of more than one unit: instability ruins the goal striving process of any system.

The presented essay is based on linear algebra and system dynamics. The specific background is multiple control of linear systems. For detailed information about the social unit, see Starkermann (1990).

3 THE MODEL OF THE DEMOCRACY

For the sake of brevity, and in order to remain within a comprehensible frame, only the unconscious information exchange will be considered. Fig. 2 depicts the model for four units.

It is acknowledged in psychology that the unconscious exchange of information (manifested in the attitude units have toward each other) is multiply times stronger in the effect than the conscious exchange. Therefore, our consideration is limited to unconscious information exchange. UC is symbolically the block of the total unconscious of the system. In the parallel-paper, two extreme behavioural patterns of attitude are explained and investigated, the aggressive and the devotional.

The model, Fig.2, shows schematically four units with multilateral unconscious information exchange via the S_{ik} and S_{ki} (i,k = 1, 2, 3, 4) in a devotional state. The totally aggressive system is characterized by $S_{ik} > 0$ and $S_{ki} > 0$. The generalized devotional system is characterized by $S_{ik} < 0$ and $S_{ki} > 0$. It is obvious that in reality an enormous amount of variations can occur. The following investigation is restricted in its concern with a) the volitions G_i and b) the speed of information exchange, the velocity V. The volition G_i for a system with n units, (n = 1 to 8), is the volition at the stability limit of a particular system. This limit is a critical measure which is convenient for comparing the performance of systems. The velocity is defined as the invers of the oscillating time, when the system of n units oscillates at the stability limit.

The devotional and unconscious relationship will now be demonstrated for one to eight units, i.e., n = 1 to n = 8. It can be observed in technical, and in social systems that, with increasing numbers of units which have interactions, the stability becomes more and more critical. As a consequence, the volitions of the individual units have to be reduced, and the amount of information exchanged should go down.

Once the volitions approach zero with increasing numbers of units, the goal achievements go to zero as well, but the speed of action - as will be seen - stays relatively high. This means that with the output being zero, the velocity of exchanged information is still considerable. Such activity is then merely the internal flow of information. Such information is clearly not anymore in the interest of goal approach.

Two features will demonstrate these facts, namely
a) The volitions of the individual unit, and
b) The velocity of acting

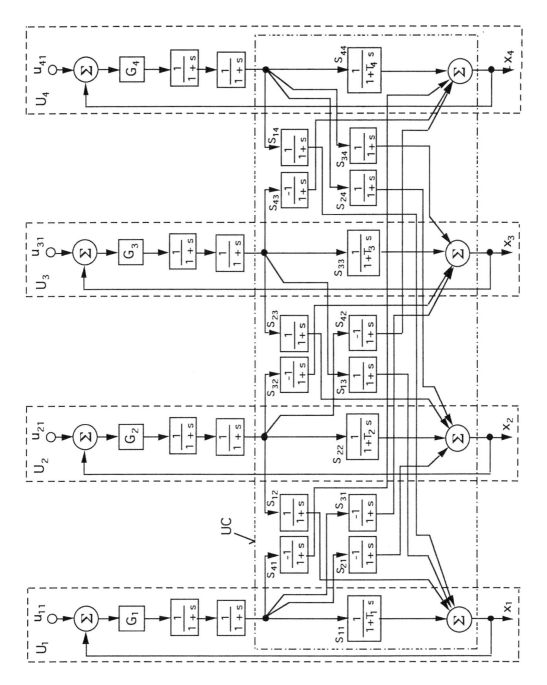

Fig. 2: The model of the democracy. Analog of four multilaterally connected units in a devotional unconscious relationship, i.e., $S_{ik} = -1$ for $i>k$, and $S_{ik} = +1$ for $i<k$. For the explanation of the term "devotional", see Starkermann (1992).

as a growing number of units involved. These two results are subject to two system parameters, α and β: α) if there is bilateral and multilateral communication,

β) if the speed of communication varies; slow, medium, and fast.
The results is:
The denser the information exchange and the faster it is, the worse is the performance of the democracy.

4 THE INVESTIGATION

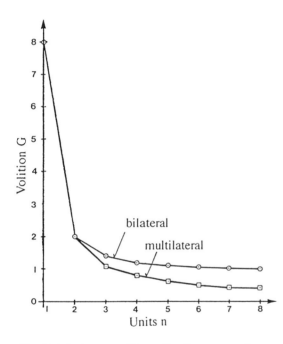

Fig. 3: Maximum volitions G_i of systems of n units each. Bilateral and multilateral information exchange. In every system of n units, all G_i's are equal, i.e., $G_1 = G_2 = \ldots \ldots G_n$.

Fig. 3 illustrates the decline of the permissible volitions for growing systems with units n from 1 to 8. There are two curves. One is for bilateral information exchange, i.e., each unit exchanges information only with the two neighbouring units. The other curve depicts volitions for multilateral connections, i.e., each unit is information-wise connected to all other units. It can be seen that the more there is information flowing around, the more the system is endangered to destabilize and the more the volitions for achieving the own goal have to be reduced. The G_1 for n = 1 is 8, whereas the G_i's for n = 8 are - in bilateral connections - only 1, that is eight times less, and in multilateral connection, the G_i's are about 0.4, what is 20 times less. There is the saying by Friedrich Schiller in William Tell: "Der Starke ist am mächtigsten allein."

Fig. 4 shows the speed of acting (in any regard). The figure is corresponding with Fig. 3. The findings reflect well the reality. The single unit (n = 1) acts the fastest. The measure on the ordinate is the inverse of the oscillating time of the system, $2\pi/\sqrt{3} = 3.6$, which is 0.276. This value can be calculated with the characteristic equation of the unit:

$$\frac{G_1}{(1+s)^3} + 1 = 0$$

With s = iω, the equation results in G = 8 and $\omega = \sqrt{3}$. The oscillating time is $2\pi/\omega$. The value 8 can be seen in Fig. 3. The more mutual information flow by attitude-information-exchange occurs, the slower the system works toward its goals u_{n1}. See the English proverb: He travels fastest who travels alone. But in Fig. 3 it seems that the limit of the volitions in the multilateral information exchange tends toward zero, the death of the democracy, whereas the speed approaches a limit which is by far not zero. The zero achieving system still shows activity although no accomplishment any-

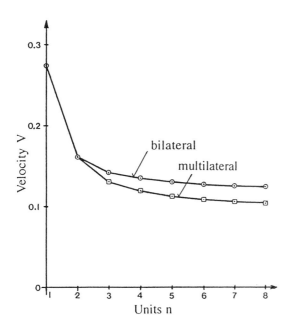

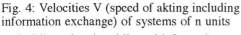

Fig. 4: Velocities V (speed of akting including information exchange) of systems of n units

each. Bilateral and multilateral information exchange. The velocitiy is the invers of the oscillation time at the stability limit of the individual system with its n units.

more, because the volitions go to zero. The unconscious behaviour among the units goes on! If the attitude is hate, hate goes on, although the system is in a ruined state of achievements!

With Fig. 5 the question can be answered, whether delay of information exchange is advantageous. The answer is **yes**. (Better, indeed, it is to stop information completely.) Three curves show measurements for multilateral systems with T_{ii} (of the eigen-transfer function S_{11}) = 0.5, 1, and 2. The time constant T_{ik} and T_{ki}, respectively, of the information transfers S_{ik} and S_{ki} is 1, i.e., the transfer functions are $S_{ik} = -1/(1+s)$, $(i>k)$; and $S_{ki} = 1/(1+s)$, $(i<k)$. By making the eigen-transfer of the units $T_{ii} = 0.5$, the transfer to the other units via $T_{ik} = 1$ is slower than the eigen-transfer. With $T_{ii} = 1$, the mutual transfer is as fast as the own one, and with $T_{ii} = 2$, the mutual transfer is faster than the eigen-transfer. Thus, it can be seen that the slower the transfer **with respect to the own**, the higher the volition can be before the systems run into instability. Higher volitions mean better gaol achievements. Due to the multilateral concept, it seems that with n = ∞, the volitions tend to zero again.

Fig. 6, finally illustrates corresponding results: The slower the information transfer, the faster the systems reach their goals; and the faster the transfer, the slower the systems reach their goals.

Life-praxis shows that the model is responding correctly .

CONCLUSION

Each living system consists of self-control of its units and of attitudes toward other units of the sys-tem. The basic philosophy is: The more the system's units are exchanging information among themselves, less is the individual's volition for reaching the own goal and less fast the system performs its goal approach. This fact holds true whether a technical plant, or a society of human beings constitute a system. Decentralization is the healing cure, or in accordance to Louis XI (11423-83): Divide et impera. Cut information channels which consume internal energy and time and force the autonomous individuals, or small systems, to do their work.

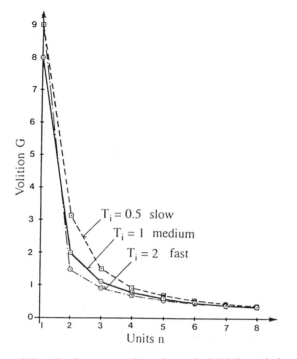

Fig. 5: Maximum volition G_i of systems of n units each. Multilateral slow, medium, and fast information exchange.

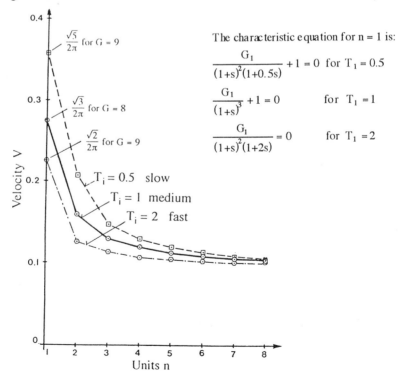

Fig. 6: Velocities V of systems of n units. Slow, medium, and fast information exchange.

REFERENCES

Starkermann, R. (1990). "A Mathematical Model for the Social Behaviour of the Individual", Preprints of the 11–th IFAC Congress, Tallinn, USSR, Vol. 12, pp. 56–61.

Starkermann, R. (1992). "The Natural Law of Love and Hate", Preprints of the (IFAC) SWIIS '92 Workshop on: Supplementary Ways of Increasing International Stability, Toronto, Canada, pp. 96-102.

A Nonlinear Stochastic Process for Quality Growth

Chris P. Tsokos, Department of Mathematics, University of South Florida, Tampa, FL 33620-5700

Abstract

The aim of the present study is to utilize the nonhomogeneous Poisson or Weibull process to characterize the failure times of a system or product through the design and development stages. Such modeling for failure data of a system is referred to as reliability growth analysis. A complete formulation of the Weibull process for reliability growth analysis is given. Some of the recent developments in the subject area are also stated. Utilizing some real data on system development, we illustrate the usefulness of the methods presented.

1 Introduction

The objective of the present study is to introduce a nonlinear stochastic process that can be effectively used in the development and testing process of a product, the main thrust being to improve the quality of the product by increasing its reliability. This approach to product improvement is referred to as reliability growth. That is, the reliability of a certain system or product will improve with time as design flaws of the product are identified and corrected through a redesigning processing and retesting. In 1964, Duane, [3], investigated empirically the phenomenon of reliability growth through the testing of aircraft engines. He noticed a linear relationship between the logarithm of the number of times the system failed and the logarithm of the operating time. Crow, [2], proposed an effective analytical characterization of reliability growth. There are a number of scientists and engineers who have worked on the subject area with attempts to improve the theoretical and application aspects of reliability growth. Here we shall discuss some recent developments and introduce some important analytical constraints associated with the stochastic process which characterize the probabilistic behavior of reliability growth analysis.

Consider the case in which we wish to find the probability of achieving n failures in the time interval $(0, t)$, that is

$$P\left(x = n;\, t\right) = \frac{e^{-\int_0^t v(x)dx} \left\{\int_0^t v\left(x\right)dx\right\}^n}{n!}. \qquad (1.1)$$

When the failure intensity function, $v\left(t\right) = \lambda$, expression (1.1) reduces into a homogeneous Poisson process,

$$P(x = n; \, t) = \frac{e^{-\lambda t} (\lambda t)^n}{n!}, \, 0 < n. \tag{1.2}$$

Crow, [2], proposed for tracking reliability growth that the failure intensity function be

$$v(t) = \frac{\beta}{\theta} \left(\frac{t}{\theta}\right)^{\beta-1} \tag{1.3}$$

where β and θ are called the shape and scale parameter, respectively. Thus we have

$$P(x = n; \, t) = \frac{e^{-\int\limits_{0}^{t} \left(\frac{\beta}{\theta}\right)\left(\frac{x}{\theta}\right)^{\beta-1} dx} \left\{\int\limits_{0}^{t} \left(\frac{\beta}{\theta}\right) \left(\frac{x}{\theta}\right)^{\beta-1} dx\right\}^n}{n!}$$

$$= \frac{1}{n!} e^{-\frac{t^\beta}{\theta^\beta}} \left(\frac{t}{\theta}\right)^{n\beta}. \tag{1.4}$$

Expression (1.4) is called a nonhomogeneous Poisson process or a Weibull process. This process plays an integral part in the analysis and modeling of product reliability growth. The Weibull process is also used for predicting the failure behavior of a product. That is, it is quite important to be able to predict the next failure time after some failures have already occurred. The time difference between the expected next failure time and current failure time is called the Mean Time Between Failures, $MTBF$. The relationship between $MTBF$ and the failure intensity function, $v(t)$, is quite important in reliability growth analysis of a product. $v(t)$ will provide the expected number of failures in $(t_n, \, t_{n+1})$, a unit of time, whereas $\frac{1}{v(t)}$ will give us the expected time of the next failure given that the nth failure occurred at time t_n. Many scientists, [1], [2], [6], in reliability growth modeling have used the reciprocal of the intensity function to be equal to $MTBF$. Qiao and Tsokos, [5], have shown that such a characterization is not true in general.

In what follows, we shall be concerned with the estimating of the parameters inherent in the Weibull process and analytically identifying the relationship between $MTBF$ and $\frac{1}{v(t)}$, Qiao and Tsokos, [5].

2   The Weibull Process

Let $T_1 < T_2 < \cdots < T_n$ be the first n ordered failure times characterized by the Weibull Process. The time to the first failure time T_1 is Weibull distributed with shape and scale parameter β, θ, respectively, that is, the probability density function is

$$f_1(t) = \frac{\beta}{\theta} \left(\frac{t}{\theta}\right)^{\beta-1} e^{-\left(\frac{t}{\theta}\right)^\beta}.$$

Let $f_i(t \,|t_1, \, \cdots, \, t_{i-1})$ denote the conditional probability distribution function of T_i given $T_{i-1} = t_{i-1}, \, \cdots, T_1 = t_1$. Then the truncated Weibull distribution is given by

$$f_i(t \,|t_1, \, \cdots, \, t_{i-1}) = \frac{\beta}{\theta} \left(\frac{t}{\theta}\right)^{\beta-1} \exp\left[-\left(\frac{t}{\theta}\right)^\beta + \left(\frac{t_{i-1}}{\theta}\right)^\beta\right], \, (t > t_{i-1}). \tag{2.1}$$

Thus, the cumulative distribution function of (2.1) is

$$F_i\left(t\,|t_1,\,\cdots,\,t_{i-1}\right) = 1 - \exp\left[-\left(\frac{t}{\theta}\right)^{\beta} + \left(\frac{t_{i-1}}{\theta}\right)^{\beta}\right].$$

The likelihood function for $T_1 = t_1,\,\cdots,\,T_n = t_n$ is:

$$L\left(\beta,\,\theta\right) = \prod_{i=1}^{n} f_i\left(t_i\,|t_1,\,\cdots,\,t_{i-1}\right)$$

$$= \left(\frac{\beta}{\theta}\right)^{n} \exp\left[-\left(\frac{t_n}{\theta}\right)^{\beta}\right] \prod_{i=1}^{n}\left(\frac{t_i}{\theta}\right)^{\beta-1}. \tag{2.2}$$

It follows that

$$\log\,L\left(\beta,\,\theta\right) = n\log\,\beta - n\beta\log\,\theta - \left(\frac{t_n}{\theta}\right)^{\beta} + (\beta-1)\sum_{i=1}^{n}\log\,t_i,$$

and

$$\begin{cases} \frac{\partial\log\,L(\beta,\,\theta)}{\partial\beta} = \frac{n}{\beta} - n\log\,\theta - \left(\frac{t_n}{\theta}\right)^{\beta}\log\,\frac{t_n}{\theta} + \sum_{i=1}^{n}\log\,t_i = 0, \\ \frac{\partial\log\,L(\beta,\,\theta)}{\partial\theta} = -n\frac{\beta}{\theta} + \beta t_n^{\beta}\theta^{-\beta-1} = 0. \end{cases} \tag{2.3}$$

Solving the equations (2.3), we can obtain the maximum likelihood estimate (MLE) for the parameter β and θ, at t_n. That is,

$$\begin{cases} \widehat{\beta} = \dfrac{n}{\sum\limits_{i=1}^{n-1}\log\left(\frac{t_n}{t_i}\right)}, \\ \widehat{\theta} = \dfrac{t_n}{n^{\frac{1}{\beta}}}. \end{cases} \tag{2.4}$$

Thus, we can use these estimates to estimate the failure intensity function of the Weibull process

$$\widehat{v}_n = \frac{\widehat{\beta}}{\widehat{\theta}}\left(\frac{t_n}{\widehat{\theta}}\right)^{\widehat{\beta}-1} \tag{2.5}$$

and

$$\frac{1}{\widehat{v}_n} = \frac{\theta}{\widehat{\beta}}\left(\frac{\widehat{\theta}}{t_n}\right)^{\widehat{\beta}-1}. \tag{2.6}$$

The aim of the development of the present model is structured so that the reliability of a product will improve with time as design flaws are repaired and removed. Thus, we would want to have v_n, the failure intensity, to be decreasing with time, which means $\beta < 1$. Furthermore, we are only interested in the reliability growth model. In what follows we will consider how to deal with $MTBF$ estimation for the case when $\beta < 1$.

3   Mean Time Between Failures of a Product

The Mean Time Between Failures, or $MTBF$, is the expected interval length from the current failure time, say $T_n = t_n$, to the next failure time, $T_{n+1} = t_{n+1}$. We shall use $MTBF_n$ to denote the $MTBF$ at current state $T_n = t_n$. Then it follows that

$$MTBF_n = \int_{t_n}^{\infty} t f_{n+1}(t\,|t_1, \cdots, t_n)\, dt - t_n, \tag{3.1}$$

where $\int_{t_n}^{\infty} t f_{n+1}(t\,|t_1, \cdots, t_n)\, dt$ is the expected $(n+1)$th failure time under the condition $T_n = t_n$.

For the Weibull process, we have

$$MTBF_n = \int_{t_n}^{\infty} t \frac{\beta}{\theta} \left(\frac{t}{\theta}\right)^{\beta-1} e^{-\left(\frac{t}{\theta}\right)^{\beta} + \left(\frac{t_n}{\theta}\right)^{\beta}} dt - t_n. \tag{3.2}$$

Note that when the shape parameter of the process $\beta = 1$, we have

$$MTBF_n = \int_{t_n}^{\infty} \frac{t}{\theta} e^{-\frac{t}{\theta} + \frac{t_n}{\theta}} dt - t_n$$

$$= \theta.$$

Also, the reciprocal of the intensity function results in

$$\frac{1}{v_n} = \theta.$$

Thus, in the case $\beta = 1$, we have

$$MTBF_n = \frac{1}{v_n}.$$

In view of our previous remarks concerning the relationship of $MTBF$ and $\frac{1}{v_n}$, it is clear that these two concepts are equal iff $\beta = 1$. Thus, in such a case the failure intensity function is driven only by the behavior of the scale parameter. Furthermore, obtaining effective estimates of the intensity function will totally depend on having robust estimates of θ.

In general, for any β, we have

$$MTBF_n = \int_{t_n}^{\infty} \frac{\beta}{\theta^{\beta}} t^{\beta} e^{-\left(\frac{t}{\theta}\right)^{\beta} + \left(\frac{t_n}{\theta}\right)^{\beta}} dt - t_n$$

$$= \theta \beta e^{\left(\frac{t_n}{\theta}\right)^{\beta}} \int_{\frac{t_n}{\theta}}^{\infty} t^{\beta} e^{-t^{\beta}} dt - t_n$$

$$= \theta e^{\left(\frac{t_n}{\theta}\right)^{\beta}} \left[\frac{t_n}{\theta} e^{-\left(\frac{t_n}{\theta}\right)^{\beta}} + \int_{\frac{t_n}{\theta}}^{\infty} e^{-t^{\beta}} dt \right] - t_n$$

$$= \frac{\theta}{\beta} e^{\left(\frac{t_n}{\theta}\right)^{\beta}} \int\limits_{\left(\frac{t_n}{\theta}\right)^{\beta}}^{\infty} e^{-t} t^{\frac{1}{\beta}-1} dt. \tag{3.3}$$

Also, using the negative reciprocal of the failure intensity function we have

$$\frac{1}{v_n} = \frac{\theta}{\beta} \left(\frac{\theta}{t_n}\right)^{\beta-1}$$

$$= \frac{\theta}{\beta} e^{\left(\frac{t_n}{\theta}\right)^{\beta}} \left(\frac{\theta}{t_n}\right)^{\beta-1} \int\limits_{\left(\frac{t_n}{\theta}\right)^{\beta}}^{\infty} e^{-t} dt.$$

Moreover, if $\beta > 1$, then $\frac{1}{\beta} - 1 < 0$, and

$$\int\limits_{\left(\frac{t_n}{\theta}\right)^{\beta}}^{\infty} e^{-t} t^{\frac{1}{\beta}-1} dt < \int\limits_{\left(\frac{t_n}{\theta}\right)^{\beta}}^{\infty} e^{-t} \left(\left(\frac{t_n}{\theta}\right)^{\beta}\right)^{\frac{1}{\beta}-1} dt$$

$$= \left(\frac{\theta}{t_n}\right)^{\beta-1} \int\limits_{\left(\frac{t_n}{\theta}\right)^{\beta}}^{\infty} e^{-t} dt.$$

Similarly, if $\beta < 1$, then $\frac{1}{\beta} - 1 > 0$, and

$$\int\limits_{\left(\frac{t_n}{\theta}\right)^{\beta}}^{\infty} e^{-t} t^{\frac{1}{\beta}-1} dt > \left(\frac{\theta}{t_n}\right)^{\beta-1} \int\limits_{\left(\frac{t_n}{\theta}\right)^{\beta}}^{\infty} e^{-t} dt.$$

Thus, we can conclude that the relationship of $MTBF$ and the reciprocal of the intensity function can be summarized as follows:

$$MTBF_n \begin{cases} = \frac{1}{v_n} & \text{if } \beta = 1 \\ > \frac{1}{v_n} & \text{if } \beta < 1 \\ < \frac{1}{v_n} & \text{if } \beta > 1. \end{cases} \tag{3.4}$$

Additional details can be found in Qiao and Tsokos, [5]. However, it must be stated that published works on the subject area which assumed $MTBF_n$ is equal to $\frac{1}{v_n}$ must be revisited.

4   Computations of MTBF and v_n

In order to achieve the aim of increasing the reliability of product development using the nonhomogeneous Poisson process, we must have the failure intensity function decreasing. We can realize such a behavior of $\frac{1}{v_n}$ is by having $\beta < 1$. Thus, we shall proceed to develop analytical relationships of $\frac{1}{v_n}$ and $MTBF_n$ when the shape parameter is less than one.

Let $\alpha = \dfrac{1}{\beta}$ and $h = \left(\dfrac{t_n}{\theta}\right)^{\beta}$, then $\alpha > 1$ and v_n and $\dfrac{1}{v_n}$ can be given by

$$v_n = \frac{\beta t_n^{\beta-1}}{\theta^{\beta}} = \frac{1}{\alpha\theta}h^{1-\alpha}$$

and

$$\frac{1}{v_n} = \alpha\theta h^{\alpha-1}. \tag{4.1}$$

Furthermore, we can write $MTBF_n$ as

$$MTBF_n = \theta e^h \alpha \int\limits_{h}^{\infty} e^{-t} t^{\alpha-1} dt.$$

To study the behavior of $MTBF_n$ and $\dfrac{1}{v_n}$ for $\beta < 1$, we need to partition the shape parameter into two parts, $0 < \beta \le \dfrac{1}{2}$ and $\dfrac{1}{2} < \beta < 1$.

Part 1: Shape parameter $0 < \beta \le \dfrac{1}{2}$. For $0 < \beta \le \dfrac{1}{2}$, we have $\alpha \ge 2$. Assume $\alpha = m + 1 + \delta$, where $m \ge 1$ is an integer and $\delta \in [0, 1)$.

Note a special case is when $\delta = 0$. In this case, the $MTBF_n$ can be found analytically as follows:

$$MTBF_n = \theta e^h \alpha e^{-h} \left[h^m + (\alpha - 1) h^{m-1} + \cdots \right.$$

$$+ (\alpha - 1)(\alpha - 2)\cdots(\alpha - m + 1) h$$

$$\left. + (\alpha - 1)(\alpha - 2)\cdots(\alpha - m) \right]$$

$$= \alpha\theta h^m \left[1 + (\alpha - 1)\frac{1}{h} + \cdots + \frac{(\alpha - 1)\cdots(\alpha - m)}{h^m} \right]$$

$$= \frac{1}{v_n} \left[1 + (\alpha - 1)\frac{1}{h} + \cdots + \frac{(\alpha - 1)\cdots(\alpha - m)}{h^m} \right].$$

From this expression, it is clear that $MTBF_n$ is quite different from $\dfrac{1}{v_n}$. The difference is given by

$$\frac{1}{v_n} \left[(\alpha - 1)\frac{1}{h} + \cdots + \frac{(\alpha - 1)\cdots(\alpha - m)}{h^m} \right],$$

which is quite significant for certain parameters.

For a general δ, we can write

$$MTBF_n = \theta\alpha h^{\delta+m} \left[1 + \frac{\alpha - 1}{h} + \cdots + \frac{(\alpha - 1)(\alpha - 2)\cdots(\alpha - m + 1)}{h^{m-1}} \right.$$

$$+\frac{(\alpha-1)(\alpha-2)\cdots(\alpha-m)}{h^m}e^h h^{-\delta}\int_h^\infty t^{\alpha-m-1}e^{-t}dt\Bigg]$$

$$=\frac{1}{v_n}\Bigg[1+\frac{\alpha-1}{h}+\cdots+\frac{(\alpha-1)(\alpha-2)\cdots(\alpha-m+1)}{h^{m-1}}$$

$$+\frac{(\alpha-1)(\alpha-2)\cdots(\alpha-m)}{h^m}e^h h^{-\delta}\int_h^\infty t^{\alpha-m-1}e^{-t}dt\Bigg] \tag{4.2}$$

$$=\frac{1}{v_n}\Bigg[1+\frac{\alpha-1}{h}+\cdots+\frac{(\alpha-1)(\alpha-2)\cdots(\alpha-m+1)}{h^{m-1}}$$

$$+\frac{(\alpha-1)(\alpha-2)\cdots(\alpha-m)}{h^m}$$

$$+\frac{(\alpha-1)(\alpha-2)\cdots(\alpha-m-1)}{h^m}e^h h^{-\delta}\int_h^\infty t^{\alpha-m-2}e^{-t}dt\Bigg]. \tag{4.3}$$

Note that for $\alpha=m+1+\delta$, we have

$$\alpha-m-1=\delta\geq 0$$

and

$$\alpha-m-2=\delta-1<0.$$

Also, we can conclude that

$$e^h h^{-\delta}\int_h^\infty t^{\alpha-m-1}e^{-t}dt$$

$$\geq e^h h^{-\delta}h^{\alpha-m-1}\int_h^\infty e^{-t}dt$$

$$=1.$$

Similarly, we have

$$e^h h^{-\delta}\int_h^\infty t^{\alpha-m-2}e^{-t}dt$$

$$< e^h h^{-\delta}h^{\alpha-m-2}\int_h^\infty e^{-t}dt$$

$$=\frac{1}{h}.$$

Thus, from equation (4.2) we can write

$$MTBF_n \geq \frac{1}{v_n}\left[1 + \frac{\alpha - 1}{h} + \cdots + \frac{(\alpha - 1)(\alpha - 2)\cdots(\alpha - m + 1)}{h^{m-1}}\right.$$

$$\left. + \frac{(\alpha - 1)(\alpha - 2)\cdots(\alpha - m)}{h^m}\right]$$

$$\geq \frac{1}{v_n}\left[1 + \frac{\alpha - 1}{h}\right].$$

On the other hand, from expression (4.3) we have

$$MTBF_n \leq \frac{1}{v_n}\left[1 + \frac{\alpha - 1}{h} + \cdots + \frac{(\alpha - 1)(\alpha - 2)\cdots(\alpha - m)}{h^m}\right.$$

$$\left. + \frac{(\alpha - 1)(\alpha - 2)\cdots(\alpha - m - 1)}{h^{m+1}}\right]$$

$$\leq \frac{1}{v_n}\left[1 + \frac{\alpha - 1}{h} + \left(\frac{\alpha - 1}{h}\right)^2 + \cdots + \left(\frac{\alpha - 1}{h}\right)^m\right]$$

$$\leq \frac{1}{v_n}\frac{1}{1 - \frac{\alpha - 1}{h}}.$$

Therefore, we can conclude that for $0 < \beta \leq \frac{1}{2}$, $MTBF_n$ is bounded from below by $\frac{1}{v_n}\left(1 + \frac{\alpha - 1}{h}\right)$ and above by $\frac{1}{v_n}\frac{1}{1 - \frac{\alpha - 1}{h}}$, respectively. That is,

$$\frac{1}{v_n}\left(1 + \frac{\alpha - 1}{h}\right) \leq MTBF(T_n) \leq \frac{1}{v_n}\frac{1}{1 - \frac{\alpha - 1}{h}}$$

where

$$\alpha = \frac{1}{\beta}, \quad h = \left(\frac{t_n}{\theta}\right)^\beta$$

and

$$\frac{1}{v_n} = \frac{\theta}{\beta}\left(\frac{\theta}{t_n}\right)^{\beta - 1}.$$

Thus, using the maximum likelihood estimates of the shape and scale parameter, we can obtain estimates of $MTBF_n$, that is,

$$\widehat{MTBF}_n = \frac{1}{2}\left[\frac{1}{\hat{v}_n}\left(1 + \frac{\hat{\alpha} - 1}{\hat{h}}\right) + \frac{1}{\hat{v}_n}\frac{1}{1 - \frac{\hat{\alpha} - 1}{\hat{h}}}\right].$$

Part 2: Shape parameter $\frac{1}{2} < \beta < 1$. In this part for $\frac{1}{2} < \beta < 1$ we have $1 \leq \alpha < 2$. Assume $\alpha = 1 + \delta$, where $0 \leq \delta < 1$.

Recall that $\frac{1}{v_n} = \alpha\theta h^{\alpha - 1}$ and thus, we can write

$$MTBF_n = \theta e^h \alpha \int_h^\infty e^{-t} t^\delta dt$$

$$\geq \alpha \theta h^\delta e^{-h} \int_h^\infty e^{-t} dt$$

$$= \alpha \theta h^\delta$$

$$= \frac{1}{v_n}.$$

At the same time, we also have the following result

$$MTBF_n = \alpha \theta e^h \left[h^\delta e^{-h} + \delta \int_h^\infty h^{\delta-1} e^{-x} dx \right]$$

$$\leq \alpha \theta e^h \left[h^\delta e^{-h} + \delta \int_h^\infty h^{\delta-1} e^{-t} dt \right]$$

$$\leq \frac{1}{v_n} \left[1 + \frac{\delta}{h} \right].$$

Therefore, $MTBF_n$ is bounded from below and above by:

$$\frac{1}{v_n} \leq MTBF_n \leq \frac{1}{v_n} \left[1 + \frac{\delta}{h} \right]$$

where $\delta = \frac{1}{\beta} - 1 \in [0, 1)$. Again, we can employ the maximum likelihood estimates of β and θ given above to obtain estimates of $MTBF_n$ and v_n. Qiao and Tsokos, [5], have proposed that $MTBF_n$ can best be estimated by

$$\widehat{MTBF}_n = \frac{1}{2\hat{v}_n} \left[2 + \frac{\hat{\delta}}{\hat{h}} \right].$$

In summary, for $0 < \beta \leq \frac{1}{2}$, we have

$$\frac{1}{v_n} \left(1 + \frac{\alpha-1}{h} \right) \leq MTBF_n \leq \frac{1}{v_n} \frac{1}{1 - \frac{\alpha-1}{h}}$$

and we propose that $MTBF_n$ be estimated by

$$\widehat{MTBF}_n = \frac{1}{\hat{v}_n} \left[\frac{\hat{\alpha}-1}{\hat{h}} + \frac{1}{1 - \frac{\hat{\alpha}-1}{\hat{h}}} \right].$$

In the case where $\frac{1}{2} < \beta < 1$, we have

$$\frac{1}{v_n} \leq MTBF_n \leq \frac{1}{v_n} \left(1 + \frac{\alpha-1}{h} \right)$$

and we propose that $MTBF_n$ be estimated by

$$\widehat{MTBF}_n = \frac{1}{2\widehat{v}_n}\left(2 + \frac{\widehat{\alpha}-1}{\widehat{h}}\right).$$

where

$$\alpha = \frac{1}{\beta}, \ h = \left(\frac{t_n}{\theta}\right)^\beta, \ \widehat{\alpha} = \frac{1}{\widehat{\beta}}, \ \widehat{h} = \left(\frac{t_n}{\widehat{\theta}}\right)^{\widehat{\beta}}$$

and

$$\frac{1}{v_n} = \frac{\theta}{\beta}\left(\frac{\theta}{t_n}\right)^{\beta-1}, \ \frac{1}{\widehat{v}_n} = \frac{\widehat{\theta}}{\widehat{\beta}}\left(\frac{\widehat{\theta}}{t_n}\right)^{\widehat{\beta}-1}.$$

The maximum likelihood estimates used above have been recently improved upon by Higgins and Tsokos, [4], Tsokos and Rao, [7], and Qiao and Tsokos, [5]. The MLE of both shape and scale parameters are quite biased and the estimates can be improved substantially by correcting for biasedness. Furthermore, the MLE of the scale parameter depends only on t_n and thus the estimate of θ can become unstable.

5 Application

Crow, [2], obtained the following failure data during the design and development of a certain product.

Table 5.1. Crow's, [2], Failure Data

.7	3.7	13.2	17.6	54.5	99.2	112.2	120.9	151.0	163.0
174.5	191.6	282.8	355.2	486.3	490.5	513.3	558.4	678.1	668.0
785.9	887.0	1010.7	1029.1	1034.4	1136.1	1178.9	1259.7	1297.9	1419.7
1571.7	1629.8	1702.4	1928.9	2072.3	2525.2	2928.5	3016.4	3181.0	3256.3

The data represents failure times of the system, that is, the system was of age $t_1 = .7$ units of time when the first failure occurred, of age $t_2 = 3.7$ when the second failure occurred, etc. It was at age $t_{40} = 3256.3$ units of time when the system experienced the 40th failure. One of the aims in the analysis of this data is to predict the time at which the 41st failure would occur.

Crow, [2], obtained the following estimates

$$\widehat{\beta} = \frac{n}{\sum_{i=1}^n \log\left(\frac{t_n}{t_i}\right)} = \frac{40}{\sum_{i=1}^{40}\log\left(\frac{3256.3}{t_i}\right)} \approx .49$$

and

$$\widehat{\theta} = \frac{t_n}{n^{1/\widehat{\beta}}} = \frac{3256.3}{40^{1/.49}} \approx 1.7461.$$

Also,

$$\widehat{h} = \left(\frac{t_n}{\widehat{\theta}}\right)^{\widehat{\beta}} = \left(\frac{3256.3}{1.7461}\right)^{.49} \approx 40.25,$$

$$\widehat{v}_n = \frac{\widehat{\beta}t_n^{\widehat{\beta}-1}}{\widehat{\theta}^{\widehat{\beta}}} = \frac{(.49)\,(3256.3)^{-.51}}{1.7461^{.49}} \approx .006$$

and

$$\frac{1}{\widehat{v}_n} \approx 166.$$

Thus, Crow, [2], concluded that the next failure time of the system would be 166 ages from the 40th failure time.

Using the methods developed by Qiao and Tsokos, [5], for $0 < \beta \leq \frac{1}{2}$, we can calculate the lower and upper bounds estimates of $MTBF_n$ as given below:

$$\text{lower bound } = \frac{1}{\widehat{v}_n}\left(1 + \frac{\widehat{\alpha}-1}{\widehat{h}}\right)$$

$$= 166\left(1 + \frac{1.0408 - 1}{40.25}\right) \approx 170.235,$$

$$\text{upper bound } = \frac{1}{\widehat{v}_n}\frac{1}{1 - \frac{\widehat{\alpha}-1}{\widehat{h}}}$$

$$= 166\frac{1}{1 - \frac{1.0408}{40.25}} \approx 170.350.$$

Thus, our estimate of the true $MTBF_n$ satisfies the following inequality

$$170.235 \leq MTBF_n \leq 170.350.$$

We can proceed to obtain an effective estimate of $MTBF_n$, that is,

$$\widehat{MTBF}_n = \frac{\text{lower } + \text{ upper}}{2} = 170.293.$$

We then can conclude that the next failure time will be around 170.293 ages from the 40th failure times. Note that Crow's estimate is not within the estimated bounds. Thus, we expect the system will fail at the age of 3426.3 units of time.

References

[1] F.W. Breyfogle, Statistical Methods for Testing, Development and Manufacturing, John Wiley & Sons, New York, NY (1991).

[2] L.H. Crow, Tracking Reliability Growth, Proceedings of the Twentieth Conference on the Design of Experiments, *ARO Report 75-2*, US Army Research Office, Research Triangle Park, NC (1975), pp. 741–754.

[3] J.T. Duane, Learning Curve Approach to Reliability Monitoring, *IEEE Transactions on Aerospace* (April 1964), Vol. 2, pp. 563–566.

[4] J.J. Higgins and C.P. Tsokos, A Quasi-Bayes Estimate of the Failure Intensity of a Reliability-Growth Model, *IEEE Transactions on Reliability* (1981), 30, pp. 471–475.

[5] H. Qiao and C.P. Tsokos, Analysis of the Mean Time Between Failures for the Nonhomogeneous Poisson Process, to appear.

[6] P.A. Tobias and D.C. Trindade, Applied Reliability, Van Nostrand Reinhold Company, New York, NY (1986).

[7] C.P. Tsokos and A.N.V. Rao, Estimation of Failure Intensity for the Weibull Process, *IEEE Transactions of Reliability*, to appear.

An Extension of the Method of Quasilinearization
for Reaction–Diffusion Equations

A.S. VATSALA, Department of Mathematics, P.O. Box 41010 University
of Southwestern Louisiana, Lafayette, LA 70504

1   INTRODUCTION.

It is well known that the method of quasilinearization offers an approach to obtain approximate solutions to nonlinear differential equations. See Bellman (1973), Bellman (1965) and Chan (1974). Also the method enables one to obtain lower or upper bounds for the solutions. However this is possible when the forcing term $f(t, u)$ is uniformly convex or uniformly concave for each $t \in [0, T]$. Recently the method has been extended by Lakshmikantham (1993) to obtain two sided bounds for the solution of scalar first order ordinary differential equations. In this paper we are extending the method of quasilinearization to obtain two sided bounds for the solution of the scalar reaction diffusion equations. The method offers monotone sequences which converge quadratically to the solution of reaction diffusion equations.

2   EXISTENCE AND COMPARISON RESULTS.

In this section we recall known existence and comparison results, see Ladde (1985), Lakshmikantham (1968) Walter (1970), and Pao (1992). Though these results are known in more general form, we merely state them in a form suitable to develop our main result. For this purpose consider the reaction diffusion systems with initial and boundary conditions of the form

$$
\begin{array}{rcll}
\mathcal{L}u_k & = & f_k(t, x, u) & in \ Q_T \\
Bu_k & = & \phi_k & on \ \Gamma_T \\
u_k(0, x) & = & u_{0k}(x) & in \ \bar{\Omega}
\end{array}
\tag{2.1}
$$

where $u = (u_1, u_2) \in R^2$, Ω is a bounded domain in R^m with boundary $\partial\Omega \in C^{2+\alpha}$ and closure $\bar{\Omega}$, $Q_T = (0, T] \times \Omega$, $\Gamma_T = (0, T) \times \partial\Omega$, $\bar{Q}_T = [0, T] \times \bar{\Omega}$, $\bar{\Gamma}_T = [0, T] \times \partial\Omega$, $T > 0$. Here $\mathcal{L}$ is a second order differential operator defined by

$$\mathcal{L} = \frac{\partial}{\partial t} - L \tag{2.2}$$

where

$$L = \sum_{i,j=1}^{m} a_{ij}(t, x) \frac{\partial^2}{\partial x_i \partial x_j} + \sum_{i=1}^{m} b_i(t, x) \frac{\partial}{\partial x_i} \tag{2.3}$$

and B is the boundary operator given by

$$Bu_k = p(t, x)u_k + q(t, x)\frac{du_k}{d\gamma}$$

where $\frac{du_k}{d\gamma}$ denotes the normal derivative of u_k and $\gamma(t, x)$ is the unit outward normal vector field on $\partial\Omega$ for $t \in [0, T]$.

Without further mention here and in the next section all vectorial inequalities mean that the same inequalities hold between their corresponding components.

We list the following assumptions for convenience.

(A_0) (i) For each $i, j = 1, \cdots m$, a_{ij}, $b_j \in C^{\frac{\alpha}{2}, \alpha}[\bar{Q}_T, R]$ and $\mathcal{L}$ is strictly uniformly parabolic in $\bar{Q}_T$;

(ii) $p, q \in C^{1+\frac{\alpha}{2}, 1+\alpha}[\bar{\Gamma}_T, R]$, $p(t, x) > 0$, $q(t, x) \geq 0$ on Γ_T;

(iii) $f \in C^{\frac{\alpha}{2}, \alpha}[[0, T] \times \bar{\Omega} \times R^2, R^2]$, that is $f(t, x, u)$ is Hölder continuous in t and (x, u) with exponent $\frac{\alpha}{2}$ and α respectively;

(iv) $\phi_k \in C^{1+\frac{\alpha}{2}, 1+\alpha}[\bar{\Gamma}_T, R]$ and $u_{0k}(x) \in C^{2+\alpha}[\bar{\Omega}, R]$ for $k = 1, 2$;

(v) The initial boundary value problem (2.1) satisfies the compatibility condition of order $[\frac{(\alpha + 1)}{2}]$. See Ladde (1985) for definition.

(A_1) $\alpha, \beta \in C^{1,2}[\bar{Q}_T, R^2]$ such that $\alpha(t, x) \leq \beta(t, x)$ on $\bar{Q}_T$ and

$$
\begin{array}{llll}
\mathcal{L}\alpha_1 \leq f_1(t, x, \alpha_1, \beta_2) & \text{in } Q_T, & \alpha_k(0, x) \leq u_{0k}(x) & \text{on } \bar{\Omega} \\
\mathcal{L}\alpha_2 \leq f_2(t, x, \beta_1, \alpha_2) & \text{in } Q_T, & B\alpha_k(t, x) \leq \phi_k & \text{on } \Gamma_T \\
\mathcal{L}\beta_1 \geq f_1(t, x, \beta_1, \alpha_2) & \text{in } Q_T, & \beta_k(0, x) \geq u_{0k}(x) & \text{in } \bar{\Omega} \\
\mathcal{L}\beta_2 \geq f_2(t, x, \alpha_1, \beta_2) & \text{in } Q_T, & B\beta_k(t, x) \geq \phi_k & \text{on } \Gamma_T.
\end{array}
$$

We now recall an existence theorem which we need to develop our main result. We note that this is a special case of Theorem 4.46 in Ladde (1985) and Theorem 9.3, chapter 8 of Pao (1992).

THEOREM 2.1   Suppose $f(t, x, u)$ in (2.1) is quasimonotone nonincreasing in u and assumptions (A_0) and (A_1) hold. Then the initial boundary value problem possesses a solution $u \in C^{1+\frac{\alpha}{2}, 2+\alpha}[\bar{Q}_T, R^2]$ such that $\alpha(t, x) \leq u(t, x) \leq \beta(t, x)$.

Next we present a comparison theorem which is needed for our main result. This is a special case of Theorem 10.6.3 in Lakshmikantham (1968).

THEOREM 2.2   Suppose that
  (i) $m \in C^{1,2}[\bar{Q}_T, R^2_+]$ such that $\mathcal{L}m \leq f(t, x, m)$ where $f(t, x, u) \in C[Q_T \times R^2, R^2]$ where the operator $\mathcal{L}$ is parabolic.
  (ii) $g \in C[[0, T] \times R^2_+, R^2]$, $g(t, y)$ is quasimonotone nondecreasing in y for each $t \in [0, T]$, $r(t, 0, y_0) \geq 0$ is the maximal solution of the differential system

$$y' = g(t, y), \ y(0) = y_0 \geq 0$$

existing for $t \geq 0$ and

$$f(t, x, z) \leq g(t, z), \ z \geq 0$$

  (iii) $m(0, x) \leq r(0, 0, y_0)$ for $x \in \bar{\Omega}$.
  Then $m(t, x) \leq r(t, 0, y_0)$ on $\bar{Q}_T$.

3   MAIN RESULT.

Consider the nonlinear scalar diffusion equation of the form

$$
\begin{aligned}
\mathcal{L}u = \quad & \frac{\partial u}{\partial t} - Lu = F(t, x, u) \quad && on \ Q_T \\
& u(0, x) = u_0(x) \quad && on \ \bar{\Omega} \\
& Bu(t, x) = \phi(x) \quad && on \ \Gamma_T
\end{aligned}
\tag{3.1}
$$

where Ω, $\partial\Omega$, Q_T, Γ_T, B, $\mathcal{L}$, L are as in section 2.1 satisfying assumptions (i) (ii) of A_0.

DEFINITION 3.1   Suppose that F admits a decomposition

$$F(t, x, u) = f(t, x, u) + g(t, x, u)$$

and $v_0, w_0 \in C^{1,2}[\bar{Q}_T, R]$ such that $v_0(t, x) \leq w_0(t, x)$ on $\bar{Q}_T$.
  Then we say that v_0, w_0 are coupled lower and upper solutions of (3.1), if

$$
\begin{aligned}
\mathcal{L}v_0 & \leq f(t, x, v_0) + g(t, x, w_0) \\
v_0(0, x) & \leq u_0(x), Bv_0(t, x) \leq \phi(x)
\end{aligned}
\tag{3.2}
$$

$$
\begin{aligned}
\mathcal{L}w_0 & \geq f(t, x, w_0) + g(t, x, v_0) \\
w_0(0, x) & \geq u_0(x), Bw_0(t, x) \geq \phi(x).
\end{aligned}
\tag{3.3}
$$

We define the closed set

$$\Lambda = [(t, x, u) : v_0(t, x) \leq u \leq w_0(t, x), \ (t, x) \in \bar{Q}_T].$$

We are now in a position to prove our main result.

THEOREM 3.1   Assume that
(A^*_1) (i) $F(t, x, u) = f(t, x, u) + g(t, x, u)$ be such that $f_u, g_u \in C^{\frac{\alpha}{2}, \alpha}[Q_T \times R, R]$ and $f(t, x, u)$ and $g(t, x, u)$ are uniformly strictly convex and concave in u respectively for $(t, x) \in \bar{Q}_T$ and $f_u(t, x, u) \geq 0$ and $g_u(t, x, u) \leq (w_0(t, x) - v_0(t, x))g_{uu}(t, x, u)$ on Λ.
  (ii) $\phi \in C^{1 + \frac{\alpha}{2}, 1 + \alpha}[\bar{\Gamma}_T, R]$ and $u_0 \in C^{2 + \alpha}[\bar{\Omega}, R]$.

(iii) The initial boundary value problem (3.1) satisfies the compatibility condition of order $[\frac{(\alpha+1)}{2}]$.

(iv) v_0, $w_0 \in C^{1,2}[\bar{Q}_T, R]$ such that $v_0(t,x) \leq w_0(t,x)$ on $\bar{Q}_T$ are coupled lower and upper solutions of (3.1) on $\bar{Q}_T$.

Then there exists monotone sequences $\{v_n(t,x)\}$, $\{w_n(t,x)\}$ that converge uniformly to the unique solution of (3.1) on $\bar{Q}_T$. Moreover the convergence of these sequences is quadratic.

Proof.   It is easy to observe that for any $\eta, \mu \in \Lambda$, that $f(t,x,u)$ and $g(t,x,u)$ satisfy the following inequalities

$$f(t,x,\mu) - f_u(t,x,\eta)\mu \geq f(t,x,\eta) - f_u(t,x,\eta)\eta \tag{3.4}$$

$$g(t,x,\eta) - g_u(t,x,\eta)\eta \geq g(t,x,\mu) - g_u(t,x,\eta)\mu. \tag{3.5}$$

Now we define the reaction diffusion systems for any $\eta, \mu \in \Lambda$ such that $\eta(t,x) \leq \mu(t,x)$. This will be later needed to define the sequences.

Consider the reaction diffusion system

$$\begin{aligned}\mathcal{L}u_1 &= H_1(t,x,P(u_1),P(u_2),\eta,\mu)\\ u_1(0,x) &= u_0(x),\ Bu_1(t,x) = \phi(x)\end{aligned} \tag{3.6}$$

$$\begin{aligned}\mathcal{L}u_2 &= H_2(t,x,P(u_1),P(u_2),\eta,\mu)\\ u_2(0,x) &= u_0(x),\ Bu_2(t,x) = \phi(x)\end{aligned} \tag{3.7}$$

where

$$\begin{aligned}H_1(t,x,u_1,u_2,\eta,\mu) &= f(t,x,\eta) + f_u(t,x,\eta)(u_1-\eta)\\ &+ g(t,x,\mu) + g_u(t,x,u_2)(u_2-\mu)\end{aligned} \tag{3.8}$$

$$\begin{aligned}H_2(t,x,u_1,u_2,\eta,\mu) &= f(t,x,\mu) + f_u(t,x,u_2)(u_2-\eta)\\ &+ g(t,x,\eta) + g_u(t,x,\eta)(u_1-\eta)\end{aligned} \tag{3.9}$$

where

$$P(u_i(t,x)) = \max[\eta(t,x), \min[u(t,x),\mu(t,x)]] \text{ for } i = 1,2. \tag{3.10}$$

It is easy to observe from the hypothesis that $H(t,x,u) \equiv (H_1(t,x,u), H_2(t,x,u))$ is quasimonotone nonincreasing in u on Λ with $v_0 = \eta$, and $w_0 = \mu$. Suppose η, μ are coupled lower and upper solutions of (3.1), then we can prove $((\eta,\eta),(\mu,\mu))$ are coupled lower and upper solutions of the system (3.6) and (3.7). That is we have to prove

$$\mathcal{L}\eta \leq f(t,x,\eta) + g(t,x,\mu) \leq H_1(t,x,\eta,\mu,\eta,\mu) \tag{3.11}$$

$$\mathcal{L}\eta \leq f(t,x,\eta) + g(t,x,\mu) \leq H_2(t,x,\mu,\eta,\eta,\mu) \tag{3.12}$$

$$\mathcal{L}\mu \geq f(t,x,\mu) + g(t,x,\eta) \geq H_1(t,x,\mu,\eta,\eta,\mu) \tag{3.13}$$

$$\mathcal{L}\mu \geq f(t,x,\mu) + g(t,x,\eta) \geq H_2(t,x,\eta,\mu,\eta,\mu). \tag{3.14}$$

It is easy to see that (3.11) and (3.14) follows from (3.8) and (3.9). Here we prove (3.12), and (3.13) follows on the same lines.

To prove (3.12) consider

$$\begin{aligned}f(t,x,\eta) + g(t,x,\mu) &- H_2(t,x,\mu,\eta,\eta,\mu)\\ &= f(t,x,\eta) - [f(t,x,\mu) + f_u(t,x,\eta)(\eta-\mu)]\\ &\quad + g(t,x,\mu) - [g(t,x,\eta) + g_u(t,x,\eta)(\mu-\eta)]\\ &\leq -f_u(t,x,\sigma)(\mu-\eta) + f_u(t,x,\eta)(\mu-\eta),\ \eta < \sigma < \mu\\ &= -f_{uu}(t,x,\sigma_1)(\sigma-\eta)(\mu-\eta) \leq 0,\ \eta < \sigma_1 < \sigma < \mu,\end{aligned}$$

since $f_{uu}(t, x, \sigma_1) \geq 0$ from hypothesis. This proves the system $(3.6) - (3.7)$ satisfies all the hypothesis of Theorem 2.1, with $((\eta, \eta), (\mu, \mu))$ as coupled lower and upper solutions of (3.6), (3.7). This proves $u_1(t, x)$, $u_2(t, x)$ exists on Q_T such that $\eta(t, x) \leq u_1(t, x) \leq \mu(t, x)$ and $\eta(t, x) \leq u_2(t, x) \leq \mu(t, x)$. Further since $H(t, x, u)$ satisfies Lipschitz Condition on Λ, $u_1(t, x)$ and $u_2(t, x)$ are unique. Also there is no loss of generality in replacing $P(u_1)$ and $P(u_2)$ by $u_1(t, x)$ and $u_2(t, x)$ respectively in (3.6), (3.7). Also we prove $u_1(t, x) \leq u_2(t, x)$ and $u_1(t, x)$ and $u_2(t, x)$ are coupled lower and upper solutions of (3.1). Set $\alpha(t, x) = u_1(t, x) - u_2(t, x)$, then

$$
\begin{aligned}
\mathcal{L}\alpha(t, x) &= f(t, x, \eta) + f_u(t, x, \eta)(u_1 - \eta) + g(t, x, \mu) + g_u(t, x, u_2)(u_2 - \mu) \\
&\quad -[f(t, x, \mu) + f_u(t, x, u_2)(u_2 - \eta) + g(t, x, \eta) + g_u(t, x, \eta)(u_1 - \eta)] \\
&\leq f(t, x, u_1) + g(t, x, u_2) - [f(t, x, u_2) + g(t, x, u_1)] \\
&= [f_u(t, x, \xi_1) - g_u(t, x, \xi_2)]\alpha(t, x)
\end{aligned}
$$

$$
\alpha(0, x) \leq 0, \quad B\alpha(t, x) \leq 0.
$$

This proves $\alpha(t, x) \leq 0$ by the comparison theorem in Ladde (1985). Also it is easy to see from (3.3), (3.4) that

$$
\mathcal{L}u_1(t, x) \leq f(t, x, u_1) + g(t, x, u_2), u_1(0, x) = u_0(x), Bu_1(t, x) = \phi(x) \tag{3.15}
$$

$$
\mathcal{L}u_2(t, x) \geq f(t, x, u_2) + g(t, x, u_1), u_2(0, x) = u_0(x), Bu_2(t, x) = \phi(x) \tag{3.16}
$$

Since $\eta(t, x) \leq u_1(t, x) \leq u_2(t, x) \leq \mu(t, x)$.

Now we define the sequences $\{v_n(t, x)\}$, $\{w_n(t, x)\}$ as follows

$$
\mathcal{L}v_n = H_1(t, x, v_n, w_n, v_{n-1}, w_{n-1}), v_n(0, x) = u_0(x), Bv_n(t, x) = \phi(x)
$$

$$
\mathcal{L}w_n = H_2(t, x, v_n, w_n, v_{n-1}, w_{n-1}), w_n(0, x) = u_0(x), Bw_n(t, x) = \phi(x).
$$

If we start with v_0, w_0 as coupled lower and upper solutions of (3.1), then it is easy to see from (3.6), (3.7), (3.15), (3.16) that there exists unique $v_1(t, x)$, $w_1(t, x)$ such that $v_0(t, x) \leq v_1(t, x) \leq w_1(t, x) \leq w_0(t, x)$ and that v_1, w_1 are coupled lower and upper solutions of (3.1). Now using mathematical induction, it is easy to see that

$$
v_0 \leq v_1 \leq \cdots \leq v_n \leq w_n \leq w_{n-1} \leq \cdots \leq w_0.
$$

We can now use known standard arguments as in Ladde (1985), and Pao (1992) to show that $v_n(t, x) \to v(t, x)$, $w_n(t, x) \to w(t, x)$ uniformly on Q_T and satisfy

$$
\mathcal{L}v(t, x) = f(t, x, v) + g(t, x, w), v(0, x) = u_0(x), Bv(t, x) = \phi(x) \tag{3.17}
$$

$$
\mathcal{L}w(t, x) = f(t, x, w) + g(t, x, v), w(0, x) = u_0(x), Bw(t, x) = \phi(x). \tag{3.18}
$$

Since f, g are Lipschitzian on Λ, it is clear that

$$
v(t, x) \equiv w(t, x).
$$

Consequently we have $v(t, x) = w(t, x) = u(t, x)$ is the unique solution of (3.1) to which the sequences $v_n(t, x)$, $w_n(t, x)$ converge.

Finally we show $v_n(t, x)$ and $w_n(t, x)$ converges to u quadratically. To do this let $\alpha_n(t, x) = u - v_n(t, x)$, $\beta_n(t, x) = w_n(t, x) - u$ on Q_T so that $\alpha_n(0, x) = 0$, $B\alpha_n(t, x) = 0$, $\beta_n(0, x) = 0$, $B\beta_n(t, x) = 0$.

Also from hypothesis we can show

$$
\begin{aligned}
\mathcal{L}\alpha_n(t,x) \leq\ & f_u(t,x,v_{n-1})\alpha_n - g_u(t,x,v_{n-1})\beta_n \\
& + f_{uu}(t,x,\xi_1)\alpha_{n-1}^2 - g_{uu}(t,x,\sigma_1)\beta_{n-1}^2
\end{aligned}
\tag{3.19}
$$

where $v_{n-1} < \xi_1 < \xi < u$ and $u < \sigma < \sigma_1 < \beta_{n-1}$.

Similarly,

$$
\begin{aligned}
\mathcal{L}\beta_n(t,x) \leq\ & f_u(t,x,w_{n-1})\alpha_n - g_u(t,x,v_{n-1})\beta_n \\
& + f_{uu}(t,x,z_1)\alpha_{n-1}^2 - g_{uu}(t,x,\zeta_1)\beta_{n-1}^2
\end{aligned}
\tag{3.20}
$$

where $u < \zeta < \zeta_1 < w_{n-1}$, $v_n < z < z_1 < u$.

We can rewrite system (3.19), (3.20) vectorially as

$$
\mathcal{L}r_n \leq Ar_n + Br_{n-1}^2
\tag{3.21}
$$

where $r_n = \begin{pmatrix} u - v_n \\ w_n - u \end{pmatrix}$, $A = \begin{pmatrix} a_1 & a_2 \\ a_2 & a_1 \end{pmatrix}$, $B = \begin{pmatrix} b_1 & b_2 \\ b_2 & b_1 \end{pmatrix}$ and $f_u(t,x,u) \leq a_1$, $-g_u(t,x,v) \leq a_2$, $f_{uu}(t,x,u) \leq b_1$, $-g_{uu}(t,x,v) \leq b_2$ on Q_T. Now using the Comparison Theorem 2.1 and noting that $r_n \geq 0$ and computing the solution of the corresponding ordinary differential equation, we get

$$
0 \leq r_n(t,x) \leq \int_0^t e^{A(t-s)} B \max_{Q_T} r_{n-1}^2(s,x)ds.
$$

This in turn proves

$$
\begin{pmatrix} \max\limits_{Q_T} |u(t,x) - v_n(t,x)| \\ \max\limits_{Q_T} |w_n(t,x) - u(t,x)| \end{pmatrix} \leq A^{-1}e^{AT}B \begin{pmatrix} \max\limits_{Q_T} |u(t,x) - v_{n-1}(t,x)|^2 \\ \max\limits_{Q_T} |w_{n-1}(t,x) - u(t,x)|^2 \end{pmatrix}.
\tag{3.22}
$$

The estimate (3.22) holds provided A^{-1} exists, that is if $a_1 \neq a_2$. If $a_1 = a_2$, there is no loss of generality in assuming $b_1 = b_2$. In this case adding the two inequalities and using the scalar version of Theorem 2.1, we can obtain

$$
\begin{aligned}
& \max_{Q_T}[|u(t,x) - v_n(t,x)| + |w_n(t,x) - u(t,x)|] \\
& \leq \frac{b}{a}e^{2aT}[\max_{Q_T}[|u(t,x) - v_{n-1}(t,x)|^2 + |w_{n-1}(t,x) - u(t,x)|^2].
\end{aligned}
\tag{3.23}
$$

The estimates (3.22) and (3.23) proves the quadratic convergence of the sequences $\{v_n(t,x)\}$, $\{w_n(t,x)\}$. This completes the proof.

References

[1] Bellman, R. (1973). Methods of Nonlinear Analysis, Vol II, Academic Press, New York.

[2] Bellman, R., and Kalaba, R. (1965). Quasilinearization and Nonlinear Boundary Value Problems, American Elsvier, New York.

[3] Chan, C. Y. (January 1974). Positive Solutions for Nonlinear Parabolic Second Initial Boundary Value Problem, Quarterly of Applied Mathematics.

[4] Ladde, G. S., Lakshmikantham, V. and Vatsala, A. S. (1985). Monotone Iterative Techniques for Nonlinear Differential Equations, Pitman, Boston.

[5] Lakshmikantham, V. (1993). An Extension of the Method of Quasilinearization. (to appear in Journal of Optimization, theory and applications.)

[6] Lakshmikantham, V, and Leela, S. (1968). Differential and Integral inequalities, Vol I, II, Academic Press, New York.

[7] Pao, C. V. (1992). Nonlinear Parabolic and Elliptic Equations, Plenum Press, New York.

[8] Walter, W. (1970). Differential and Integral Inequalities, Springer-Verlag, New York.

Geometric Methods in Population Dynamics

M. L. ZEEMAN University of Texas at San Antonio, San Antonio, TX 78249-0664

1 INTRODUCTION

Consider a community of n interacting species. The growth rate of the i-th species is generally viewed as being in some sense "proportional" to its population size x_i; where the proportionality factor, known as the *per capita* growth rate, reflects the interaction of the species. Thus we have a system of ordinary differential equations

$$\dot{x}_i = x_i N_i(x), \qquad i = 1, \ldots, n \tag{1}$$

where the dot denotes differentiation with respect to time; $x_i(t)$ is the population size at time t and the vector $x = (x_1, \ldots, x_n)$ lies in the closed positive cone $\mathbf{R}^n_+$. We denote the interior of the positive come by $\text{Int}\mathbf{R}^n_+$.

For distinct i and j, $\text{sign}(\frac{\partial N_i}{\partial x_j})$ and $\text{sign}(\frac{\partial N_j}{\partial x_i})$ reflect the relationship between the i-th and j-th species. If both quantities are positive, then the growth of each species promotes the growth of the other. That is: they cooperate. If both quantities are negative, the species compete. Finally, if the quantities are of opposite signs, then the two species have a predator-prey relationship. The matrix $DN = (\frac{\partial N_i}{\partial x_j})$ is known as the *community matrix* of the system.

When the *per capita* growth rates N_i are affine, equations (1) form the classical Lotka-Volterra system

$$\dot{x}_i = F_i(x) = x_i(b_i - (Ax)_i), \quad i = 1 \ldots, n \tag{2}$$

which was independently introduced by Lotka and Volterra in the 1920's. Here A is an $n \times n$ matrix. We shall assume that all the entries of A are strictly positive to ensure that system (2) is competitive.

The two-dimensional Lotka-Volterra systems are well understood: If the two species cooperate or compete there are no periodic orbits, and all bounded trajectories of the flow converge to a fixed point. The same results hold when the species have a predator-prey relationship, except for certain degenerate cases when the interior of the positive cone is foliated by concentric periodic orbits surrounding a fixed point. These results are discussed in most elementary texts on ecology, and are based on a geometric analysis of the nullclines of the system (the sets on which one component of the vector field vanishes). For example, see [17, 10, 16, 19].

Less is known about the dynamics of n-dimensional Lotka-Volterra systems for $n > 2$. In three dimensions, there is fairly extensive treatment of the matrix analysis required to determine the dynamics at a fixed point from its linearisation [6, 8, 9, 24], but this type of analysis is notoriously difficult to generalise, and provides only local information. In [29] a high dimensional generalisation of the classical nullcline analysis was developed, and used to partially classify the dynamical behaviour in three-dimensional systems, and to predict the occurrence of Hopf bifurcations, and hence periodic orbits. However, this method cannot provide a complete classification of the periodic behaviour, even in three dimensions.

There has been much work on producing or ruling out periodicity in systems with various kinds of symmetry [1, 11, 12, 13, 15, 18, 23, 25], another direction of study has been to prove survival or extinction results by imposing various inequalities on the parameters [2, 3, 21, 30]. Numerical studies have also been made to investigate the periodicity and even chaos in particular examples [5]. However, we are still far from a general theory with which to predict the long term behaviour of an arbitrary given Lotka Volterra system.

In this paper we survey some of the geometric results that have recently been developed to study competitive Lotka-Volterra systems. In each of §§2-5 we briefly describe a geometric tool, together with some of its applications.

2 MONOTONICITY AND THE CARRYING SIMPLEX

The condition of mutual competition ensures that system (2) is a competitive monotone system, meaning that the standard partial ordering on $\mathbf{R}^n$ is preserved by the time-reversed flow of F. Thus we can apply a theorem of M. W. Hirsch (theorem 1.7 [14]) as follows:

Define the carrying simplex, Σ, of system (2) to be the boundary of the basin of repulsion of the origin. To be precise, we define $R(0) = \{x \in \mathbf{R}^n_+ : \alpha(x) = 0\}$, and $\Sigma = \partial R(0) \setminus R(0)$, where $\alpha(x)$ denotes the alpha limit set of x and $\partial R(0)$ denotes the boundary of $R(0)$ taken in $\mathbf{R}^n_+$. We remove $R(0)$ from $\partial R(0)$ to avoid topological awkwardness at the coordinate planes.

THEOREM 2.1 (Hirsch) Given system (2), every trajectory in $\mathbf{R}^n_+ \setminus \{0\}$ is asymptotic to one in Σ, and Σ is an invariant Lipschitz submanifold homeomorphic to the unit simplex in $\mathbf{R}^n_+$ by radial projection.

This means that the dynamics on $\mathbf{R}^n_+$ of an n-dimensional autonomous competitive Lotka Volterra system are actually determined by the dynamics on the carrying simplex, an invariant surface of co-dimension one.

Further analysis of the carrying simplex leads to some interesting questions. For example, Hirsch's theorem guarantees that Σ is at least Lipschitz, while Brunovski [7] and Mierczynski [20] have given conditions under which Σ is at least C^1. Whether the carrying simplex is always C^1 for a competitive system remains an open question to date. However, example 3.1 in §3 shows that a weakening of the competition hypothesis in theorem 2.1 can lead to a carrying simplex which is only Lipschitz, or worse.

In [31] the concept of the carrying simplex is generalised to nonautonomous competitive Lotka-Volterra systems as follows. Assume for the moment that the coefficients in system (2) are continuous functions of t, bounded above and below by strictly positive constants. To fix notation, let $a_{ij}^u, b_i^u, a_{ij}^l, b_i^l$ be the least upper bounds and greatest lower bounds of $a_{ij}(t), b_i(t)$ respectively. We define the *upper system* of our nonautonomous system (2) by

$$\dot{x}_i = x_i \left(b_i^u - \sum_{j=1}^n a_{ij}^l x_j \right), \qquad i = 1, \ldots, n$$

and the *lower system* by

$$\dot{x}_i = x_i \left(b_i^l - \sum_{j=1}^n a_{ij}^u x_j \right), \qquad i = 1, \ldots, n.$$

Each of the upper and lower systems are autonomous, and hence have carrying simplices Σ^u and Σ^l by theorem 2.1. We define the *thickened carrying simplex* of nonautonomous system (2) to be the region in $\mathbf{R}_+^n$ between Σ^u and Σ^l, and prove in [31] that:

THEOREM 2.2 The thickened carrying simplex is a compact globally attracting set for nonautonomous system (2).

3   NULLCLINE ANALYSIS

One way to geometrically capture the algebraic simplicity of the autonomous Lotka-Volterra equations is by the nullclines of the system. The i-th *nullcline* of system (2) is the set in $\mathbf{R}_+^n$ on which $\dot{x}_i = 0$. It is given by $\{x_i = 0\} \cup N_i$ where N_i is the hyperplane $b_i = \sum_{j=1}^n a_{ij} x_j$, which has positive axial intercepts $\frac{b_i}{a_{ij}}$.

In two dimensions, the nullclines are lines, and the classical two dimensional results are obtained by a geometric analysis of the configuration of these lines.

In [29] a high dimensional generalisation of the classical nullcline analysis was developed, and used in conjunction with the carrying simplex to partially classify the dynamical behaviour in three-dimensional systems, and to predict the occurrence of Hopf bifurcations, and hence periodic orbits.

Here, we digress for a moment to discuss an example which illustrates the relationship between the nullclines and the carrying simplex in the two-dimensional case, and how a weakening of the competition hypothesis in theorem 2.1 can lead to a carrying simplex which is only Lipschitz, or worse.

EXAMPLE 3.1 Consider the two-dimensional Lotka-Volterra system

$$\begin{aligned} \dot{x} &= x((2+\delta) - 2x - \delta y) \\ \dot{y} &= y((2+\epsilon) - \epsilon x - 2y) \end{aligned} \tag{3}$$

which has an attracting fixed point at $P = (1,1)$ for every value of δ, ϵ such that $\delta\epsilon < 4$.

When $\delta, \epsilon > 0$, system (3) is competitive, the N_i have strictly positive normals, and Σ is a C^1 curve, tangent to the weak stable eigenspace of DF_P at P. See figure 1(a).

When $\delta, \epsilon = 0$, the nullclines are parallel to the coordinate axes. This forces Σ to coincide with the nullclines, and hence it is Lipschitz, but not C^1. See figure 1(b).

When $\delta > 0, \epsilon < 0$ system (3) is a predator-prey system, and DF_P has a complex conjugate pair of eigenvalues. Hence trajectories spiral in towards P, so that Σ (the boundary of the basin of repulsion of the origin) has few, if any, of the properties guaranteed by Hirsch's theorem in the competitive case. In particular, it is neither a Lipschitz submanifold, nor homeomorphic to the unit simplex under radial projection. See figure 1(c).

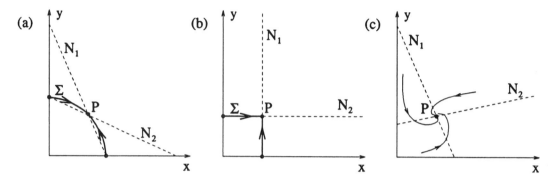

Figure 1: The nullclines and carrying simplex, Σ of system (3): (a) when $\delta, \epsilon > 0$; (b) when $\delta, \epsilon = 0$, and (c) when $\delta > 0, \epsilon < 0$.

We now return to competitive Lotka-Volterra systems of arbitrary finite dimension.

As the dimension of the system increases, so does the amount of dynamical information that cannot be deduced from the nullcline analysis. However, in all dimensions this approach can be useful for proving extinction results, by inductively applying the following lemma:

LEMMA 3.2 Given system (2), if there exist i, j such that N_i lies entirely above N_j, then species x_j is driven to extinction.

Here, we say N_i lies entirely above N_j if N_i is contained in the unbounded component of $\mathbf{R}_+^n \setminus N_j$. See [21] for a proof.

From this lemma, simple algebraic criteria can be deduced that guarantee the extinction of any number of particular species within the system. Moreover, lemma 3.2, and hence the associated extinction results, also hold for nonautonomous systems as follows.

Once again, assume for the moment that the coefficients in system (2) are continuous functions of t, bounded above and below by strictly positive constants. With notation as in §2, for $i = 1, \ldots, n$ let N_i^u, N_i^l be the i-th nullclines of the upper and lower systems respectively. Now define the i-th *thickened nullcline*, denoted N_i, to be the region in $\mathbf{R}_+^n$ between N_i^u and N_i^l. Then lemma 3.2 holds as stated for the thickened nullclines of nonautonomous system (2).

4   SPLIT LIAPUNOV FUNCTIONS

This and the next section describe tools that are designed to capture the algebraic simplicity of the autonomous equations in new ways, in order to deduce some of the dynamical information that nullcline analysis cannot provide as the dimension of the system increases.

Consider the case when system (2) has a fixed point at $P \in \text{Int}\mathbf{R}_+^n$. Let H be any hyperplane through P, let h be a normal to H, and define $\alpha = hA^{-1}$, so

$$h = \alpha A = \sum_1^n \alpha_i A_i,$$

where A_i is the i-th row of A. Now define

$$V = \prod_1^n x_i^{\alpha_i}$$

Then the proof of the following lemma (see [28]) is straightforward after system (2) has been re-written as

$$\dot{x}_i = x_i A_i \cdot (P - x)$$

LEMMA 4.1

$$
\begin{aligned}
\dot{V}(x) &> 0, \text{ when } h \cdot (P - x) > 0 \\
\dot{V}(x) &= 0, \text{ when } h \cdot (P - x) = 0 \\
\dot{V}(x) &< 0, \text{ when } h \cdot (P - x) < 0
\end{aligned}
$$

In other words, V is a *split Liapunov function*: monotone increasing along orbits to one side of H, and monotone decreasing along orbits to the other side of H. Therefore, if we know that an orbit lies entirely to one side of H we can apply V as a Liapunov-like function to deduce that the orbit is non-recurrent. Here, we call an orbit *recurrent* if each point in the orbit belongs to its own ω-limit set. For example, a periodic orbit is recurrent.

Applying lemma 4.1 to the case when H is the tangent hyperplane to the carrying simplex at P yields:

THEOREM 4.2 If the carrying simplex lies all to one side of H then there is no non-trivial recurrence, and the global dynamics in $\text{Int}\mathbf{R}_+^n$ are determined by the local dynamics at P.

DEFINITION 4.3 We call a C^1 hypersurface M in R^n convex if, for all $m \in M$, the surface M lies entirely in one of the closed half spaces determined by the hyperplane T tangent to M at m. We call M strictly convex if, for all $m \in M$, $T \cap M = m$.

COROLLARY 4.4 If the carrying simplex is strictly convex then there is no non-trivial recurrence, and the global dynamics in $\text{Int}\mathbf{R}_+^n$ are determined by the local dynamics at P.

COROLLARY 4.5 Any periodic (or recurrent) orbit of system (1) must intersect every hyperplane through P.

One interesting aspect of these results is that they relate the global geometric structure of the carrying simplex Σ to the global dynamics of the system. The disadvantage is that global geometric information about Σ is hard to obtain.

§5 describes a tool that deduces global geometric information about Σ from local information about $\dot{x}$ at P and on $\partial\Sigma$.

5   CONVEXITY AND THE PINK LEMMA

Once again, consider the case when system (2) has a fixed point at $P \in \text{IntR}_+^n$. Let H be a hyperplane in $\mathbf{R}^n$ with unit normal h, and define $Q = F \cdot h \mid_H$ (the restriction to H of the component $F \cdot h$ of F normal to H).

PINK LEMMA 5.1 Q is quadratic.

The proof of the pink lemma [1] is immediate, as Q is the composition of a linear map (projection) with the quadratic map F. In the two-dimensional case, Q is a quadratic function of one variable, and can be used (see [27]) to prove:

THEOREM 5.2 If system (2) is two-dimensional, then the carrying simplex is convex.

In the n-dimensional case, the pink lemma can be applied to the case when H is the tangent hyperplane to Σ at P, and used in conjunction with the split Liapunov function of §4 to relax the global hypothesis of theorem 4.2 to hypotheses about Q at P and about $\dot{x}$ on $\partial\Sigma$. See [28] for a proof.

THEOREM 5.3 If P repels on Σ, the boundary $\partial\Sigma$ attracts on Σ, and Q is definite, then Σ lies all to one side of H, so the basin of repulsion of P is the interior of Σ.

As stated, theorem 5.3 is an extinction result. An analogous result holds for the case when P attracts on Σ and $\partial\Sigma$ repels on Σ, provided that we add a mild hypothesis about the relative strengths of the eigenvalues at P. Then we can conclude that P is globally attracting, and all strictly positive initial conditions lead to stable coexistence of all n species.

---

[1] So named because it was first proved on a tiny piece of pink paper

QUESTIONS. The tools described here, and particularly those of the last two sections have by no means been fully explored, or even fully developed. There are many questions raised. For example, so far these tools have been used mainly to rule out recurrence in high dimensional systems. Can they be used to study the case when we do have recurrence? The pink lemma was very easy to use in the two-dimensional case, but is harder to work with in higher dimensions. Can the idea be simplified further to make it more useful in higher dimensions? Can we tie more of these results together by relating geometric properties of Q to the nullcline analysis?

REFERENCES

[1] Ahmad, S. (1988) On Almost Periodic Solutions of the Competing Species Problems. *Proceedings of the American Mathematical Society,* **102**: vol. 4.

[2] Ahmad, S. On the Nonautonomous Volterra-Lotka Competition Equations. *Proceedings of the American Mathematical Society,* To appear.

[3] Ahmad, S. and Lazer, A. C. One Species Extinction in an Autonomous Competition Model. *Proceedings of the World Congress on Nonlinear Analysis,* To appear.

[4] Ahmad, S. and Lazer, A. C. On the Nonautonomous N-Competing Species Problem. *Journal of Mathematical Analysis and Applications,* To appear.

[5] Arneodo, A., Coullet, P., Peyraud, J. and Tresser, C. (1982) Strange Attractors in Volterra Equations for Species in Competition. *Journal of Mathematical Biology* **14**, 153–157.

[6] Bahl, C. A. and Cain, B. E. (1977) The Inertia of Diagonal Multiples of 3×3 Real Matrices. *Linear Algebra and its Applications* **18**, 267–280.

[7] Brunovski, P. (1992) Controlling Non-Uniqueness of Local Invariant Manifolds. *Preprint.*

[8] Clark, C. E. and Hallam, T. G. (1982) The Community Matrix in Three Species Community Models. *Journal of Mathematical Biology* **16**, 25–31.

[9] Cross, G. W. (1978) Three Types of Matrix Stability. *Linear Algebra and its Applications* **20**, 253–263.

[10] Freedman, H. I. (1980) *Deterministic Mathematical Models in Population Ecology.* (Marcel Dekker Inc., New York).

[11] Gilpin, M. E. (1975) Limit Cycles in Competition Communities. *American Naturalist* **109**, 51–60.

[12] Goh, B. S. (1977) Global Stability in Many-Species Interactions. *American Naturalist* **111**, 135–143.

[13] Gopalsamy, K. (1985) Globally Asymptotic Stability in a Periodic Lotka-Volterra System. *Journal of Mathematical Analysis and Applications*, **159**: 44–50.

[14] Hirsch, M. W. (1988) Systems of Differential Equations that are Competitive or Cooperative. III: Competing Species. *Nonlinearity*, **1**: 51–71.

[15] Hofbauer, J. (1981) On the Occurrence of Limit Cycles in the Volterra-Lotka equation. *Nonlinear Analysis, Theory, Methods and Applications* **5**, 1003–1007.

[16] Hofbauer, J. and Sigmund, K. (1988) *The Theory of Evolution and Dynamical Systems*. Cambridge University Press, Cambridge.

[17] Lotka, A. J. (1956) *Elements of Mathematical Biology*. (Dover Publications, Inc., New York).

[18] MacArthur, R. (1970) Species Packing and Competitive Equilibrium for Many Species. *Theoretical Population Biology* **1**, 1–11.

[19] May, R. M. (1975) *Stability and Complexity in Model Ecosystems*. Princeton University Press, Princeton.

[20] Mierczynski, J. (1993) The C^1 Property of Carrying Simplices for a Class of Competitive Systems of Ordinary Differential Equations. *Journal of Differential Equations*, To appear.

[21] Montes de Oca, F. and Zeeman, M. L. Extinction in Nonautonomous Competitive Lotka-Volterra Systems. To appear.

[22] Montes de Oca, F. and Zeeman, M. L. Balancing Survival and Extinction in Nonautonomous Competitive Lotka-Volterra Systems. In preparation.

[23] Resigno, A. (1968) The Struggle for Life II: Three Competitors. *Bulletin of Mathematical Biophysics* **30**, 291–297.

[24] Strobeck, C. (1973) N-Species Competition. *Ecology* **54**, 650–654.

[25] Tineo, A. and Alvarez, C. (1991) A Different Consideration about the Globally Asymptotically Stable Solution of the Periodic n-Competing Species Problem. *Journal of Mathematical Analysis and Applications*, **159**: 44–50.

[26] Tineo, A. (1992) On the Asymptotic Behaviour of some Population Models. *Journal of Mathematical Analysis and Applications*, **167**: 516–529.

[27] Zeeman, E. C. and Zeeman, M. L. (1993) On the convexity of carrying simplices in competitive Lotka-Volterra systems. *Differential Equations, Dynamical Systems and Control Science*. Marcel Dekker, Inc., New York.

[28] Zeeman, E. C. and Zeeman, M. L. Ruling out recurrence in competitive Lotka-Volterra systems. *To appear.*

[29] Zeeman, M. L. (1993) Hopf Bifurcations in Competitive Three-Dimensional Lotka-Volterra Systems. *Dynamics and Stability of Systems*, **8**: 189–217.

[30] Zeeman, M. L. Extinction in Competitive Lotka-Volterra Systems. *Proceedings of the American Mathematical Society,* To appear.

[31] Zeeman, M. L. Thickened Carrying Simplices in Nonautonomous Competitive Lotka-Volterra Systems. To appear.

Uniform Asymptotic Stability in Functional Differential Equations with Infinite Delay

Bo Zhang Department of Mathematics and Computer Science
Fayetteville State University, Fayetteville, NC 28301

1 INTRODUCTION

In this paper we consider a system of functional
differential equations with infinite delay

$$x'(t) = F(t, x_t), \quad x \in R^n, \tag{1}$$

and obtain a Liapunov-type stability theorem. Our work
provides a unified approach to the stability theory for
infinite delay systems.

Let $R = (-\infty, +\infty)$, $R^+ = [0, +\infty)$, $R^- = (-\infty, 0]$. $|\cdot|$ denotes
the Euclidean norm on R^n. Let BC be the space of bounded
continuous functions $\phi: R^- \to R^n$ with the supremum norm $\|\cdot\|$.
If x is a continuous function defined on $(-\infty, A)$ with $A \in R$,

and if $t \in (-\infty, A)$ is a fixed number, then $x_t: R^- \to R^n$ is defined by $x_t(s) = x(t+s)$ for all $s \in R^-$. Define $C_H = \{ \phi \in BC: \|\phi\| < H \}$ for some $0 < H \leq +\infty$. We assume that $F: R \times C_H \to R^n$ with $F(t,0) = 0$.

In (1), $x'(t)$ denotes the right-hand derivative of x at t. A function $x: (-\infty, A) \to R^n$ is called a solution of (1) through $(\sigma, \phi) \in (-\infty, A) \times C_H$ if x is differentiable and satisfies (1) on $[\sigma, A)$ with $x_\sigma = \phi$. We denote by $x(\sigma, \phi)$ a solution of (1) with $x_\sigma = \phi$. The value of $x(\sigma, \phi)$ at t will be $x(t) = x(t, \sigma, \phi)$. The remainder of our conditions on F will be indirect, but we cite results ensuring their fulfillment. The basic assumption is that

> for each $(\sigma, \phi) \in R \times C_H$ there is a solution   (2)
> $x(\sigma, \phi)$ of (1) defined on $[\sigma, +\infty)$.

Sawano [14] asks that

(H_1) if $x: (-\infty, A) \to R^n$ is bounded continuous, then $F(t, x_t)$
        is measurable in $t \in (-\infty, A)$,

(H_2) for any bounded set V in BC there exists a function
        $m(t) = m_V(t)$ locally integrable on R such that
        $|F(t, \phi)| \leq m(t)$ for every $\phi \in V$,

(H_3) $F(t, \phi)$ is continuous in ϕ for each $t \in R$.

He then shows that for each $(\sigma, \phi) \in R \times C_H$, (1) has a solution $x = x(\sigma, \phi)$ of Carthéodory type defined on $0 \leq t < \alpha$ for some $\alpha > 0$ and if $x(\sigma, \phi)$ is noncontinuable beyond α, then

$$\lim_{t \to \alpha^-} \sup |x(t, \sigma, \phi)| = +\infty.$$

It is clear that if $F(t, x_t)$ is continuous in $t \in (-\infty, A)$ whenever $x: (-\infty, A) \to R^n$ is bounded and continuous, then the solution $x(\sigma,\phi)$ is continuously differentiable on $[\sigma,A)$.

Let $V(t,\phi)$ be a continuous functional defined on $R \times C_H$. The upper right-hand derivative of V along solutions of (1) is defined by

$$V_{(1)}{}'(t,\phi) = \lim_{\delta \to 0^+} \sup \{V(t+\delta,x_{t+\delta}(\sigma,\phi)) - V(t,\phi)\}/\delta.$$

DEFINITION 1 The zero solution of (1) is uniformly stable (US) if for each $\epsilon > 0$ there exists $\delta > 0$ such that $[(t_0,\phi) \in R \times C_H, \|\phi\| < \delta, t \geq t_0]$ imply that $|x(t,t_0,\phi)| < \epsilon$.

DEFINITION 2. The zero solution of (1) is uniformly asymptotically stable (UAS) if it is uniformly stable and there exists $\delta > 0$ such that for each $\epsilon > 0$ there exists $T > 0$ such that $[(t_0,\phi) \in R \times C_H, \|\phi\| < \delta, t \geq t_0 + T]$ imply $|x(t,t_0,\phi)| < \epsilon$.

DEFINITION 3 $W: R^+ \to R^+$ is called a wedge if W is continuous and strictly increasing with $W(0) = 0$. Throughout the paper W, W_j ($j = 0,1,2,\cdots$) will denote wedges.

DEFINITION 4 A continuous function $G: R^+ \to R^+$ is convex downward if $G[(t+s)/2] \leq [G(t) + G(s)]/2$ for all $t, s \in R^+$.

Jensen's inequality will be used in this paper. For reference, we refer to [1] and [13].

Jensen's Inequality. Let W be convex downward and let $f,p: [a,b] \to R^+$ be continuous with $\int_a^b p(s)ds > 0$, then

$$\int_a^b p(s)ds \; W\left[\int_a^b p(s)f(s)ds / \int_a^b p(s)ds\right] \leq \int_a^b p(s)W(f(s))ds.$$

LEMMA 1. Let W_1 be a wedge. Then for any $L>0$ there exists a convex downward wedge W_0 such that

$$W_0(r) \leq W_1(r) \text{ for all } r \in [0,L].$$

In fact, $W_0(r) = \int_0^r W_1(s)ds/L$ will suffice. For any continuous function $\phi: [a,b] \to R^n$, we define

$$\|\phi\|[a,b] = \sup \{|\phi(s)|: a \leq s \leq b\}.$$

In order to put the problem into its historical context we consider the ordinary differential equation

$$x'(t) = f(t,x(t)) \tag{3}$$

where $f: R \times R^n \to R^n$ is continuous. The following result is well-known (see [3], p. 261).

THEOREM A. Let $V: R \times R^n \to R^+$ be continuous such that

(i) $W_1(|x|) \leq V(t,x) \leq W_2(|x|)$,

(ii) $V_{(3)}'(t,x) \leq - W_3(|x|)$.

Then the zero solution of (3) is uniformly asymptotically stable.

Extending Theorem A to functional differential equations has been the subject of extensive investigations for many years. For results on equations with finite delay, we refer to Burton [2], Hale [6], Krasovskii [12], Yoshizawa [16], and Zhang [17]. In [4], Burton and Zhang extended Theorem A to (1) for unbounded delay. Their result may be stated as follows.

THEOREM B. ([4]) Suppose that there exists a continuous functional $V: R \times BC \to R^+$ and $\Phi: R^+ \to R^+$ with $\Phi \in L^1[0,+\infty)$

such that

(i) $W_1(|\phi(0)|) \leq V(t,\phi) \leq W_2(|\phi(0)|) + W_3\left[\int_{-\infty}^{0} \Phi(-s)W_4(|\phi(s)|)ds\right]$,

(ii) $V_{(1)}'(t,\phi) \leq -W_5(|\phi(0)|)$.

Then the zero solution of (1) is UAS.

Hering [7] generalized Theorem A to (1) on the space C_g which generated much interest (see [3] and [5]). We recall some definitions.

$G = \{$ g: $R^- \rightarrow [1,+\infty)$: g is continuous, non-increasing,

$$g(0)=1, \text{ and } g(r) \rightarrow +\infty \text{ as } r \rightarrow -\infty \}.$$

For any $g \in G$, we define

$C_g = \{ \phi: R^- \rightarrow R^n: \phi \text{ is continuous and } |\phi|_g < +\infty \}$ (4)

where $|\phi|_g = \sup_{s \leq 0} |\phi(s)|/g(s)$. Then $(C_g, |\cdot|_g)$ is a Banach

space. The following result may be found in the recent work of Hering [7-9].

THEOREM C.   Suppose that there exists a continuous functional V: $R \times C_g \rightarrow R^+$, $g^o \in G$ with $g<g^o$, and a positive constant $\gamma>0$ such that

(i) $W_1(|\phi(0)|) \leq V(t,\phi) \leq W_2(|\phi(0)|) + W_3(|\phi|_{g^o})$,

(ii) $V_{(1)}'(t,\phi) \leq - W_4(|\phi(0)|)$,

(iii) $W_1(r) - W_3(r) > 0$ for $r \in (0,\gamma)$.

Then the zero solution of (1) is g-UAS.

Finally, we recall the recent counter example of Kato [11] which suggests that conditions

$W_1(|\phi(0)|) \leq V(t,\phi) \leq W_2(\|\phi\|)$, $V_{(1)}'(t,\phi) \leq - W_3(|\phi(0)|)$

will not guarantee the uniform asymptotic stability and some extra condition on V must be added.   In this paper, we will

focus on the space BC and use the ideas from [4] and [7] to
fully extend Theorem B. The Liapunov functional obtained
here has a large upper bound and can be applied to highly
perturbed systems.

2.   The Main Result.

When (1) has a unbounded delay, an example of Seifert [15]
shows that if UAS is expected, then (1) must have some
type of fading memory. It is also believed that in order
to prove that the zero solution of (1) is UAS or solutions of
(1) are uniformly ultimately bounded using a Liapunov
functional V, then the upper bound of V must have a fading
memory with respect to the norm on the space of initial
functions (see [4], [8], and [18]). We introduce the
following definition.

DEFINITION 5. A semi-norm $|\cdot|_B$ on BC is said to have a
fading memory with respect to $\|\cdot\|$ if $|\phi|_B \leq \|\phi\|$ for all $\phi \in$
BC and if for each $\epsilon > 0$ and $D > 0$ there exists a $h > 0$ such that

$$|\phi|_B \leq \max [\ \|\phi(\cdot)\|^{[-\sigma,0]}, \ \epsilon \ ]$$

whenever $\sigma \geq h$ and $\|\phi(\cdot)\|^{(-\infty,-\sigma]} \leq D$.

EXAMPLE 1. $|\cdot|_g$ has a fading memory with respect to $\|\cdot\|$.

In fact, if $\phi \in$ BC, then $|\phi|_g = \sup\limits_{s \leq 0} |\phi(s)|/g(s) \leq \|\phi\|$.
Moreover, for any $\epsilon > 0$ and $D > 0$ there exists a constant $h > 0$
such that $D < \epsilon g(-h)$. Thus, if $\sigma \geq h$ and $\|\phi(\cdot)\|^{(-\infty,-\sigma]} \leq D$,
then

$$|\phi|_g = \max \left[\sup_{-\sigma \leq s \leq 0} \frac{|\phi(s)|}{g(s)}, \quad \sup_{s \leq -\sigma} \frac{|\phi(s)|}{g(s)} \right]$$

$$\leq \max \left[\|\phi(\cdot)\|^{[-\sigma,0]}, \frac{D}{g(-h)} \right] \leq \max \left[\|\phi(\cdot)\|^{[-\sigma,0]}, \epsilon \right].$$

Thus $|\cdot|_g$ has a fading memory with respect to $\|\cdot\|$.

EXAMPLE 2. Let $\alpha: R^- \to R^+$ be continuous with $\int_{-\infty}^{0} \alpha(s)ds \leq \frac{1}{2}$.

Then $|\cdot|_\alpha$ introduced in [10], $|\phi|_\alpha = \int_{-\infty}^{0} \alpha(s)\|\phi(\cdot)\|^{[s,0]}ds$,

has a fading memory with respect to $\|\cdot\|$.

It is clear that $|\phi|_\alpha \leq \|\phi\|$ for all $\phi \in BC$. For any $\epsilon > 0$

and $D > 0$, there exists $h > 0$ such that $2D\int_{-\infty}^{-h} \alpha(s)ds < \epsilon$. If $\sigma \geq h$

and $\|\phi(\cdot)\|^{(-\infty,-\sigma]} \leq D$, then

$$|\phi|_\alpha = \int_{-\infty}^{0} \alpha(s)\|\phi(\cdot)\|^{[s,0]}ds$$

$$= \int_{-\sigma}^{0} \alpha(s)\|\phi(\cdot)\|^{[s,0]}ds + \int_{-\infty}^{-\sigma} \alpha(s)\|\phi(\cdot)\|^{[s,0]}ds$$

$$\leq \int_{-\sigma}^{0} \alpha(s)\|\phi(\cdot)\|^{[s,0]}ds$$

$$+ \max\left[\|\phi(\cdot)\|^{[-\infty,-\sigma]}, \|\phi(\cdot)\|^{[-\sigma,0]}\right] \int_{-\infty}^{-\sigma} \alpha(s)ds$$

$$\leq \int_{-\sigma}^{0} \alpha(s)ds \, \|\phi(\cdot)\|^{[-\sigma,0]} + \max\left[D, \|\phi(\cdot)\|^{[\sigma,0]}\right] \int_{-\infty}^{-\sigma} \alpha(s)ds$$

$$\leq \max \left[\|\phi(\cdot)\|^{[-\sigma,0]}, \epsilon \right].$$

Thus, $|\cdot|_\alpha$ has a fading memory with respect to $\|\cdot\|$.

THEOREM 1. Suppose that there exists a continuous

functional $V: R \times C_H \to R^+$, a semi-norm $|\cdot|_B$ having a fading memory with respect to $\|\cdot\|$, a continuous function $\Phi: R^+ \to R^+$ with $\Phi \in L^1[0,+\infty)$, and a positive constant γ such that

(i) $W_1(|\phi(0)|) \leq V(t,\phi)$

$$\leq W_2\left[|\phi(0)| + \int_{-\infty}^{0} \Phi(-s)W(|\phi(s)|)ds\right] + W_3(|\phi|_B),$$

(ii) $V_{(1)}'(t,\phi) \leq -W_4(|\phi(0)|)$,

(iii) $W_1(r) - W_3(r) > 0$ for all $r \in (0,\gamma)$.

Then the zero solution of (1) is UAS.

Proof.   Let $J = \int_{0}^{+\infty} \Phi(s)ds$.   For any $(t,\phi) \in R \times C_H$, we have

$$V(t,\phi) \leq W_2\left[|\phi(0)| + \int_{-\infty}^{0} \Phi(-s)W(|\phi(s)|)ds\right] + W_3(|\phi|_B)$$

$$\leq W_2\left[|\phi(0)| + W(\|\phi\|)J\right] + W_3(\|\phi\|) \leq W_2^*(\|\phi\|)$$

for some wedge W_2^*.   For any $\epsilon > 0$, there exists $\delta > 0$ $(\delta < H)$ such that $W_2^*(\delta) < W_1(\epsilon)$.   Let $x(t_0,\phi)$ be a solution of (1) with $\|\phi\| < \delta$. Then for $t \geq t_0$ we have

$W_1(|x(t)|) \leq V(t,x_t) \leq V(t_0,\phi) \leq W_2^*(\|\phi\|) < W_2^*(\delta) < W_1(\epsilon)$.
This implies that $|x(t)| < \epsilon$ for $t \geq t_0$ and the zero solution of (1) is US.

Next find $\delta > 0$ of uniform stability for $\epsilon = \min\{1,\gamma\}$, where γ is given in (iii).   Without loss of generality, we assume that $1 < \gamma$.   To complete the proof, we must show that for each $\epsilon > 0$ there exists $T > 0$ such that $[(t_0,\phi) \in R \times C_H,$ $\|\phi\| \leq \delta,$ $t \geq t_0 + T]$ imply $|x(t,t_0,\phi)| < \epsilon$.

Let $\epsilon > 0$ be given and find a constant M with $0 < M < 1$ such that $W_2(3M) + W_3(M) < W_1(\epsilon)$.   By condition (iii) there exists $\sigma > 0$ such that $0 < \sigma < M$ and $W_1(r) - W_3(r) \geq \sigma + W_2(3\sigma)$ for $r \in [M,1]$.

Since W_1 is uniformly continuous on $[\sigma, 1]$, there exists constant m such that $0 < m < M - \sigma$ and

$$W_1(r) - W_1(r-m) < \sigma \text{ for all } [M, 1]. \tag{5}$$

This implies that

$$W_1(r-m) - W_3(r) > W_1(r) - \sigma - W_3(r) \tag{6}$$

$$\geq \sigma + W_2(3\sigma) - \sigma = W_2(3\sigma) \text{ for } r \in [M, 1].$$

Since $|\cdot|_B$ has a fading memory with respect to $\|\cdot\|$, for $D = 1 + \delta$ and $\sigma > 0$ defined above there exists $h > 0$ such that

$$|\psi|_B \leq \max [\|\psi\|^{[-h,0]}, \sigma]$$

whenever $\psi \in BC$ and $\|\psi\|^{(-\infty, -h]} \leq D$. We choose $h > 0$ so large that

$$W(D) \int_{-\infty}^{-h} \Phi(-s) ds < \sigma.$$

Let $x(t) = x(t, t_0, \phi)$ with $\|\phi\| \leq \delta$. It follows that $\|x_t\| \leq D$ and $|x_t|_B \leq \max [\|x\|^{[t-h,t]}, \sigma]$ for any $t \in R$. Moreover, for any $\tau \geq t_0$ and $t \geq \tau$ we have

$$V(t, x_t) \leq V(\tau, x_\tau) - \int_\tau^t W_4(|x(s)|) ds \leq W_2^*(D) - \int_\tau^t W_4(|x(s)|) ds.$$

This implies that there exists a constant $L > 0$ depending on D such that for each $\tau \geq t_0$ there exists $t^* \in [\tau, \tau + L]$ with $|x(t^*)| < \sigma$. Consequently, we can find a sequence $\{t_n\}$ such that

$$t_{n-1} + h \leq t_n \leq t_{n-1} + h + L, \text{ and} \tag{7}$$

$$|x(t_n)| < \sigma \text{ for } n = 1, 2, \cdots.$$

For any $t \geq t_0 + h$, we have

$$V(t,x_t) \leq W_2\Big[|x(t)| + \int_{t-h}^{t} \Phi(t-s)W(|x(s)|)ds$$

$$+ \int_{-\infty}^{t-h} \Phi(t-s)W(|x(s)|)ds\Big] + W_3\Big[\max[\|x\|^{[t-h,t]}, \sigma] \Big]$$

$$\leq W_2\Big[|x(t)| + \int_{t-h}^{t} \Phi(t-s)W(|x(s)|)ds + W(D)\int_{-\infty}^{-h} \Phi(-s)ds\Big]$$

$$+ \max \Big[W_3(\|x\|^{[t-h,t]}), W_3(\sigma) \Big]$$

$$\leq W_2\Big[|x(t)| + \int_{t-h}^{t} \Phi(t-s)W(|x(s)|)ds + \sigma \Big]$$

$$+ \max \Big[W_3(\|x\|^{[t-h,t]}), W_3(\sigma) \Big].$$

Thus,

$$V(t_n,x_{t_n}) \leq W_2\Big[2\sigma + \int_{t_n-h}^{t_n} \Phi(t_n-s)W(|x(s)|)ds \Big]$$

$$+ \max \Big[W_3(\|x\|^{[t_n-h,t_n]}), W_3(\sigma) \Big].$$

Notice also that $|x(t)| \leq 1$ for all $t \geq t_0$. By Lemma 1, there exists a convex downward wedge W_5 such that $W_5(r) \leq W_4(W^{-1}(r))$ for $0 \leq r \leq W(1)$. Thus, for $t \geq t_0$ we have

$$V_{(1)}'(t,x_t) \leq - W_4(|x(t)|) = - W_5\Big[W(|x(t)|)\Big]. \qquad (8)$$

Let $Q = \sup \{\Phi(s): 0 \leq s \leq h\}$ and $K > 0$ be an integer such that

$$W_2^*(D) - (K-1)hW_5(\sigma/hQ) < 0.$$

For any integer n and $t \geq t_{n+K}$, integrate (8) from t_n to t and use Jensen's inequality to obtain

$$V(t,x_t) \leq V(t_n,x_{t_n}) - \int_{t_n}^{t} W_5\Big[W(|x(s)|)\Big]ds$$

$$\leq W_2^*(D) \ - \sum_{j=n+1}^{n+K} \int_{t_j-h}^{t_j} W_5\Big[W(|x(s)|)\Big]ds \leq W_2^*(D)$$

$$-\sum_{j=n+1}^{n+K} hW_5\Bigg[\int_{t_j-h}^{t_j} \frac{1}{h}W(|x(s)|)ds\Bigg].$$

We now claim that there exists an integer i, n+1≤i≤n+K, such that

$$Q \int_{t_i-h}^{t_i} W(|x(s)|)ds < \sigma. \tag{9}$$

In fact, suppose $Q \int_{t_j-h}^{t_j} W(|x(s)|)ds \geq \sigma$ for all j, n+1≤j≤n+K. Then

$$V(t,x_t) \leq W_2^*(D)-\sum_{j=n+1}^{n+K} hW_5(\sigma/hQ) \leq W_2^*(D) - (K-1)hW_5(\sigma/hQ) < 0,$$

a contradiction. Thus (9) holds. This together with (7) yields that there exists a subsequence $\{s_n\}$ of $\{t_n\}$ such that

$$Q \int_{s_n-h}^{s_n} W(|x(s)|)ds < \sigma. \tag{10}$$

Moreover, $s_{n-1} + h \leq s_n \leq s_{n-1} + h + KL$ for n =1,2,···, where $s_o = t_o$. Thus,

$$V(s_n,x_{s_n}) \leq W_2(3\sigma) + \max\Big[\ W_3(\|x\|^{[s_n-h,s_n]}),\ W_3(\sigma)\ \Big].$$

Let $I_j = [s_j-h,\ s_j]$. On each I_j we have either

(A) $\|x\|^{[s_j-h,s_j]} \leq M$ or

(B) $|x(\tau_j)| > M$ for some $\tau_j \in I_j$.

If (A) holds, then for $t \geq s_j$ we have

$$W_1(|x(t)|) \leq V(t,x_t) \leq V(s_j,x_{s_j}) \leq W_2(3\sigma) + W_3(M)$$

$$\leq W_2(3M) + W_3(M) < W_1(\epsilon).$$

This implies that $|x(t)| < \epsilon$ for $t \geq s_j$. Now suppose that (B) holds. Let $M_j = \|x(\cdot)\|^{[s_j-h, s_j]}$. We will show that $|x(t)| < M_j - m$ for all $t \geq s_j$, where m is given in (5). Indeed, if there exists $t^* \geq s_j$ such that $|x(t^*)| = M_j - m$, then

$$W_1(M_j-m) = W_1(|x(t^*)|) \leq V(t^*, x_{t^*}) \leq V(s_j, x_{s_j})$$

$$\leq W_2(3\sigma) + \max\left[\, W_3(\|x\|^{[s_j-h, s_j]}),\, W_3(\sigma)\,\right] \leq W_2(3\sigma) + W_3(M_j)$$

which contradicts (6). Thus,

$$|x(t)| < M_j - m \quad \text{for all } t \geq s_j.$$

Now choose the first positive integer N such that $1 - Nm \leq M$. If (B) holds on I_j for $j = 1, 2, \cdots N$, then for $t \geq s_N$ we have

$$|x(t, t_o, \phi)| < M_N - m < M_N - 2m < \cdots < 1 - Nm \leq M.$$

Thus (A) must occur on some I_j with $j \leq N+1$. That is

$$|x(t, t_o, \phi)| < \epsilon \quad \text{for all } t \geq s_{N+1} \geq s_j.$$

Notice that

$$s_{N+1} \leq t_o + (N+1)(h+KL) =: t_o + T.$$

Therefore, $|x(t, t_o, \phi)| < \epsilon$ for $t \geq t_o + T$ and the proof is complete.

COROLLARY 1. Suppose that there exists a continuous functional $V: R \times C_H \to R^+$, a function $g \in G$, a continuous function $\Phi: R^+ \to R^+$ with $\Phi \in L^1[0, +\infty)$, and a positive constant γ such that

(i) $W_1(|\phi(0)|) \leq V(t, \phi)$

$$\leq W_2(|\phi(0)| + \int_{-\infty}^{0} \Phi(-s)W(|\phi(s)|)ds) + W_3(|\phi|_g),$$

(ii) $V_{(1)}'(t, \phi) \leq - W_4(|\phi(0)|),$

(iii) $W_1(r) - W_3(r) > 0$ for all $r \in (0,\gamma)$.

Then the zero solution of (1) is UAS.

Due to limitation of the length of the paper, readers are refered to [17-18] for more applications of Theorem 1 and related Liapunov functionals.

REFERENCES

1. L. C. Becker, T. A. Burton, and S. Zhang, Functional differential equations and Jensen's inequality, J. Math. Anal. Appl., 138: 137-156 (1989).

2. T. A. Burton, Uniform asymptotic stability in functional differential equations, Proc. Amer. Math. Soc. 68: 195-199 (1978).

3. T. A. Burton, Stability and Periodic Solutions of Ordinary and Functional Differential Equations, Academic Press, Orlando, (1985).

4. T. A. Burton and S. Zhang. Unified boundedness, periodicity, and stability in ordinary and functional differential equations, Anal. Mat. Pur. Appl. CXLV: 129-158 (1986).

5. C. Corduneanu, Integral Equations and Stability of Feedback Systems, Academic Press, New York, (1973).

6. J. K. Hale, Theory of Functional Differential Equations, Springer-Verlag, New York (1977).

7. R. H. Hering, Boundedness and Stability in Functional Differential Equations, Ph.D. Dissertation, Southern Illinois University, (1988).

8. R. H. Hering, Boundedness and periodic solutions in infinite delay systems, J. Math. Anal. Appl., 163: 521-535 (1992).

9. R. H. Hering, Uniform asymptotic stability in infinite delay systems, J. Math. Anal. Appl., to appear.

10. Q. Huang and K. Wang, Space C_h, boundedness and periodic solutions of FDE with infinite delay, Scientia Sinica, Series A, Vol. XXX: 807-818 (1987).

11. J. Kato, A conjecture in Liapunov method for functional

differential equations, Preprint.

12. N. N. Krasovskii, Stability of Motion, Stanford Univ. Press, Stanford, CA, (1963).

13. I. P. Natanson, Theory Functions of a Real Variable, Vol.II, Ungar, New York, 1960.

14. K. Sawano, Exponential asymptotic stability for functional differential equations with infinite delay, Tohoku Math. J., 31: 471-486 (1978).

15. G. Seifert, Liapunov-Razumikhin conditions for stability and boundedness of functional differential equations of Volterra type, J. Differential Equations 14: 424-430 (1973).

16. T. Yoshizawa, Stability Theory by Liapunov's Second Method, Math. Soc. Japan, Tokyo, (1966).

17. Bo Zhang, A stability theorem in functional differential equations, Differential and Integral Equations, to appear.

18. Bo Zhang, Boundedness in functional differential equations, Nonlinear Analysis, to appear.

Index

Admissibility   89

Algorithm   152
    Monotone Iterative   57
    Symplectic   152

Ascoli's Theorem   44, 50, 54

Attractors   83

Belousov-Zhabetinski System   79

Bifurcation Diagram   77

Blowup   189, 190
    Rate   191

Boundary Value Problem   57, 127

Capillary Problem   139

Carrying Capacity   26, 31

Chaos   67

Chemical Oscillator   67

Chemical Systems   67

Coexistence   101

Comparison of Even Order Elliptic
    Equations   159

Comparison Principle   2, 138, 144, 227, 237
    Method   93, 277, 303, 332

Competitive System   109

Completely Continuous Operator   130, 133

Conservation Matrix   68

Contraction Mapping   8, 49

Control   88, 217, 269, 303

Convection Dominated Parabolic
    Equations   58

Convexity   1, 8

Degree Theory   19

Democracy   369

Differential Equations
    Functional   35, 349
    Impulsive   199, 227
    Integro   26
    Parabolic   277
    Uncertain   217

Dissipative System   102

Dynamical System   69, 102, 176, 209
    Large-Scale   117
    Semi   102

Eigenvalue Problem   293

Elliptic Dirichlet Problems   240

Equation
    Delay   25, 91
    Diffusive Logistic   109
    Drift Diffusion   18
    Fading Memory   35
    Fixed Point   129
    Heat   189
    Integral   35, 88
    Logistic   28

Parabolic   57

Uniformly Elliptic   139

Feedback   217

Fixed Point Index   127

Fixed Point Theorems of Cone Expansion or
  Compression Type   128

Free Boundary Problems   1

Gauss Curvature Bounds   137

Generalized Lipshitz Condition   88

Global Attractor   102

Gray-Scott Network   73

Green's Function   138

H-Graphs   137

Hamiltonian System   151

Hopf Bifurcations   79

Hurwitz Determinant   76

Infinite Delay Equations   91

Instability   57, 69

Invariance Principle   203

Linearizing About Equilibria   35

Linear Eigenvalue Problem   105

Linear Integro Differential Systems   96

Lotka Volterra System   109, 339

Lyapunov Function   37, 117, 342

Multiple   104

Perturbed   117

Vector   117

Lyapunov Functional   92, 351

Maximal and Minimal Solutions   134

Maximum Principle   6, 21, 109, 241

Mixed Boundary Conditions   60

Mixed Quasomonatone Property   61

Monotone Flows   176, 209

Monotone Iterative Technique   57

Multiple Positive Solutions   133

Neural Networks   203

Nodally Oscillatory   159

Omega Limit Set   33

Order Preserving   102, 209

Oscillations   67

P-Capacitary Potential   18

P-Laplacian   18

Persistence (Permanence)   101

Uniform   102

Poincaré-Bendixson Theorem   33

Poisson Hamiltonian   153

Population Diversities   26

Population Dynamics   25, 101, 339

Quasilinearization   331

Quasimonotone   119

Quasisolutions   65

Reaction-Diffusion Models   101

Reaction Diffusion System   101, 277, 33

Saddle Node Bifurcations   78

Schaefer's Fixed Point Theorem   35

Second Order Elliptic Equations   137

Semiconductor System   17

Skew Product Flow   103

Solutions